LEARN
FROM
THE MASTERS!

Cover image courtesy of Smithsonian Institution libraries, Photo number 86-2236. The image is the cover of *Arithmeticae Practicae Methodus Facilis,* by Frisius Gemma, published in 1550. The book was a gift from Dr. Bern Dibner to the Smithsonian libraries.

The photos of Johannes Kepler, René Descartes, and Gottfried Wilhelm Leibniz are from *A History of Mathematics,* Second Edition, Carl B. Boyer, Revised by Uta C. Merzbach, pp. 324, 334, and 399, respectively, copyright 1991 by John Wiley and Sons. Reprinted by permission of John Wiley and Sons, Inc.

Cover design by Adriana Barbieri.

Typeset in ten point New Times Roman by Beverly J. Ruedi.

Technical drawings by Amy Fabbri using FreeHand.

© *1995 by*
The Mathematical Association of America (Incorporated)
Library of Congress Catalog Card Number 95-78461

ISBN 0-88385-703-0

Printed in the United States of America

Current Printing (last digit):
10 9 8 7 6 5 4 3 2 1

LEARN FROM THE MASTERS!

Edited by

Frank Swetz

John Fauvel

Otto Bekken

Bengt Johansson

Victor Katz

Published by
THE MATHEMATICAL ASSOCIATION OF AMERICA

CLASSROOM RESOURCE MATERIALS

Published by
THE MATHEMATICAL ASSOCIATION OF AMERICA

———

CLASSROOM RESOURCE MATERIALS

This series provides supplementary material for students and their teachers—laboratory exercises, projects, historical information, textbooks with unusual approaches for presenting mathematical ideas, career information, and much more.

Proofs Without Words, by Roger Nelsen
A Radical Approach to Real Analysis, by David Bressoud
She Does Math! Real-Life Problems from Women on the Job, edited by Marla Parker
Learn from the Masters! edited by Frank Swetz, John Fauvel, Otto Bekken, Bengt Johansson, and Victor Katz

Mathematical Association of America
1529 Eighteenth Street, NW
Washington, DC 20036
1-800-331-1MAA   FAX: 1-202-265-2384

PREFACE

In August of 1988 following the close of ICME6, an international conference/workshop on the History of Mathematics was held in Kristiansand, Norway. The "Kristiansand Conference" was conceived and organized by Otto Bekken of Agder College, Norway and Bengt Johansson of Göteborg University, Sweden. Financial support for the conference was principally supplied by ABACUS Publishers, Mölndal, Sweden. The invited participants, mathematicians and mathematics educators from twelve countries spanning four continents, were drawn together by their mutual interest in the history of mathematics and, particularly, its use as a pedagogical tool in the teaching of mathematics. Conference organizers planned an environment that allowed for "a sharing of our interests to improve the teaching of mathematics by the use of materials from the history of mathematics." Obscure research talks were to be avoided and presentations were to stress the classroom use of ideas, topics, concrete materials and problems from the history of mathematics. In brief, this week-long gathering was intended as an exchange of ideas and sharpening of thoughts on the historical enrichment of the mathematics curriculum.

Gimlekollen Mediasenter in the hills above Kristiansand supplied a sylvan setting where the participants lived and worked together for a week. In this cloistered atmosphere, the presentations, discussions and activities combined to form a truly consciousness-raising and rewarding experience. It is hoped that some of the spirit of the Kristiansand Conference is maintained by this collection of papers, the conference proceedings, and will be conveyed to the readers.

At one time, Niels Henrik Abel (1802–1829), Norway's greatest mathematician, lived near Kristiansand. Although Abel's life was tragically short, he left a bountiful legacy for mathematics. One hundred sixty years after Abel spent his last productive summer at Froland, a country estate a short distance from Kristiansand, this conference took place. Thus it was fitting that the participants paid their respect to this great mathematician and interesting historical figure. A group pilgrimage was undertaken to Froland and Abel's grave site. (See the Tribute to Abel.) In one of Abel's notebooks as a note in the margin, he observed:

> It appears to me that if one wants to make progress in mathematics one should study the masters.

Following this advice and as a further tribute to Niels Henrik Abel, we title this work *Learn from the Masters*.

Frank Swetz
Harrisburg, PA
August, 1994

CONTENTS

PART I

History in School Mathematics

While many would agree that the teaching of mathematics at all levels can be enriched by historical reflection, perhaps that consensus is even stronger when directed at the secondary school level. At this level, historical enrichment can have a profound effect! For it is at the secondary or high school level that students first experience the power of mathematics and begin to realize the wide scope of its applications and possibilities. Hopefully, this cognitive impact can be stimulating, resulting in an anticipation and enthusiasm for a deepening of mathematical knowledge, but, disconcertingly, it can also be intimidating, especially for a student who has lacked obvious structure in his or her mathematical learning. It is in this latter instance, particularly, that the history of mathematics can supply a structure of understanding relating reasons with results. History can provide a logic between the definition of a mathematical concept and its application or, more historically correct, between the application and the definition-theory of a concept. The following discussions seek to identify and clarify some techniques and pedagogical approaches for using the history of mathematics in secondary teaching.

Shmuel Avital makes a broad case for the learning/teaching benefits resulting from historical awareness. Building on his teacher education experiences in Israel, Avital is a strong proponent for teachers' studies on and about the history of mathematics.

Phillip Jones examines several concepts related to the general idea of "number" and describes their evolution. He then notes how a familiarity with this evolution lends itself to understanding how mathematics and mathematical research work.

Frank Swetz contributes three papers to this section: the first advocates the use of actual historical problems in classroom instruction; the second documents the development of trigonometry as it could be revealed to students; the third considers the history of a particular topic in mathematical modeling. All of these contributions also help to illustrate the multicultural aspects of the development of mathematics, yet another dimension that can be used to make mathematics learning "interesting."

Are logarithms a dead topic? Not according to John Fauvel who shows that through historical considerations there is a lot of mathematical life still left in this topic.

Victor Katz also discusses logarithms and supports their importance as a mathematical topic. He reviews the historical evolution of the concept, concluding that a Napieran approach to teaching logarithms may be well suited to today's classrooms.

Jan van Maanen shows how a fourteenth century legal problem can supply a bounty of worthwhile mathematical learning experiences; how seventeenth century drawing instruments lead to modern problems on conic sections and how the struggles of Torricelli, Sluse and Huygens in attempting to understand the mathematical properties of infinitely extended solids can help students today. Van Maanen's discussions further emphasize the rich variety of historical materials available for classroom use.

Thus, at the school level, whether one is interested in improving calculation, geometric reasoning, analytic thinking or trigonometry applications, the history of mathematics can assist in these efforts.

History of Mathematics Can Help Improve Instruction and Learning

Shmuel Avital

I have often been surprised that mathematics, the quintessence of truth, should have found admirers so few and so languid. Frequent consideration and minute scrutiny have at length unravelled the course: viz THAT THOUGH REASON IS FEASTED, IMAGINATION IS STARVED: while reason is luxuriating in its proper paradise, imagination is wearily travelling on a dreary desert.

[Samuel Taylor Coleridge, 4, p. 21]

Mathematics as taught in school is perceived by most secondary school students as a subject devoid of history. The teacher becomes the source of all that has to be learned on the subject and his task is to convey that knowledge to the student. Usually in the instructional process the understanding of the process of mathematical creation and of the age-old grappling with mathematical problems are completely lost. Mathematics to most students is a closed subject, located in the mind of the teacher who decides whether answers are correct or not. This situation is particularly harmful to mathematics teaching, more than to teaching in most of the other sciences. Mathematics is by nature a cumulative subject; most of what was created millennia ago—both content and processes—is still valid today. Exposing students to some of this development has the potential to enliven the subject and to humanize it for them.

If we want to change the present situation we have to do it through the teachers. Little allowance is usually made in the secondary school curriculum for the history of mathematics. The best that teachers can do to change this situation is to enrich their own instruction by means of appropriate references to the historical development of mathematics. It is our task to educate mathematics teachers in the use of history in order to make their instruction more meaningful to students. The question before us, then, is how to achieve this objective. Research in pre- and in-service teacher education has shown that teachers tend to teach their students in the manner they, themselves, were taught. See [8 pp. 178, 180–181]. Hence to achieve our goal we have to expose prospective teachers to approaches that are directly applicable to their own teaching. This quest leads us to analyze the historical development of mathematics, searching for ideas of high pedagogical value. We have to isolate problems encountered in the teaching and learning of mathematics in secondary schools, and then point out developments in the history of mathematics which can help the teacher better understand and cope with these problems. I shall discuss how a knowledge and understanding of the historical development of mathematics can contribute in four specific areas of instruction and learning. These areas are:

1. Acquiring insight into students' learning difficulties.
2. Improving modes of instruction.
3. Incorporating problem solving and problem posing in instruction.

4. Drawing attention to emotional and affective factors in the creation and learning of mathematics.

Examine History to Understand Learning Difficulties

In the history of mathematics, we encounter many developments that were slow in coming to be accepted. Quite often, we find that mathematicians made important developmental advances in their discipline, which were, nevertheless, disregarded by later colleagues. I believe one is justified in conjecturing that secondary school students face learning difficulties in areas similar to those we encounter in the historical development. I shall list three examples:

1. The acceptance of negative numbers.   Those of us who have taught in secondary schools, or are involved in teacher education, are aware of the difficulties students encounter in understanding and performing operations with negative numbers. Students have difficulties in recognizing that $a - b$ is equivalent to $a + (-b)$, that $(-a)(-b) = ab$, that $-a$ need not represent a negative number, etc. In general, we find that in the common perception of the concept "number" the negative numbers take a place secondary to the positive ones, or even to the number zero. Go into a class of 15- to 16-year-olds and ask: "Given a number, can you add another number to it and yet get a result smaller than the given number?" Suggest that they jot down their answers, so that you don't get the answers of the few "geniuses." Try it: studying students' responses is a fascinating and fruitful activity.

The historical development of the number concept reveals a widespread opposition to the introduction of negative numbers. In most early work on the solution of equations, negative roots were not considered. Hindu mathematicians in the sixth century, and Chinese mathematicians even before that, worked out all the rules for operations with negative numbers. Still, three centuries later we do not find any of this in the work of Arab mathematicians, although they were definitely aware of the Hindus' contribution. Much later, in the seventeenth century, the French mathematician Blaise Pascal (1623–1662) saw no need for the introduction of negative numbers. Even in the nineteenth century, the English mathematician Augustus de Morgan (1806–1871) considered numbers less than zero inconceivable. See [9, pp. 252, 593]. The negative numbers gained some acceptance together with the complex numbers in the sixteenth century. When Cardan (1501–1576) published the algebraic solution to cubic equations it turned out that to find the solution of equations which have three distinct real roots we have to use complex numbers, namely, square roots of negative numbers. It is only in the nineteenth century that the negative numbers achieved a fully recognized position in the number system. A study conducted at the Weizmann Institute of Science in Israel gives some support to a conjecture that exposing teachers to the historical development of negative numbers can help them improve their understanding of students' difficulties in this topic at school. See [1, p. 10].

2. The use of symbolic notation.   The situation is similar with regard to the acceptance and use of symbolic notation in mathematics. Difficulty in developing facility with the meaning and use of symbols is one of the major obstacles in the learning of algebra in school. Two factors seem to be involved:

(i)   The similarity of some symbols which are completely different in meaning. Consider the visual similarity of the following: $2+2$, 22, $2 \cdot 2$, 2^2; or a_2, $2a$, a^2; or the distinction between constants and variables in the equation $ax + by = c$, etc. No wonder that some students, when faced with $\cos x / \cos 2x$ cancel $\cos$ and x and get the answer $\frac{1}{2}$.

(ii) The huge amount of information some symbols encompass. Consider for instance the expression $|x - 2|$. Worse yet, when trying to simplify and find the value of $x^2/|x|$, the learner has to bear in mind that even though numerator and denominator are both positive, x itself may take on a negative value.

We all know the tortuous path taken by the historical development of mathematical symbolism. The Greek mathematician Diophantus (c. 250) used syncopated symbols, that is shortened words, for operations. He even used a symbol for the unknown. None of these symbols is to be found in Arabic writings of the eighth and ninth centuries, although the authors knew of Diophantus's and Hindu work. Babylonians used Sumerian words in the form of pictorial symbols to represent unknowns in a problem [9, p. 10]. So did Diophantus [5, p. 42]. None of these is found in the work of the Arabic mathematician Al-Khwarizmi (c. 825) whose books, in their Latin translations, influenced the mathematics of medieval times. The book *Hisab Al-jabr W'al Muqabalah*, from which the term algebra was derived, does not contain a single mathematical symbol.

The difficulties in the historical adoption of mathematical symbolism can also be seen in the sluggishness in acceptance of the use of letters to denote classes of numbers, such as coefficients, as noted by the mathematician Morris Kline when speaking about Greek mathematics:

> There was no recognition of the enormous contribution that letters could make in increasing the effectiveness and generality of algebraic methodology. [9, p. 144]

A great contribution in the use of letters for different classes of numbers was made by the French mathematician François Viète (1540–1603) who symbolically distinguished between constants and variables by suggesting the use of consonants for the former and vowels for the latter. But it was not until the time of Descartes (1596–1660) that a consistent system of representation was generally accepted.

Understanding the historical difficulties in the introduction of suitable symbolism will help to convince teachers that much patience is needed in helping students grasp the difference in meaning between coefficients and variables. Particular efforts should be made to develop students' ability to understand the use of a reasonable symbolism in algebraic proofs.

3. Rigor and abstraction. A third point in attempts to understand students' learning difficulties, by looking at the historical development, concerns the levels of rigor and abstraction. The use of rigor and abstraction requires a maturity, based mainly on experience in the area in which the rigor is to be applied, and not just age. As abstraction and rigor have to be based upon accumulated experience, even adults should be gradually led to the abstract via the familiar when exposed to a new theory. The New Math movement of 1960–1970, and in particular the School Mathematics Study Group (SMSG) in the US failed in their attempts at reform through a lack of understanding that a precise rigorous language simply cannot be imposed upon young people who have not accumulated enough experience in the given area. One of the greatest mistakes of the curricular changes that took place at that time was the introduction of operational set-theory (not just the concept of "set," which can be conceptually very helpful).

Historical studies show that the natural numbers were the last among various number sets, even after the quaternions and Cayley numbers, to receive a rigorous axiomatic structure. The

Zermelo-Fraenkel axioms for set theory came even later. Indeed the conception of rigor itself changes with time.

History Can Teach Us How To Teach

A basic problem with reading published articles in mathematics is that in almost all publications the author ignores the background, the grappling, and the failures that led to the ideas he or she writes about. The published form of "Definition, Theorem, Proof, Corollary etc." leaves the reader asking: "Heavens, how did he/she get the idea for these definitions and theorems?" The confusion is much greater when we teach in this way.

Start with specific examples or with an intuitive approach in almost all courses. Very often in undergraduate courses, and sometimes even secondary school textbooks, one finds that the presentation of a topic begins with a generalization with specific examples following afterwards. When exposed to any new material, a learner needs a frame of reference to which to relate this material. Such frames are provided by familiar material. A generalization that comes before examples is usually lost on us, and to many the loss may be so great that the subsequent examples cannot make up for it.

Many historical developments went from specific examples to generalizations. Egyptian and Babylonian mathematics deal almost entirely with specific examples. Here is one example as quoted in [5, p. 71]: I have added up the area and the side of my square: 45′. (The Babylonians used the sexagesimal system in which this means 45/60). The solutions to these problems were always of prescriptive nature: "do this, do that."

Van der Waerden in his book *Science Awakening* could find only two examples of generalization in his examination of Babylonian mathematics. See [12, pp. 73, 74]. The American mathematician and philosopher R. L. Wilder, states:

> There is little question that Greek mathematics represented a natural evolution from Babylonian mathematics. . . . It is quite possible that it was this wedding of the Greek philosophical bent and the Babylonian science of numbers that caused what is so often termed a gap, but which was actually only a leap to a higher level. [13, pp. 150–151]

Greek mathematics, which was the first to introduce general statements (theorems), seems to have been an evolutionary-revolutionary continuation of the Babylonian tradition. I believe that, similarly to what happened in history, instruction which moves from specific examples to theoretical generalizations will help students make the leap to the higher level.

On the other hand the Greek heritage may, to some extent, be responsible for the educational transgressions pointed out at the beginning of this section. Greek mathematicians, particularly Euclid and Archimedes, use smooth and polished formulations in which the initial grappling with the material has been obliterated. It is therefore beneficial to expose students and prospective teachers to Archimedes' (287–212 B.C.) letter to Eratosthenes (c. 276–c.196 B.C.), usually referred to as *The Method*, in which Archimedes describes how he arrived at some of his theorems. Let us read the master.

> I thought fit to write out for you and explain in detail in the same book the peculiarity of a certain method, by which it will be possible for you to get a start to enable you to investigate some of the problems in mathematics by means of mechanics. This

procedure is, I am persuaded, no less useful even for the proof of the theorems themselves; for certain things first became clear to me by a mechanical method, although they had to be demonstrated by geometry afterwards because their investigation by the said method did not furnish an actual demonstration. But it is of course easier, when we have previously acquired by the method some knowledge of the questions, to supply the proof than it is to find it without any previous knowledge. [8, p. 13]

For Archimedes, an approach via mechanics led to what we may consider to be an intuitive proof. We can never know the amount of experience learners possess. So the best way to convey a proof is first to give an intuitive overview of the proof and then present the proof. After one generates ideas in a heuristic way, the student realizes the objective and then he/she is ready for a rigorous proof. It is then a good idea to summarize, using an intuitive overview.

Finally, only an historical perspective can help our student grasp the importance of the fact that, unlike the natural sciences, mathematics is a cumulative subject. Results created thousands of years ago are still valid today. For example, one can best understand the revolution in mathematics which established the validity of non-Euclidean geometry only after being exposed to the long historical struggle to prove the fifth postulate. By sending students to read about this history, we supply them with a profound understanding that mathematics grows from within, the new being an extension of the old.

Use Problems from the Past

Teachers may ask "Where do I find the time to teach history?" The best answer is: "You do not need any extra time." Just give an historical problem directly related to the topic you are teaching; tell where it comes from; and send the students to read up its history on their own. So, when dealing with simple fractions tell students about Egyptian fractions; refer to that "Old School Book," the *Rhind Papyrus*; and give them the problem of writing 3/5 as a sum of Egyptian fractions—unit fractions with different denominators. In an upper grade, when dealing with algebraic proofs, let students try to prove the validity of Sylvester's algorithm for changing a given fraction into a sum of Egyptian fractions. (See exercise 3.) When dealing with mathematical induction show them the Babylonian summation:

$$1^2 + 2^2 + 3^2 + \cdots + 10^2 = \left[\frac{1}{3} + \left(\frac{2}{3}\right) \cdot 10\right](1 + 2 + 3 + \cdots + 10)$$

and let students formulate a generalization and prove it. When you teach congruence of triangles, let students discover Thales' method for finding the distance of ships from the shore; when you deal with solutions of algebraic equations, let students explore what types of cubic equation the Babylonians could solve by using their tables for $n^2 + n^3$. History can follow the curriculum from topic to topic. Some approaches to historical problems not only enrich instruction, but actually show ways that are educationally better than the modern ones. The Egyptian hieroglyphic symbols for addition and subtraction were Λ for plus and Λ for minus. The symbols represent moving legs. (The Egyptians, like the Arabs and Hebrews, wrote from right to left.) Elementary school teachers like the idea of using these, with changed directions, as symbols to accompany the usual signs "+" and "−". These symbols can enliven the introduction of negative numbers to twelve- and thirteen-year-old students.

Once, in an in-service course for heads of mathematics departments, I asked an Arab student to present some ideas from Al-Khwarizmi. The student had a photocopy of a fourteenth century text of *Al'jabr W'al Muqabalah*. He presented Al-Khwarizmi's approach to the solution of quadratic equations. The drawing in Figure 1 shows how Al-Khwarizmi would represent the solution of the equation $x^2 + 4x = 12$.

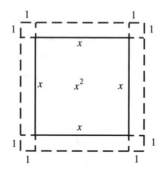

FIGURE 1

Another teacher in the same course mentioned that she was, just at that time, teaching this topic and would try to use this approach with her students. She returned a week later, and told us: "For the first time I felt students truly understood the solution by 'completing the square'." To give one more example: nine- to ten-year-old children have difficulties with the "carrying" process in multiplication, much more so than in addition. The drawing in Figure 2 shows an example of the Hindu system of writing the multiplication $32 \times 43 = 1376$. There is no carrying in the multiplication, as every product is written out in full.

Elementary school teachers think this method may help in teaching children the multiplication of numbers written with more than one digit.

1. Educate to pose problems.   An important activity in doing mathematics is *asking questions*. Each achievement, each creation, has its origin in questions which, when answered, are usually followed by more questions. From studying history we can see that most questions arise as a direct continuation of what has gone before: "If I can do this, can't I do something similar after

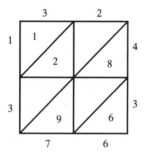

FIGURE 2

some items have been changed?" We cannot know who was the first to pose the famous problems of antiquity: trisection of angles, duplication of a cube, and quadrature of a circle, nor how they were prompted. We may conjecture that these three problems that deal with construction with Platonic tools, a straightedge and compass, originated as an outcome of things the Greeks could do. On this interpretation the problems were:

(i) We know how to bisect an angle with these tools, but can we also trisect it with the same tools?

(ii) We know how to double the area of a square; what about doubling the volume of a cube?

(iii) We know how to square the area of any convex polygon; what about squaring the area of a circle?

Our conjecture can be extended to a fourth problem that also originated in ancient Greece. We know how to inscribe in a circle equilateral triangles, squares, pentagons, and hexagons, but what about inscribing a heptagon?

The same questions can and should be asked in any classroom, but you rarely find secondary school students who have ever been exposed to them. The problems are directly related to areas that are part of the curriculum. There is no need to spend extra time on them in class; just lead students to raise the questions and then send them to read on their own about the problems.

An important educational principle is that we have to educate our students to ask valid questions. Pólya, in his model for problem solving given in the book *How To Solve It* [11], lists as the fourth step in a solution process "Looking back." In this step, he suggests that after solving a problem the solver should look back and ask himself some questions related to the problem that has been solved. One of these questions is: "What other questions can I ask?" See [11, p. xvii]. Teachers know that learners ignore this step. When a student succeeds in solving a problem he/she underlines the answer and is finished. We conjecture that by exposing students to the enormous role problems have played in the historical development of mathematics, with patience and in an appropriate classroom climate, we shall start getting pertinent questions.

2. Include problems with the answer "It cannot be done". Another important aspect in learning from history about the posing of problems concerns the possible answers to these problems. The answer to many problems of antiquity was: "It cannot be done." This was the answer to the three classical problems of antiquity, to the construction of the heptagon with Platonic tools, to the problem of deriving the fifth postulate from the other axioms, and to many others. This is not true for usual textbook problems, most of which have what we may call a "positive" answer. At most, we can find problems of the nature: "Prove that this or that is impossible." Such a formulation kills the spirit of the problem. It took humanity 2000 years, and lots of grappling, to show that the parallel postulate must remain a postulate. It took about three hundred years to show that quintic equations cannot generally be solved by radicals. We can and should educate students to conjecture "It cannot be done" and then try to prove this conjecture. For example, ask nine- and ten-year-olds to search for an integral square with an even number of divisors; eleven-year-olds to obtain the sum 45 by adding 8 numbers taken from the set 1,3,4,7,11 allowing the use of any number more than once; twelve-or thirteen-year-olds to place seven points in the plane, no three being on the same line, and connect each of the points with exactly three of the others. (Tell them that connecting lines may intersect, etc.) In each case, mix those problems with others that can be done, and let them search before they conjecture "it cannot be done" and seek to prove it.

3. Generate exploratory problems as well. Most textbook problems are of the closed form ("solve this," "find that"), requiring one specific answer. However, many historical problems are of an exploratory nature, in which the search for a proof or a solution has generated outstanding mathematics. Such problems include, for instance, Fermat's last theorem, that the equation $x^n + y^n = z^n$, with x, y, z positive integers has no solutions for integral $n > 2$, and the prime number theorem, that there exists an elementary function which represents the number of primes not exceeding a given number. In school there should at least be some problems of an exploratory nature, similar to historical ones, which provide more opportunities for students' exploration. Some examples of suitable school problems are: "How does a product of two positive numbers change when a number is subtracted from one factor and added to the other?" "What reduced fractions p/q can be written as a sum of two unit fractions?" In such problems, every child can do something, while the more able ones can explore in greater depth.

4. Discuss open problems. Earlier, I mentioned the wrong impression many students obtain that mathematics is a closed subject—the teacher knows everything. A good way to eradicate this impression is to convey to students some of the open problems that they can understand. The only open problems we usually find in textbooks are the Goldbach conjecture that any even integer greater than 4 can be represented as a sum of two prime numbers, and Fermat's last theorem (which is no longer open). There are many more open problems that secondary school students can easily understand. For instance: "Are there an infinite number of twin primes?" "Are there any odd perfect numbers?" "Can we define functions which will generate only primes?" "Is Euler's constant a rational or irrational number?" "Is π^e a transcendental number?" See [9, p. 450].

Mathematics is a Human Creation: Mathematicians Have Emotions

Mathematics as taught in school, and also to a large extent at the university, has the reputation of being a "dull drill" subject. Primarily, mathematicians themselves are responsible for this false image. The polished style of publications eliminates the human side of the grappling, of the perseverance, of the ups and downs experienced on the way to final achievement. We can enliven the subject by including, indeed emphasizing, this human factor. The story of Archimedes running naked through the streets of Syracuse is known to some, but the story of Hamilton's (1805–1865) perseverance for years in trying to extend the known number sets and define algebras, first of ordered triples and then of ordered quadruples, preserving the operational rules, is educationally much more enlightening. See [6, pp. 389–390].

For centuries, great mathematicians made repeated attempts to solve the problem of the fifth postulate; to find a solution of the quintic by radicals; or to find a proof of Fermat's last theorem. These stories, well told, will bring mathematics closer to the heart of the student. We have to give our students some feeling for the pains of failure in mathematics. For instance, look at the senior Bolyai writing to his son who informed him of his intention to attack the problem of the fifth postulate:

I have travelled all the reefs of the infernal dead sea, and have always come back with a broken mast and a torn sail. [10, p. 33]

We can also expose students to the elation of success: the younger Bolyai's reaction when he succeeded in developing his non-Euclidean geometry:

Out of nothing I have created a new and wonderful world. [10, p. 31]

To break the image of mathematics as a "boring" subject we can add color and enliven it by considering its human side and from time to time expose students to anecdotes from the lives of mathematicians. In some cases, such anecdotes can even help solve educational problems. The story of the struggle of the young Evariste Galois (1811–1832) can help a young creative child with adaptation problems and perhaps overcome disciplinary difficulties at school. The story of the great Norwegian mathematician, Niels Henrik Abel (1802–1829) can encourage a young gifted poor student to fulfill his potential in spite of hardships. See [2, pp. 555–557; 638–642].

Summary: Teaching Mathematics as a Part of Human Culture

Language and mathematics developed side by side. Language has day by day reinforcement, and thus becomes an integral part of human life under all conditions. Mathematics does not have such reinforcement and therefore requires intensive schooling. It is our task to organize this schooling in such a way that it conveys to our students the basic attributes of mathematics as a part of human culture. We can achieve this goal by relating the topics we teach to their historical developments. As we have seen:

(a) The historical development can teach us about possible learning difficulties.
(b) It can help us improve our teaching by trying to follow the process of creation in mathematics.
(c) It can induce us to create a classroom climate of search and investigation and not just of conveying information.
(d) It will lead us to use exercises in which there is a search progressing to a goal which can be reached through the accumulation of data.
(e) It will teach us to include problems to which the answer is "It cannot be done."
(f) Exposing our students to open problems will show them that mathematics is an open subject in which grappling with problems can be an exciting activity.
(g) It will help us to humanize the subject by exposing our students to the affective aspect of doing mathematics.

For Further Consideration

1. The mathematician Carl Fredrich Gauss (1777–1855) defined a subset of the set of complex numbers, usually referred to as Gaussian integers. This subset contained a smaller subset referred to as Gaussian primes. Formulate a conjecture about what may have induced Gauss to define these subsets.

2. The partial sums of the series $1 + 2 + 3 + \cdots$ are usually referred to as triangular numbers. In the seventeenth century, the mathematician Gottfried Wilhelm Leibniz (1646–1716) was challenged to develop a rule for the sums of the reciprocals of these partial sums. Use an approach mentioned in this article to develop a conjecture about the required rule.

3. The mathematician James Joseph Sylvester (1814–1897) developed the following algorithm to write any positive simple fraction p/q as a sum of Egyptian fractions (fractions with numerator 1 and different denominators):

(i) Subtract from p/q the largest Egyptian fraction which is not greater than p/q.

(ii) Do the same to the remainder, and so on.

Prove that this algorithm always leads to a finite sum of Egyptian fractions.

4. The name of the Greek mathematician Diophantus is attached to polynomial equations in which there are more variables than equations. Discuss why it would be desirable to introduce the solution of linear equations of this type, before teaching the solution of sets of two linear equations with two variables each.

5. Geometry underwent great upheaval in the first half of the nineteenth century. Discuss in what way this upheaval may be in keeping with the spirit of the time.

Bibliography

1. Arcavi, Abraham *et al*: "Maybe a Mathematics Teacher Can Profit from the Study of the History of Mathematics," *For the Learning of Mathematics* 3 (1982), 30–37.
2. Boyer, Carl B.: *A History of Mathematics*. New York: Wiley, 1968.
3. Chace, A. R. *et al*: *The Rhind Mathematical Papyrus*. Oberlin, Oh.: Mathematical Association of America, 1927. (Reprinted by National Council of Teachers of Mathematics, 1979.)
4. Coleridge, Ernst H.: *The Complete Poetical Works of Samuel Taylor Coleridge*. Oxford: Oxford University Press, 1967.
5. Dedron, P. and Itard, J.: *Mathematics and Mathematicians*. Milton Keynes: Open University Press, 1978. (Original French edition published in 1959.)
6. Eves, Howard: *An Introduction to the History of Mathematics*, 4th ed. Philadelphia, Saunders College Publishing, 1976.
7. Heath, T. L.: *The Works of Archimedes*. New York: Dover, 1953.
8. International Commission on Mathematics Instruction: *Proceedings of the 6th International Congress on Mathematical Education*, Action Group on Pre-Service Teacher Education. Budapest: ICMI Secretariat, 1988.
9. Kline, Morris: *Mathematical Thought from Ancient to Modern Times*. New York: Oxford University Press, 1972.
10. Meschkowski, Herbert: *Non-Euclidean Geometry*. New York: Academic Press, 1965.
11. Polya, George: *How to Solve It*. Princeton: Princeton University Press, 1973.
12. van der Waerden, B. L.: *Science Awakening, I*. New York: Oxford University Press, 1961.
13. Wilder, R. L.: *Evolution of Mathematical Concepts*. New York: Wiley, 1968.

The Role in the History of Mathematics of Algorithms and Analogies

Phillip S. Jones

In 1575 Pedro Nuñez wrote, "Oh well it had been if those authors who have written in mathematics had delivered to us their inventions ... in the same manner and with the same discourse as they were found out." John Wallis quoted him approvingly in 1686 [19, p. 3]. Three hundred years later Jonathan Golan wrote, "as we all know the process of *doing mathematics* involves, in the end, concealment of one's tracks." He then went on to show his depreciation of this process and his agreement with Nuñez's view, by complimenting the journal *Linear and Multilinear Algebra* for publishing a series of letters giving a "blow-by-blow" account of an attempt to solve a problem [2, p. 106]. This same shortcoming of mathematical exposition was in evidence when Jaques Barzun, a prominent literary scholar, wrote,

> I have an impression—it amounts to a certainty—that algebra is made repellent by the unwillingness or inability of teachers to explain why. ... There is no sense of history behind the teaching, so the feeling is given that the whole system dropped down ready-made from the skies, to be used only by born jugglers. [2, p. 82]

Barzun's comment was too narrow when he restricted it to algebra and too generous when he implied that history can completely solve the problem of conveying understanding and a feeling for the nature, growth processes, and aliveness of mathematics.

Of course, the major task of the historical researcher is to determine "the facts" of mathematical growth. However, he also has a responsibility to display the setting in which new developments took place, and he should not shirk the more difficult and uncertain task of trying to determine the concealed tracks, the methods, motivations, and viewpoints of the developers. These latter artifacts are of especial importance to the users of history, the teachers and expositors who should wherever possible bring out three things:

(1) the relationships between newly developing mathematics and the philosophy and/or applications of its day;

(2) the motivations leading to a new development, not only the reasons why people did mathematics at all, but how and why they approached it as they did;

(3) the methods or thought processes, induction, deduction, experimentation, or logic which led to forward steps.

Many mathematical steps forward are the result of inductive leaps, spurred by analogy, and facilitated by effective algorithms, or prodded by criticisms and directed by conjectures, which were often incomplete and sometimes incorrect. To display these to students is not to show the feet of clay of the queen of the sciences or to reveal the skeleton in her closet, but it is to humanize her and her followers, to make their pathways more accessible and enticing even if less steep and direct.

The following illustrations of this type of exposition are simple and incomplete. The quotations are out of their complete context. This may make my proposal appear easy to achieve and trivial in importance. On the contrary, to extend it very far beyond this level becomes difficult and increasingly conjectural because of the failure of mathematicians even to show their steps, not to mention their thought processes. However, the attempt is worthwhile because it may attract the interest of some who would otherwise turn away, and because it can profitably broaden the view and understanding of those who do join us in the study and application of mathematics.

The stories of negative and complex numbers furnish the simplest illustrations. However, other elementary topics where data on these processes are available are: the theory of parallels in perspective drawing and projective geometry, the analogies between finite differences and differentials dating back to Brook Taylor (1685–1731), and the nineteenth century development of operator theory. Through all of these illustrations there run themes of disbelief and reluctance to accept new elements and new theories. This reluctance is mirrored in the use of such words as "fictitious," "imaginary," "absurd," "false," "paradoxical," and in uncertainties as to the validity of the justifications produced as proofs—or as we might view them, pseudo-proofs.

Early Concepts of Negative Numbers

Our sketch begins with Diophantus's (A.D. 250) rules for operating with *forthcomings* and *wantings*. These rules were directly equivalent to the modern rules for the product of positive and negative numbers. The product of a forthcoming and a wanting was a wanting. The product of two wantings was a forthcoming. Diophantus provided no rationalization for these rules, which he gave in a numerical context. Although some foreshadowing of negatives can be found in a few places in early Babylonian work [18, p. 61], the most likely source for Diophantus's ideas can be read into the diagrams for the "geometric algebra" of Euclid's Book II. If the line

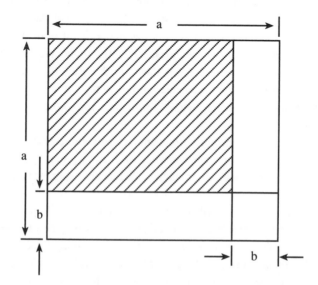

FIGURE 1

segments and rectangular areas of Proposition 7 are represented algebraically, Figure 1, a square of side a, leads to the formula $a^2 + b^2 = 2ab + (a - b)^2$. This is even more suggestive if we rewrite it in the form $(a - b)^2 = a^2 + b^2 - 2ab$. However, as we point out, the use of analogy here is the construction of algebraic-numerical models of geometric figures. We must also warn that we have used symbolic manipulation much beyond anything attributed to Diophantus or Euclid. Note also the fact that Diophantus's uses and the diagrams require the wantings to be less than the forthcomings. It seems safe to say that Diophantus had no concept of a negative quantity or number which could in any way exist independently. In fact, he characterizes the equation $4x + 20 = 0$ as "absurd," the first use observed of such almost emotionally loaded language.

Skipping to Mohammed ibn Musa al-Khwarizmi (c. 840), we find him giving the rules for operating with positive and negative numbers and then justifying them by an arithmetical analogy. Knowing that $8 \times 17 = 136$ as calculated by the usual arithmetic algorithm, he rewrote the product as $(10 - 2) \times (20 - 3)$, applied the algorithm for multiplying binomials in vertical form, and arrived at $200 - 40 - 30 + 6 = 136$. Al-Khwarizmi did not accept negative numbers as the roots of equations nor did he deal with any problem in which the final outcome would be a negative number.

This illustrates another historical concept—often the development of mathematical ideas does not follow the nice linear chronological sequence which hindsight or logic might lead one to expect. The processes of communication, skepticism and acceptance play an important role. Western Europe learned of algebra from the Arabs, especially al-Khwarizmi. Actually, the Hindu writer Brahmagupta and others had accepted negative roots for quadratics as early as the seventh century. In the thirteenth century, Fibonacci, also known as Leonardo of Pisa, suggested that a negative root might be interpreted as a loss. Cardano and Descartes in the sixteenth and seventeenth centuries, respectively, accepted negative numbers as roots of equations, but in the same centuries Recorde implicitly denied them and Harriot explicitly argued against them with considerable strength and influence as we shall see.

Cardano, in 1545, did find two roots for quadratics, but he called the second root fictitious when it was what we would have termed negative. Recorde, at essentially the same time, found and admitted the second root only when both roots were positive. René Descartes, writing in French in 1637, used *vraie* and *fausse,* true and false, as his terms for positive and negative roots, but did develop algorithms for transforming equations to change their roots by specified addends or multipliers. These processes could change true roots into false and conversely. However, he noted that none of his transformations would change imaginary roots to real.

Thomas Harriot, in his *Artis Analyticæ,* published posthumously in 1631, denied that equations could have negative roots. He claimed that what we would call the negative roots were actually the positive roots of the different equations obtained by changing the signs of the odd powers. Harriot also paid considerable attention to inequalities, and it was the concept that negative numbers were less than zero, arising in the seventeenth century, that bothered Leibniz and others for over a century. In 1712 Leibniz sent Johann Bernoulli a proof by contradiction that -1 could not be less than zero. Assuming that $1/-1 = -1/1$ and that -1 is less than zero, then in the left-hand fraction the numerator is greater than the denominator while on the right the numerator is less than the denominator. This contradicted the definition of proportion which had been accepted since Eudoxus had devised it in the fourth century B.C. to avoid the difficulties in geometry attendant upon the discovery of incommensurables.

Leibniz presented a second argument based on an initial statement that a ratio may be considered *imaginary* [sic!] when it has no logarithm. He then noted that $\log(-1/1) = \log(-1) - \log 1 = \log(-1)$. However, $\log(-1)$ is neither positive nor negative, therefore $\log(-1)$ is not really true but imaginary. It is interesting to note that Descartes had used the terms *réelle* and *imaginaire* in discussing the roots of equations which he said could be *vraie* or *fausse, réelles* or *imaginaire*. However, neither the usage of Descartes nor of Leibniz implied that these actually were new mathematical entities, imaginary numbers, but rather that these numbers were imaginary in the sense that they did not exist.

Bernoulli used integration to reply to Leibniz's second argument. Since

$$\frac{dx}{x} = \frac{-dx}{-x},$$

by integrating both sides we have $\log(x) = \log(-x)$. Hence, the graph of $\log(-x)$ is symmetric to the graph of $\log(x)$ with respect to the y axis and therefore $\log(-x)$ does exist. Here, again, we see formalism following an algorithm into new territory being used, albeit incorrectly in this case, in the struggle to justify new, tentative advances.

Negative Numbers in the Nineteenth Century

It is fascinating to read and compare the mental and linguistic gymnastics to rationalize negatives and imaginaries in the early nineteenth century gone through by algebraists such as Francis Maseres, Augustus De Morgan, George Peacock, philosophers such as Auguste Comte, and the geometer-physicist and revolutionary Lazare Carnot.

Francis Maseres published *A Dissertation on the Use of the Positive and Negative Sign in Algebra* in 1758. In a sequel published in 1800, expounding ideas attributed to Harriot, he wrote:

> The quantities called negative are such as it is impossible to find any clear idea of, being defined by Sir Isaac Newton and other algebraists, to be such quantities as are less than nothing, or as arise from the subtraction of a greater quantity from a lesser, which is an operation evidently impossible to be performed; and as to the negative roots of an equation they are in truth the real and positive roots of another equation consisting of the same terms as the first equation, but with different + and − signs prefixed to some of them, so that when writers of "algebra" talk of the negative roots of an equation, they in fact jumble two different equations together. [16, p. 286]

Lazare Carnot is best known mathematically as a geometer and for his attempt to provide an improved logical-philosophical basis for the infinitesimal calculus. The result of the latter effort was first published in 1797 but was based on a previous essay he wrote for a 1786 contest sponsored by the Berlin Academy on "how so many correct theorems have been deduced from a contradictory supposition." Carnot's concern for negatives arose from his use of them in his book *De la Correlation des Figures de Géométrie* (1801). He discussed them in "Digression sur la Nature des Quantities dites Negative" which has been reprinted with several of his geometric works. [8] He needed to give a sound basis for negative numbers since he associated changes of signs in the analogous algebraic formulas with the changing relative positions of points on a line. According to Gillespie, he rejected both the views advocated at that time: that negatives are less than zero and that they merely represent an opposite direction. He distinguished a *quantity* from an *algebraic value*; the latter he regarded as a fictitious entity introduced for purposes of calculating. He argued that negatives are not quantities, since if quantities of no magnitude may be ignored in computations, surely one could ignore those less than zero, if there were any. However, negatives can not be ignored; therefore they are not quantities [10, pp. 125, 176].

On the other hand, *values*, he argued, are only algebraic forms, quantities taken collectively with their signs; these signs indicate operations which often can not be performed (e.g., $-a$ indicates that one should subtract a from zero, which is absurd) [6, p. 2].

Carnot also viewed a negative as a magnitude governed by a sign that was wrong. This could happen when in formulating a geometric question algebraically one did not know the relative position of two points on a line and was thereby forced to make an assumption of order which might turn out to be incorrect.

Carnot's idea of "correlated figures" had as one goal "to make the formulas of any system whatever immediately applicable to another which is correlative to it." One of his simple preliminary examples was to note that if C moved along BD toward D in Figure 2 the formula $AB^2 - AC^2 = BC^2 - 2BC \cdot CD$ became $AB^2 - AC^2 = BC^2 + 2BC \cdot CD$ as C passed to the other side of D. If his ideas of correlation and use of negatives to show "inverse" quantities were employed, one derivation (and one formula) served for both cases.

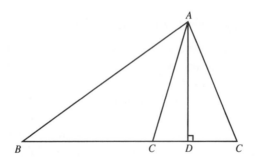

FIGURE 2

Thus, fruitful geometric analogies or applications of algebraic algorithms led Carnot to content himself with a not fully satisfactory justification of negatives and to move on to do geometry. In fact, he concluded his next book, *Géométrie de Position*, with the following paragraph:

> Although mathematical truths are all of a perfect certainty, they are not all based on the same degree of evidence. It is difficult to bring the basic notions to the desirable clarity; but one would be arrested in his progress at his first steps if, because certain fundamental principles remain enveloped in some obscurity, one refused to go on ahead. Carnot then quoted d'Alembert, *'Allez en avant, et la fois vous viendra.'* (Go forth, and faith will follow you.) [7, pp. 480–481]

Carnot's somewhat convoluted rationalizations were paralleled in a few years by the writings of the English algebraists who, perhaps even subconsciously, were struggling toward a logical, eventually axiomatic, basis for their subject. This is illustrated by a quotation from Augustus De Morgan, which may serve both to amuse and to communicate the uncertainties and error even of great mathematicians. It also points out the role of critics in stimulating the development of mathematics. In 1831 De Morgan wrote for the Society for the Diffusion of Useful Knowledge a book titled *On the Study and Difficulties of Mathematics*. On pages 71 and 72 he wrote that the teacher

> must recollect that the signs + and − are not quantities, but directions to add and subtract. Above all he must reject the definition still sometimes given of the quantity −a that it is less than nothing. ... It is astonishing that the human intellect should ever have tolerated such an absurdity as the idea of a quantity less than nothing, above all, that the notion should have outlived the belief in judicial astronomy and the existence of witches, either of which is ten thousand times more possible.

There are many historical lessons in even this sketchy account of negative numbers. These include the changing nature of and concern for proof and for structure; the shifting geographic locale of mathematical development; the role of applications and geometric models, a form of analogy; the continuity of some ancient concepts and their occasional misuse as they are uncritically applied in new situations. However, let us turn to the story of complex numbers for examples of how old algorithms and constructions when applied almost mechanically in new settings can lead to further new ideas.

Complex Numbers, Algorithms and Analogies

In his *Ars Magna* of 1545, Gerolamo Cardano proposed the following problem: "Find two numbers whose sum is ten and whose product is 40." He then remarked, "Obviously [*manifestum*] this is impossible, but nevertheless let's operate [*operabimur*]." This is an early, perhaps the earliest, use of the mathematician's "obvious," but the significant thing is that after labeling it as impossible he then said "nevertheless let's operate." His action was to apply the complete-the-square algorithm for solving quadratics, which he had just explained, to this "impossible" case which in modern symbols is $x(10 − x) = 40$. Cardano found the numbers to be $5 + \sqrt{−15}$ and $5 − \sqrt{−15}$, which he verified by adding and multiplying them. He then remarked that "this truly is sophisticated" and that if we were to go any further with it, it would require a new arithmetic which would be "as subtle as it would be useless." In other words, he drew back from constructing a new mathematical system using what to him would have been unreal elements. However, his colleague Rafael Bombelli in 1575 gave rules for operating with square roots of negatives and thereby validated Cardano's own process for solving cubic equations.

Skipping forward three centuries in this story, we come to Robert Argand who developed the geometric representation of complex numbers in an 1806 article in Gergonne's journal. He first discussed negative numbers using both mathematical and physical analogies to justify their existence and applicability, but noting that in physical situations they may be *imaginaire*. He then gave parallel arguments with reference to $\sqrt{−1}$. His key idea, represented by the diagram of Figure 3, involved the proportion $+1 : +x :: +x : −1$ and an extension of the Euclidean construction for a mean proportional. In the latter KE and KN would be mean proportionals between IK and KA. If now $KA = +1$ and $KI = −1$ we may regard KE and KN as representing $\sqrt{−1}$ and $−\sqrt{−1}$, mean proportionals between $+1$ and $−1$.

John Wallis had employed the idea that $\sqrt{−bc}$ is the mean proportional between $+b$ and $−c$ in seeking a geometric solution to quadratic equations. He had tried to rationalize negative areas using an analogy of land covered and uncovered by tides to parallel the concept of positive

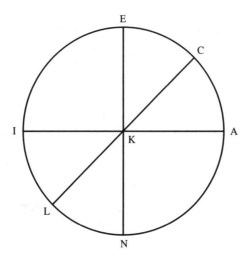

FIGURE 3

and negative numbers representing directions on a line. However, Argand's representation of complex numbers as radii of a unit circle, although related to the summation of horizontal and vertical components, was much more general and abstract. He extended the mean proportional idea, asserting "par une marche analogue" [1, p. 7] that one can insert mean proportionals between KA and KE by bisecting angle AKE to determine KC and KL.

Incidentally, Augustin Cauchy and William Rowan Hamilton both cited Argand's work with praise as they each presented alternate algebraic approaches to complex numbers. These new approaches revealed their dissatisfaction with the underlying foundation for the theory of functions of a complex variable which had been used so profitably by Euler, Cauchy himself, and many others.

Hamilton's construction of the complex numbers as couples has many ramifications for the development of modern algebra, but here we only point out the mixture of criticism and analogy which motivated Hamilton in 1823 (published 1837). He set out to give to algebra that which he felt it lacked, a sound basis such as he believed had long existed for geometry. Geometry, he felt, was firmly based on physics, on distance in particular. Thus the basis for algebra should also be a physical concept, namely, time. As a conclusion to what we could regard as an axiomatic development of a field defined by "couples" of real numbers he wrote:

> Were these definitions even altogether arbitrary, they would at least not contradict each other, nor the earlier principles of Algebra, and it would be possible to draw legitimate conclusions by rigorous mathematical reasoning, from premises thus arbitrarily assumed, but ... [as shown in an earlier essay, "Algebra as the Science of Pure Time"] these definitions are really not arbitrarily chosen, and that though others might have been assumed, no others would be equally proper. [12, p. 83]

Hamilton was a mathematician who went to an extreme to explain his modes of thought as well as his results. He made extensive use of the word "analogy." There are varied meanings for this word. In early mathematics it merely referred to a proportion, as "Napier's Analogies" in spherical trigonometry. In Hamilton's essay on "Algebra as the Science of Pure Time," if

A, B, C, D are moments, then the pairs (A, B) and (C, D) are the same only if both $A = C$ and $B = D$. However, they are analogous if $D - C = B - A$. Later, in his mathematical presentation, he began the development of powers of couples (leading to the equivalent of DeMoivre's theorem) by writing, "In like manner, if we write by analogy to the fractional powers of numbers $(c_1, c_2) = (b_1, b_2)^{u/v} \ldots$" [12, p. 93]. This use of "analogy" is close to ours: the use, in the invention and extension of mathematical systems, of similarities or parallels between new elements and/or operations and those existing in earlier mathematical systems or in the physical world. Of course, one of Hamilton's motivating drives, the plan to give algebra a firm basis in the physics of time analogous to the founding of geometry on physical distance, had already been destroyed, unbeknownst to him apparently, by the invention of non-Euclidean geometry!

Uncertainties and ambivalence as to the existence and utility of complex numbers date back to before Leibniz and continue on even after Hamilton. We noted that Descartes first used the term *imaginaire*. However, his view of imaginary number was limited. He explained that when the simultaneous solution of two equations leads to imaginary numbers one concluded that the corresponding curves did not intersect. He thus obtained real but limited information from his imaginary roots. This same ambivalence can be noted in Leibniz's comments. In the lengthy correspondence with Bernoulli, Leibniz not only defined a ratio as "imaginary" when it had no logarithm, but he also stated that such proportions as $+1 : -1 = -1 : +1$ may be used "with the same advantage and safety with which other inconceivable quantities are used." It is a little uncertain when by "imaginary" Leibniz meant elements or equations which did not exist or could not exist and when he used the term to name an existing mathematical entity. At one time, Leibniz employed complex numbers to factor $x^4 + y^4$ and used in connection with them such Latin adjectives as *elegans, mirabili, analyses miracula*. At another time he referred to a complex number as "an amphibious monster living in an ideal world in between being and not being" [14, p. 357].

In the following century De Morgan wrote at the outset of a chapter titled "Introduction of the unexplained symbol $\sqrt{-1}$":

> It is almost impossible to discredit Woodhouse's remark, 'Whether I have found a logic, by the rule of which operations with imaginary quantities are conducted, is not now the question, but surely this is evident that since they lead to right conclusions *they must have a logic!*.' [9, p. 41]

Woodhouse was probably referring to the derivation of previously known trigonometric identities via manipulation of complex numbers. De Morgan does this also, after warning the reader that "he must not think that they are demonstrated here, though they will have strong moral evidence in their favor." He promised proofs in the second book on "double algebra" [9, p. 41].

Alexander MacFarlane, in a biography of George Boole (1815–1964) commented on his use of $\sqrt{-1}$ in trigonometry by noting, "It is paradoxical to say ... we start from equations having a meaning and arrive at equations having a meaning by passing through equations which have no meaning" [15, p. 56], a comment reminiscent of the essay by Carnot cited earlier.

Somewhat the reverse of Woodhouse's viewpoint was expressed by Benjamin Peirce (1809–1880). He concluded a lecture in which he had shown $e^{\pi/2} = \sqrt[i]{i}$ by saying, "Gentlemen, that is surely true. It is absolutely paradoxical, we can't understand it, and we don't know what it means, but we have proved it, and therefore we know it must be the truth." [5, p. 6]

Analogies and Algorithms in Analysis and Geometry

There are many other areas in which one can find support for the thesis that analogies, algorithms, and the prodding of critics play an important role in stimulating and directing invention in mathematics. There are many more areas, unfortunately, where the mental steps have been concealed and, perhaps, correspondence destroyed. One obvious source is in the history of the operational calculus, a major segment of which is to be found in "The Calculus of Operations and the Rise of Abstract Algebra" by Elaine Koppelman [13]. An interested reader can refer to it to extend our examples of the use of "analogy" by tabulating the use of the term in both her writing and quotations, to add "arcane" to our collection of words for ill-understood and only partially believed mathematics [13, p. 171], to find notes on Peacock's principle of permanence of algebraic forms which seems to be closely related to our idea of analogy as well as to the ideas of Carnot cited earlier [13, p. 238], and to find Lagrange using principles which he said he did not fully understand [13, p. 172].

This excellent and extensive work could be extended backward to include Brook Taylor's use of the analogies between finite differences and fluxions and an associated juggling of symbols in his derivation of his series. It could be extended forward to include Heaviside's operational calculus and some of his amusing and rather sharp replies to critics to be found in volume II (1899) of his *Electromagnetic Theory*, and to the citation of a "principle of permanence of form" as "useful for operational application" of the calculus by Bourgin and Duffin [3, p. 489].

Another obvious place to look for fictitious elements, analogies, and permanence of form or continuity in geometry is the projective plane with its "ideal" and "imaginary" points and points at infinity. The person who began all this, Girard Desargues, was an architect whose famous theorem became known through being published as an appendix to Abraham Bosse's book on perspective, a subject whose theory had been substantially but succinctly expounded by Bosse's teacher and friend Desargues. In his case, one analogy backfired. Because among his intended readers were craftsmen such as stonecutters, Desargues invented an elaborate vocabulary which he thought would make geometric ideas more understandable to them. He used trunk (*tronc*), knot (*noeud*), branch (*rameau*), twig (*brim*), for line, point, ray, segment, etc. It is uncertain how well the craftsmen read his work, but geometers found the vocabulary confusing. Desargues' famous theorem (1636) states that if the lines joining pairs of corresponding vertices of two triangles meet at one point, then the intersection points of pairs of corresponding sides all lie on the same line. In order that the theorem be true without exception, he identified parallel lines with lines having infinitely distant intersections. Later, (1639), in a concise pamphlet on the sections of a cone, he defined the concept of an involution of points on a line. He discussed two cases, one of four points at finite distances and one in which the fourth was at an infinite distance. In this latter case he remarks,

> This is *incomprehensible* [italics mine] and seems at first to imply—since in this case the three points at finite distance give two equal parts, between which the middle point is situated—that the stump and the extreme knot are paired with the infinite distance. [17, p. 120]

Conclusion

A tongue-in-cheek article suggested that the history of science should be "X-rated" for young people because the facts of great discoveries are not always consistent with the ideal "scientific method" which it is hoped a study of science will teach. [4] No one has made a similar suggestion for the history of mathematics even though it too would deceive students were it to claim that all mathematics develops by rigorous deduction—even by deduction as accepted in its own time.

The conclusions which we have tried to illustrate are:

(1) Great and good mathematicians have said, "I don't understand," and have erred in conjectures, appraisals, and even proofs.

(2) Mathematics is a creative art, not merely a bag of algorithms nor a collection of necessary conclusions. However, the art of creating mathematics may include the use of analogies, algorithms, symbols, and assumed properties. These may all serve as a source and stimulus to creation and understanding, just as criticisms may prod reconsideration and rigorization.

(3) Communicating the humanness of great mathematicians and the methods and excitement with which mathematics is created is an important goal of instruction.

(4) Historical writing and teaching—with these goals in mind—is a major, maybe the sole, method by which these goals may be accomplished.

Bibliography

1. Argand, R.: *Essai sur une maniere de representer les quantites imaginaires dans les constructions géométriques* (New Printing). Paris: Albert Blanchard, 1971.
2. Barzun, Jacques: *Teacher in America*. New York: Doubleday and Co., 1954.
3. Bourgin, D. G. and Duffin, R. J.: "The Heaviside Operational Calculus," *American Journal of Mathematics* 59 (1937), 489.
4. Brush, Stephen G.: "Should the History of Science be Rated X?" *Science* 183 (Mar. 22, 1974), 1164.
5. Byerly, W. E.: "Reminiscences of Benjamin Peirce," *American Mathematical Monthly* 32 (1925), 5–7.
6. Carnot, Lazare: *De la correlation des figures de géométrie*. Paris: Duprat, 1801.
7. Carnot, Lazare: *Géométrie de Position*. Paris: Duprat, 1803.
8. Carnot, Lazare: "Digression sur la Nature des Quantities dites Negative," in *Memoire sur la relation qui existe entre les distances repectives de cing points quelconques pris dans l'espace, suivi d'un essai sur la theorie des transversales*. Paris: 1806.
9. De Morgan, Augustus: *Trigonometry and Double Algebra*. London: Taylor, 1849.
10. Gillespie, Charles C.: *Lazare Carnot, Savant*. Princeton: Princeton University Press, 1971.
11. Golan, Jonathan: Letter to the editor, *Notices of the American Mathematical Society* 22 (1975), 106.
12. Hamilton, William Rowan: "Theory of Conjugate Functions, or Algebraic Couples; with a Preliminary and Elementary Essay on Algebra as the Science of Pure Time," *Transactions of the Royal Irish Academy* 17 (1837), 293–422, reprinted in *The Mathematical Papers of William Rowan Hamilton*, vol. III (H. Halberstam and R. E. Ingram, eds.). Cambridge: Cambridge University Press, 1967, 3–96.
13. Koppelman, Elaine: "The Calculus of Operations and the Rise of Abstract Algebra," *Archive for History of Exact Sciences* 8 (1971-72), 155–242.
14. Leibniz, G. W.: *Werke*, Vol. V (C. I. Gerhardt, ed.). Hildesheim: Georg Olms Verlag, 1971. See Smith, D. E., *History of Mathematics*, Vol. II. (Boston: Ginn and Company, 1925), p. 264 for the original Latin and some further quotations.
15. MacFarlane, Alexander: *Lectures on Ten British Mathematicians of the Nineteenth Century*. New York: John Wiley, 1916.

16. Maseres, Francis: *Tracts on the Resolution of Affected Equations by Dr. Halley's, Mr. Raphson's, and Sir Isaac Newton's Methods of Approximating Roots*. London, 1800.

17. Taton, René: *L'Oeuvre mathématique de G. Desargues*. Paris: Presses Universitaires de France, 1951. (English version in Field, J. and Gray, J.: *The Geometrical Work of Girard Desargues*. New York: Springer, 1987.)

18. Vogel, Kurt: *Vorgriechische Mathematik II*. Hanover: Hermann Schroedel, 1959. Compare Otto Neugebauer, *The Exact Sciences in Antiquity* (Princeton: Princeton University Press, 1951), p. 108. with Neugebauer, *Mathematisch Keilschrifte-Texte*, (Berlin: Springer-Verlag, 1973), pp. 463, 474.

19. Wallis, John: *Treatise of Algebra both Historical and Practical*. London: Playford, 1685.

Archimedes

Using Problems from the History of Mathematics in Classroom Instruction

Frank J. Swetz

Many teachers believe that the history of mathematics, if incorporated into school lessons, can do much to enrich its teaching. If this enrichment is just the inclusion of more factual knowledge in an already crowded curriculum, the utility and appeal of historical materials for the classroom teacher is limited. Thus to include an historical note in a student's text on the life or work of a particular mathematician may shed an historical perspective on the content, but does it actually encourage learning or illuminate the concept being taught? The benefits of this practice can be debated.

A more direct approach to enriching mathematics instruction and the learning of mathematics through history is to have students solve some of the problems that interested early mathematicians. Such problems provide case studies of many contemporary topics encountered by students in class. They transport the reader back to the age when the problems were posed and illustrate the mathematical concerns of the period. Often, these are the same concerns that occupy modern day mathematics students. This simple realization of the continuity of mathematical concepts and processes over past centuries can help motivate learning. There can be a certain thrill and satisfaction for students in solving problems that originated centuries ago. In a sense, these problems allow the students to touch the past.

Strategies for Employing Historical Problems

In a recent issue of the *Mathematics Teacher* [12], a problem included in the "Reader Reflections" section prompted a series of follow-up discussions in later issues of the periodical. Obviously, the problem caught the attention of the readers. Interesting mathematical variants were proposed and solved by such readers as Lieske [5]. The problem was a good one because it was strikingly simple in its conception. (See Figure 1.)

> Given a right triangle with legs of length a and b and hypotenuse of length c, what is the length x of the side of the largest inscribed square utilizing the right angle as one of its vertices?

Interestingly, x is found to be the product of the legs divided by the sum of the legs, $x = ab/(a + b)$. This problem takes on even more intrigue when one learns that it was first known to be posed over 2000 years ago in China. It is the fifteenth problem in the ninth chapter of *Juizhang Suanshu* (*Nine Chapters on the Mathematical Art*) [8, p. 48]. Now, if this problem is examined in its historical context, it would also be noted that problem sixteen in this collection challenged the reader to find the radius r of the inscribed circle for a given right triangle [8, p. 49]. (See Figure 2(a).)

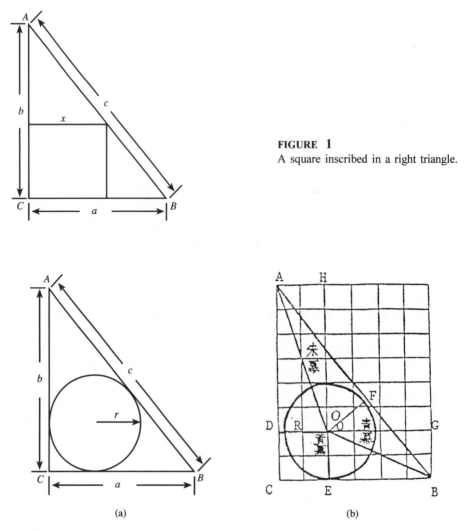

FIGURE 1
A square inscribed in a right triangle.

(a)   (b)

FIGURE 2
A variation of the problem in Figure 1 also appeared over 2000 years ago.

In a similar manner, a present-day teacher can offer this challenge to a secondary school mathematics class. Can they find the same answer for this problem that the ancient Chinese scribes found, namely,

$$r = \frac{ab}{a+b+c}.$$

The mathematical learning experience built around these problems can then be further extended by an examination and discussion of the Chinese geometric-algebraic solution scheme for the inscribed circle problem as shown in Figure 2(b). For historical interest, a facsimile of the actual Chinese solution illustration is presented including a graphical error in the circle construction. Reference lettering is a modern insertion; however, the use of a grid is a traditional Chinese procedure.

In Figure 2(b), the length of the hypotenuse AB is represented by c, the length of leg BC by a, and the length of leg CA by b. Then $2(\text{area } \triangle ABC) = ab$ or

$$ab = 4(\text{area } \triangle ADO) + 4(\text{area } \triangle BEO) + 2(\text{area rect. } DOEC).$$

These regions can be rearranged and will be equal in area to

area rect. $AHEC$ + area rect. $DGBC$ + area rect. $AHOD$ + area rect. $GBEO$.

Thus,

$$ab = br + ar + cr \qquad \text{and} \qquad r = \frac{ab}{a+b+c}.$$

Several problems in this same series are ingenious in their conception and require true perceptual and mathematical acuity on the part of the solver. To form a solution strategy, the use of a diagram is more than a suggestion, it is imperative. Problems 5 and 9 in the series are examples of this type:

A tree of height 20 feet has a circumference of 3 feet. There is an arrow-root vine which winds seven times around the tree and reaches the top. What is the length of the vine? (answer: 29 ft.) [8, p. 29]

A wooden log is encased in a wall. If we cut part of the wall away, so that only 1 inch of the log is still in the wall, the width of the exposed part measures 1 foot. What is the diameter of the log? (answer: 37″) [8, p. 36]

Comparison of solution techniques.   Historical sequences of problems from different time periods and cultures can be assembled and assigned as exercises for students to solve and compare. For example, facility in the use of the "Pythagorean Theorem" was valued in all societies:

A beam of length 30 feet stands against a wall. The upper end has slipped down a distance 6 feet. How far did the lower end move? (Babylonia, c. 1800 B.C.) (answer: 18 ft.) [10, p. 76]

The height of a wall is 10 feet. A pole of unknown length leans against the wall so that its top is even with the top of the wall. If the bottom of the pole is moved 1 foot farther from the wall, the pole will fall to the ground. What is the length of the pole? (China, 300 B.C.) (answer: 50.5 ft.) [8 p. 35]

There is a 30 foot ladder leaning against a 30 foot tower. If the foot of the ladder is 18 feet from the foot of the tower, how far from the top of the tower does it reach? (Italy, A.D. 1300) (answer: 6 ft.) [3, p. 392]

Involvement with such a series of problems exposes students to the universality of mathematical ideas and dispels the frequently held misconception of the uniqueness of discovery of certain mathematical theories. It is apparent from these problems that the relationship commonly known as the "Pythagorean Theorem" was developed and employed by many non-Greek people even before the time of Pythagoras. Later, more mathematically mature societies puzzled over indeterminate problems:

576 coins have been paid for the purchase of 78 bamboo poles. It is desired to calculate prices for large and small poles. How much is the price of each? (China, A.D. 300) (an answer: 48 small poles at 7 coins each; 30 large poles at 8 coins each) [11, p. 120]

A hundred bushels of grain are distributed among 100 persons in such a way that each man received 3 bushels, each woman 2 bushels, and each child half a bushel. How many men, women and children are there? (Europe, A.D. 775) (an answer: 20 men, 0 women and 80 children) [2, p. 251]

There were 63 equal piles of plantain fruit put together and 7 single fruits. They were divided evenly among 23 travelers. What is the number of fruits in each pile? (India, A.D. 850) (an answer: 5 fruit) [2, p. 253]

Using problems to support instruction. At times, in instructional situations, the use of historical problems reinforces and clarifies the concept being taught. For example, a discussion on the algebraic technique of "completing the square" to find the roots of a quadratic equation is enhanced by reference to actual Babylonian problems from which the concept originated, as discussed by McMillan. [6] Consider the translation of a problem from 2000 B.C.:

I have added the area and 2/3 the side of my square and it is 35/60. What is the side of my square?

If the side of the square in question is taken to be x, then the problem becomes one of solving the equation $x^2 + \frac{2}{3}x = \frac{35}{60}$. If we interpret this in the Babylonian geometric translation of the algebra, we get Figure 3. Thus by geometrically completing the square, the new area (Figure 3(b)) is seen to be $\frac{35}{60} + \frac{1}{9}$ and the algebraic solution situation becomes one of solving $(x + \frac{1}{3})^2 = \frac{35}{60} + \frac{1}{9}$. Modern students would then find that $x = \frac{1}{2}$ or $x = -\frac{7}{6}$, and the positive root agrees with the solution found by the Babylonians.

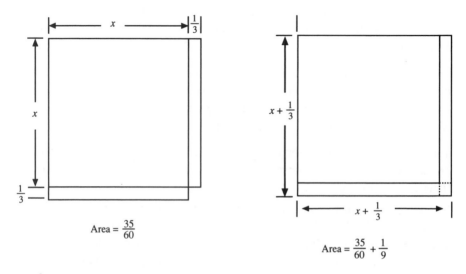

Area = $\frac{35}{60}$

Area = $\frac{35}{60} + \frac{1}{9}$

FIGURE 3
Area = $\frac{35}{60}$ Area = $\frac{35}{60} + \frac{1}{9}$

Discovery Situations

Students can be placed in the role of mathematical archaeologists and led to discoveries. Assume that a class has discussed numeration systems including the sexagesimal Babylonian cuneiform system. They are presented with a facsimile of the face of a clay tablet from 2000 B.C. now held in the Yale University Babylonian Collection (YBC 7289) [7]. (See Figure 4.) If recognized as a numeral, the symbols above the horizontal diagonal can be translated as 1, 24, 51, 10. Babylonian numerals possess a positional value. Thus, if the 1 is interpreted as one unit, the remaining numerals represent fractional components of the number represented, namely, $\frac{24}{60}$, $\frac{51}{60^2}$, and $\frac{10}{60^3}$. When the numbers are all combined, it is found that the mystery number is the sexagesimal equivalent of 1.414213 and that the cuneiform numeral below the diagonal designates the diagonal's length d in the given situation where the square has side length 30, i.e., $d = 42 + \frac{25}{60} + \frac{35}{60^2} = 42.4264$ in our number system. The $\sqrt{2}$ is approximated to the equivalent of six decimal places—impressive accuracy for 2000 B.C.! Two discoveries emanate from this example: the ancient Babylonians performed geometric constructions similar to those in use today; and they had a proficient computational technique for the extraction of square roots. Babylonian accuracy in square root extraction prompts further student investigation—just how did they do it?

Cultural, economic and sociological information can be obtained from the solutions to historical problems. For example, the height of the mast of an Egyptian ship for the historical period 250 B.C. can be found:

> If it is said to you, "Have a sailcloth made for the ships," and it is further said, "Allow 1000 cloth cubits (square cubits) for one sail and have the ratio of the height of the sail to its width as 1 to $1\frac{1}{2}$," then what is the height of the sail? (1 cubit = 50.8 cm). (an answer: 25.8 cubits) [11, p. 168]

Similarly, the size of a loaf of bread in 15th century Venice can be deduced:

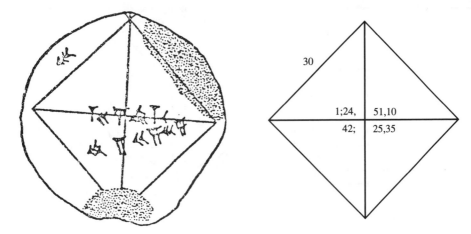

FIGURE 4
The translation of a Babylonian tablet.

> When a bushel of wheat is worth 8 lire, the bakers make a loaf of bread weighing 6 ounces; required the number of ounces in the weight of a loaf when it is worth 5 lire a bushel. (answer: $9\frac{3}{5}$ ounces) [9, p. 131]

or the hourly wages (of a 12-hour workday) for a man in post-Civil War America can be determined:

> A gentleman received $4 a day for his labor, and pays $8 a week for his board; at the expiration of 10 weeks he has saved $144; required the number of idle and working days. (answer: 14 idle days and 56 working days) [1, p. 85]

Illustrating the growth of mathematical proficiency.   Sets of study problems that demonstrate a transition in mathematical thought can be given to students. Mathematical problems from early periods in a society are usually empirically based and task oriented: taxes had to be computed, dikes and walls constructed, grain stored, and the price of merchandise reckoned. Later in the history of the society, one often finds the same problem situations modified to promote mathematical techniques. Thus, in the China of 300 B.C., land surveying concerns prompted the formulation and solution of problems involving towns and landmarks. Typical of such problems is that of a square-walled town [8, p. 55]. Several measurements outside the walls are given and the width of the town is desired (see Figure 5(a)). The computational situation gives rise to a quadratic equation, $x^2 + 34x - 71000 = 0$, for which the positive root $x = 250$ is obtained. By the thirteenth century, such problems had passed out of the realm of surveyors and became the basis for rather powerful algebraic demonstrations [4]. This transition is exemplified by the situation of the round-walled town depicted in Figure 5(b).

  Analysis of the Chinese solution processes for the round-wall town problem described in Figure 5(b) indicates the solution of a tenth degree equation, $x^{10} + 15x^8 + 72x^6 - 864x^4 -$

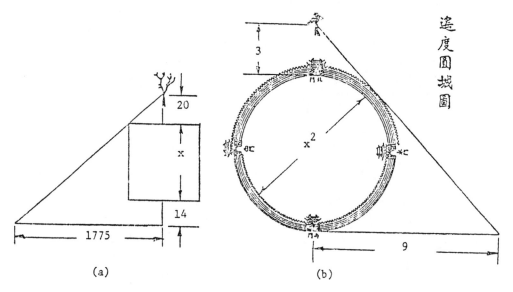

(a)   (b)

FIGURE 5
Problems of the square-walled and round-walled towns

$11,664x^2 - 34,922 = 0$. A root of $x = 3$ was found for the solution. The problem's author, Qin Jiushao, was exhibiting much mathematical bravado by letting the diameter of his town be equivalent to x^2 and demonstrating the power of his solution technique. If one lets the diameter of the town be x instead of x^2, a much simpler equation for the problem emerges: $x^4 + 6x^3 + 9x^2 - 972x - 2916 = 0$.

Similar situations also abound in European mathematical literature. Problems of partnership were a primary concern in Renaissance Italy of the fifteenth century. A problem from the *Treviso Arithmetic* of 1478 is typical:

> Three men, Tomasso, Domengo and Nicolo entered into partnership. Tomasso put in 760 ducats on the first day of January, 1472, and on the first day of April took out 200 ducats. Domengo put in 616 ducats on the first day of February, 1472, and on the first day of June took out 96 ducats. Nicolo put in 892 ducats on the first day of February, 1472, and on the first day of March took out 252 ducats. On the first day of January, 1475, they found that they had gained 3168 ducats, $13\frac{1}{2}$ grossi. Required the share of each, so that no one shall be cheated. (1 ducat = 24 grossi) (answer: Tomasso, 1061 ducats, 1g; Domengo 949 ducats, $19\frac{1}{2}$g; Nicolo 1157 ducats, 17g.) [9, p. 146]

This problem can be solved by the use of simple proportions, but a century later rather imaginative partnership problems served as a basis for complex algebraic computation. Consider an example from Cardano's *Ars Magna* (1545), chapter 26:

> Four men form an organization. The first deposits a given quantity of aurei; the second deposits the fourth power of one-tenth of the first; the third, five times the square of one-tenth the first; and the fourth, 5. Let the sum of the first and second equal the sum of the third and fourth. How much did each deposit? (answer: 7.4907, 0.3148, 2.8055 and 5 respectively) [2, p. 323]

Of course, problems with an historical allure can be assigned merely to demonstrate the mathematical capabilities and skill of our predecessors. A Babylonian tablet of 2000 B.C. requires the reader to find the radius of a circle in which is inscribed an isosceles triangle of sides 50, 50, and 60 [2, p. 82]. An Egyptian counterpart of A.D. 150 asks the reader to find the area of a circle which circumscribes an equilateral triangle whose side is 12 units [11, p. 174].

In its Babylonian context, the problem of Figure 6(a) is solved by computing the length of the altitude drawn to the 60 unit side. (The length of the altitude is 40 units.) Radii r

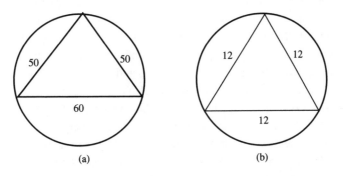

(a)   (b)

FIGURE 6
Ancient mathematicians worked problems involving inscribed triangles

are then visualized extending to the three vertices of the triangle and the relationship $r^2 = (30)^2 + (40 - r)^2$ is established. The radius r is found to be 31.25 units. The scribe solving the second century Egyptian counterpart to this problem first used the Pythagorean proposition to find the altitude of the triangle, $\sqrt{108}$ units, and then applied this result to obtain the area of the triangle, $6\sqrt{108}$ square units. Next, he employed a trapezoidal approximation for the area of the remaining circular segments, namely, $A = \frac{1}{2}(b_1 + b_2)h$ where $b_1 = h$. He realized that the length of the sagitta of the segment, which is also the height h for the trapezoidal approximation, was equal to $\frac{\sqrt{108}}{3}$. Thus, $A = \frac{1}{2}(12 + \frac{\sqrt{108}}{3})\frac{\sqrt{108}}{3}$. He then multiplied the obtained area by 3, since there are three congruent segments, and added the result to the triangle's area to obtain a total area for the circle. It appears that the Egyptian calculations were performed using the Babylonian sexagesimal system and the resulting sexagesimal fractions converted into the Egyptian format of sums of unit fractions. The scribe's computation resulted in an answer of $143 + \frac{1}{10} + \frac{1}{20}$. How accurate is this result? Is the scribe's confidence in the trapezoidal approximation for a circular segment well-founded? This approximation technique was used by several ancient peoples. (See Figure 7.)

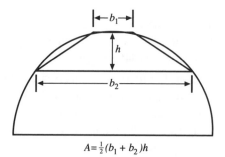

$$A = \tfrac{1}{2}(b_1 + b_2)h$$

FIGURE 7
Trapezoidal approximation of a circular segment

Conclusion

It is not uncommon when visiting a large art museum to find groups of school children accompanied by their teachers admiring and studying paintings, sculptures, and other works of art from centuries past. Their teachers or guides draw the student's attention to the artist's techniques: mastery of color, tone, interplay of light and shadow, and even the significance of the scenes described. Under this directed scrutiny, a painting or statue becomes a testimony to its creator's genius and offers some understanding of the period in which the artist lived and functioned. Learning takes place. This learning is both cognitive and affective. So, too, are the mathematics problems of history. They, in a sense, are intellectual and pedagogical works of art that testify to an expression of human genius. But, unlike the museum pieces, these problems can actually be possessed by the viewers through a participation in the solution processes. Questions originating hundreds or even thousands of years ago can be understood and answered in today's classrooms. What a dramatic realization that is!

Historical problems and problem solving, as a topic in itself, can be the focus of a lesson, but it is probably more effective if such problems are broadly dispersed within classroom drills and homework assignments. Teachers who like to specify "a problem of the week" will find that historical problems fit the task nicely. Ample supplies of historical problems can be found in some survey books on the history of mathematics. A suggested list of problem sources is included as an appendix. The seeking out and employing of historical mathematical problems in classroom instruction is a rewarding and enriching experience of which all mathematics teachers should partake.

For Further Consideration

These exercises were chosen to reinforce and expand upon the ideas expressed above. Can you see the connections?

1. Find a pair of similar mathematical problems: one from a contemporary school text and the other from several hundred years ago. How do their established solution techniques differ, if they do?

2. The following problems are from three contemporary but different ancient societies. Attempt to solve these problems. What do their contents tell us about life and the role of mathematics at that time?

> I have two fields of grain. From the first field I harvest 2/3 bushel of grain per unit area, from the second 1/2 bushel per unit area. The yield of the first field exceeds the second by 500 bushels. The total area of the two fields together is 30 square units. What is the area of each field? (Babylonia)

> A cow, a horse and a goat were in a wheat field and consumed some stalks of wheat. Damages of 5 baskets of grain were asked by the wheat field's owner. If the goat ate one-half the number of stalks eaten by the horse, and the horse ate one-half of what was eaten by the cow, how much should be paid by the owners of the goat, horse and cow, respectively? (China)

> Suppose a scribe says to you, four overseers have drawn 100 great quadruple hekats of grain and their gangs consist of 12, 8, 6, and 4 men. How much does each overseer receive? (Egypt)

3. What do you think the intent of the author was for each of the following problems?

> Given a field of width the sum of: 1, 1/2, 1/3, 1/4, 1/5, 1/6, 1/7, 1/8, 1/9, 1/10, 1/11, and 1/12 pu. It is known that the area of the field is 1 mu. What is the length of the field? (A pu is a double pace; 1 mu = 240 square pu) (*Jiuzhang suanshu*, China 300 B.C.)

> Three hundred pigs are to be killed for a feast. They are to be killed in three batches on three successive days with an odd number of pigs in each batch. How can this be accomplished? (Alcuin of York, c. A.D. 775)

There is a tree with 100 branches, each branch has 100 nests, each nest 100 eggs, each egg 100 birds. How many branches, nests, eggs and birds are there? (*Liber Abaci*, Italy, 1202)

4.   After solving the given problem, what conclusions might you reach concerning the state of mathematics in China in 300 B.C., the period when the problem originated?

Three sheafs of good crop, 2 sheafs of mediocre crop, and 1 sheaf of bad crop produce 39 dou of grain. Two sheafs of good, 3 of mediocre and 1 of bad produce 34 dou. One sheaf of good, 2 of mediocre, and 3 of bad produce 26 dou. What is the yield of a sheaf of good crop, mediocre crop and bad crop?

5.   Identify, perhaps with the help of library resources, the origins and mathematical significance of these problems:

We have a number of things, but we do not know exactly how many. If we count them by threes we have two left over. If we count them by fives we have three left over. If we count them by sevens we have two left over. How many things are there?

Fifteen Christians and fifteen Turks are on a ship which is in danger of sinking. Passengers must be sacrificed in order to save the ship. The captain is a Christian and wishes to sacrifice the Turks. How can he arrange the passengers in a circle so that in counting around, every fifteenth count should mark a Turk?

6.   In his *A History of Geometrical Methods,* Julian Coolidge discusses the topic of cultural interchange (p. 17) and cites two problems as an example. One problem is from the Chinese *Jiuzhang suanshu* of about 300 B.C. The other is from the works of the Hindu mathematician Bhaskara (c. 1150). What conclusion would you draw in the comparison of these two mathematics problems?

There grows in the middle of a pond 10 feet square a reed which projects one foot out of the water. When it is drawn down it just reaches the edge of the pond. How deep is the water? (China, 300 B.C.)

In a certain lake, swarming with red geese, the tip of a bud of a lotus was seen a span above the surface of the water. Forced by the wind it gradually advanced, and was submerged at a distance of two cubits. Compute quickly, mathematician, the depth of the pond. (India, c. 1150)

7.   In Archimedes' *Book of Lemmas* (a collection of 13 geometrical propositions), he introduced a figure that, due to its shape, has historically been known as the arbelos, or "shoemakers knife." If, in a given semicircle with radius R and diameter AB, two semicircles with radii r_1 and r_2 ($r_1 + r_2 = R$) are constructed on diameter AB so that they meet at point C on AB, then the region bounded by the three circumferences is called an arbelos. (See Figure 8.)

Prove that the length of arc AC plus the length of arc CB equals the length of arc AB.

Prove: If a perpendicular line is constructed from C intersecting the arc AB at point P, then PC is the diameter of a circle whose area equals that of the arbelos.

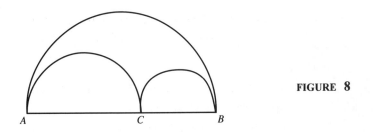

FIGURE 8

Complete the lower half of the circle with diameter AB, let the midpoint of the arc of this lower semicircle be Q, the midpoint of arc AC be M and the midpoint of arc CB be N. Prove that the area of the quadrilateral $MCNQ$ is equal to $r_1^2 + r_2^2$.

8. Archimedes' *Book of Lemmas* also contains a geometrical figure called the salinon, or "salt cellar." Construct a semicircle with diameter AB. On AB mark the distances AC and DB where $AC = DB$ and $AC + DB < AB$. Then describe semicircles, with AC and AB as diameters, on the same side of AB as the original semicircle; also describe a semicircle, with CD as the diameter, on the other side of the original semicircle. The region bounded by the semicircles is the salinon. (See Figure 9.)

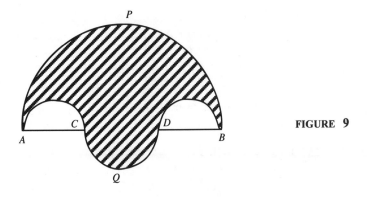

FIGURE 9

Let P be the center of arc AB and Q be the center of arc CD. Prove that the area of the salinon is equal to the area of a circle with PQ as a diameter.

Once again construct semicircle AB, construct semicircle M below AB and semicircle N above AB as shown in Figure 10. Let AP be tangent to circle N. Show that the area of the shaded region is equal to the circle which has AP as its diameter.

9. Proposition 5 of Book I of Euclid's *Elements* states:

In isosceles triangles the angles at the base are equal to one another, and, if the equal straight lines be produced further, the angles under the base will be equal to one another.

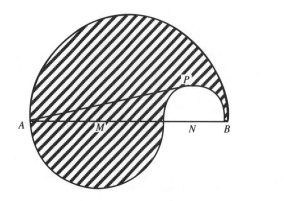

FIGURE 10

This proposition represented the usual limit of geometry instruction in the Middle Ages. Its understanding and proof formed an intellectual and mathematical "bridge" across which "fools" could not hope to pass, and therefore, it became known as the *pons asinorum*, or "bridge of fools." Figure 11 is from Isaac Barrow's edition of Euclid's *Elements* (1665). Be a mathematical detective and decipher the proof, that is, follow the proof as indicated and write out the steps in modern notation. It doesn't matter if you can't read Latin! Compare this proof of the equality of the base angles of an isosceles triangle with a modern proof.

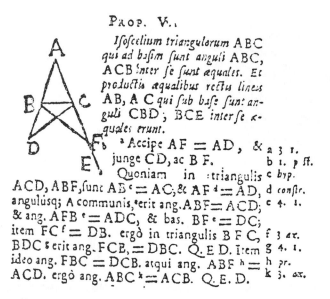

FIGURE 11
From Isaac Barrow's 1665 edition of Euclid's *Elements*.

10. The "three classical problems of antiquity" that have fascinated mathematicians for centuries are:

  1. The duplication of a given cube.

2. The trisection of an angle.

3. The quadrature of the circle.

Research these problems. Go to the library and find proofs that these problems cannot be solved using the instruments of classical geometry, i.e., the uncalibrated compass and the straightedge.

11. In 1736 the Swiss mathematician Leonhard Euler (1707–1783) solved a problem concerning a configuration of bridges in the town of Königsberg. This problem became known as the "Königsberg bridge problem." Find out what this problem was, find Euler's solution, and discuss its implications for modern day mathematics teaching.

Bibliography

1. Brooks, Edward: *The New Normal Arithmetic*. Philadelphia: Christopher Sower, 1873.
2. Burton, David M.: *The History of Mathematics: An Introduction*. Boston: Allyn and Bacon, Inc., 1985.
3. Cowley, Elizabeth Buchanan: "An Italian Manuscript." *Vassar Medieval Studies*, C. F. Fiske (ed.). New Haven: Yale University Press, 1932, 379–405.
4. Libbrecht, Ulrich: *Chinese Mathematics in the Thirteenth Century: The Shu-shu Chiu-Chang of Ch'in Chiu-shao*. Cambridge: The MIT Press, 1973.
5. Lieske, G. Spencer: "Right Triangles II," *The Mathematics Teacher* 78 (1985), 498–99.
6. McMillan, Robert: "Babylonian Quadratics," *The Mathematics Teacher* 77 (1984), 63–65.
7. Resnikoff, H. L. and Wells, R. O.: *Mathematics in Civilization*. New York: Holt, Rinehart and Winston, Inc., 1973.
8. Swetz, Frank J. and Kao, T. I.: *Was Pythagoras Chinese? An Examination of Right Triangle Theory in Ancient China*. Reston, VA: National Council of Teachers of Mathematics, 1977.
9. Swetz, Frank: *Capitalism and Arithmetic: The New Math of the Fifteenth Century*. Peru, IL: Open Court Publishing Co., 1987.
10. van der Waerden, B. L.: *Science Awakening*. New York: Wiley, 1963.
11. van der Waerden, B. L.: *Geometry and Algebra in Ancient Civilizations*. New York: Springer-Verlag, 1983.
12. Wikenfeld, Morris: "Right Triangle Relationships," *The Mathematics Teacher* 78 (1985), 12.

Suggested Problem Sources

1. Brown, R. Gene and Johnson, K.: *Paciolo on Accounting*. New York: McGraw-Hill, 1963.
2. Cardano, Girolamo: *The Great Art, or the Rules of Algebra* (Richard Witmer, trans.). Cambridge: MIT Press, 1968.
3. Chace, Arnold Buffum: *The Rhind Mathematical Papyrus*. Washington: The National Council of Teachers of Mathematics, 1978 (reprint of 1927, 1929 editions).
4. Colebrooke, Henry Thomas: *Algebra with Arithmetic and Mensuration from the Sanscrit of Brahmegupta and Bhascara*. London: John Marry, 1817 (available in reprint from University Microfilms).
5. Eves, Howard: *An Introduction to the History of Mathematics*. Philadelphia: Saunders Publishing, 1983.
6. Gies, Joseph and Francis: *Leonardo of Pisa and the New Math of the Middle Ages*. New York: Thomas Y. Crowell Co., 1969.
7. Gittleman, Arthur: *History of Mathematics*. Columbus, OH: Charles E. Merrill Publishing Co., 1975.
8. Neugebauer, O. and Sachs, A.: *Mathematical Cuneiform Texts*. New Haven: American Oriental Society, 1945.
9. Parker, R. A.: *Demotic Mathematical Papyri*. Providence: Brown University Press, 1972.
10. Sanford, Vera: *The Historical Significance of Certain Standard Problems in Algebra*. New York: Teachers College Press, 1927.

11. Smith, D. E.: "Mathematical Problems in Relation to the History of Economics and Commerce," *American Mathematical Monthly* 24 (1918), 221–223.
12. Smith, D. E.: *History of Mathematics* 2 vols. New York: Dover Publications 1958 (reprint of 1925 edition).
13. Swetz, Frank J.: *The Sea Island Mathematical Manual: Surveying and Mathematics in Ancient China*. University Park, PA: Pennsylvania State University Press, 1992.

Revisiting the History of Logarithms

John Fauvel

My father was *l'ingegné* (the engineer), with his pockets always bulging with books and known to all the pork butchers because he checked with his logarithmic ruler the multiplication for the prosciutto purchase. [Primo Levi, 12, p. 19]

The subject of logarithms, like the notorious "asses' bridge" in Euclid (*Elements* I,5) for an earlier generation, seems to mark an intellectual rite of passage: before going over there is a sense of unfathomable mystery, even danger, ahead; afterwards there is still some wonder and perplexity at just what it is one has learned. Some stumble at the hurdle and feel forever excluded, like the lame boy of Hamelin; others press on and on and still do not come to the end of what is undeniably a paradigm of the rich complexity of mathematical concerns.

All this remains true, even now that a traditional calculational justification for studying logarithms has passed into history; Primo Levi's father was from one of the last generations to go with their slide rules to the butcher's. For example, I was talking with a bright but non-mathematical friend and said that I was thinking about how to teach logarithms. "Whatever for?" he said, "Surely no one needs to learn about those any more, now that we have calculators and computers." "No, they're still important," I said confidently. As he continued to look somewhat dubious, I launched into a rather unwise discussion of what happens when you try to integrate a hyperbola, prodding his calculus memories of twenty years back.

I hope that thinking through various ways of recapturing the history immanent in logarithms, and looking at the problems that gave rise to them, may help to provide a better response next time to my friend's question, as well as providing some suggestive classroom activities for a variety of ages and contexts.

Babylonian Logarithms?

There is a wealth of accessible and interesting problems and themes on old Babylonian tablets. How one discusses these materials with students depends on their stage of development, the available time, and so on. Thus it is profitable and fun at one stage to play around with sexagesimal place-value notation, and of course with repetitive two-symbol numerals, and at a later stage with the Babylonian solutions to problems we solve with quadratic equations [17].

There is an Old Babylonian tablet (c. 1800 B.C.) which can provide a fruitful source for investigations of the pattern that in a sense underlies logarithms [14, pp. 35–36]. The numbers on one side of the tablet are arranged like this:

15	2
30	4
45	8
1	16

What can these numbers be doing? Can the students see a pattern? The next pattern inscribed on the tablet may be easier to spot:

2	1
4	2
8	3
16	4
32	5
64	6

But even so it needs some sophistication to recognize an arithmetic progression in the right-hand column, rather than a simple numerical register. There is a similar column on the tablet Plimpton 322 [2], from 1 to 15, which *does* seem to be just a line-numbering. (There is a fruitful reflection about number here—an example from later in the history makes the point also: the distinction between Napier's artificial and natural numbers lies not in the numbers, but in their place in his scheme of things.) The students who tackle this exercise will need to have had some experience with number patterns, so that they understand what they are supposed to be doing, as well as being able to recognize these and give them a name.

One advantage of taking an actual Babylonian example is that it forces the question: "What was the purpose of this?" We may not be able to provide a sure answer, but that matters less than the importance of rooting the number pattern in some real context. Neugebauer and Sachs, the translation editors of the tablet, suggested that its contents have to do with computation of interest. Presumably the idea is that if you borrow a sum and agree to double repayments for every month borrowed (i.e., 100% compound interest), the lower table shows what you would owe after so many months. This rate of interest may seem implausibly high, but it seems to be implied in article 101 of the contemporary Code of Hammurabi:

100. He shall write down the interest on the money, as much as he has obtained, and he shall reckon its days, and he shall make returns to his merchant.

101. If he does not meet with success where he goes, the agents shall double the amount of money obtained, and he shall pay it to the merchant. [11]

Nevertheless, this doesn't preclude classroom discussion of the morality of usury (see, for example [9]). Neugebauer and Sachs' discussion has other features that teachers may be interested in coming back to later as a discussion topic, once students' ideas on logarithms are more firmly rooted. They commented on the tablet as follows:

We now have an Old-Babylonian tablet which answers the question: to what power must a certain number be raised in order to yield a given number? This problem is identical with finding the logarithm to the base *a* of a given number ... The new "logarithmic" tables ... exhibit a knowledge of the basic laws of operating with exponents. In a comparison with our concept of logarithm, the only missing elements are the selection of a common base and the tabulation for constant intervals, which would be needed if the tables were to be used for practical computations in general. It is accordingly clear that the Old-Babylonian mathematicians were very close to an important discovery but failed to take the final, essential step. [14, pp. 35–6]

This style of historiography is ambitious and perhaps out of favor nowadays; the comparison with "our concept of logarithm" is certainly a heady one.

Further Comparisons of Arithmetic and Geometric Progressions

The idea for students to get used to now is of operating on two parallel patterns, and comparing the results. This might be possible a couple of years or more before one would normally be introducing the concept of "logarithm." For some students, exploring Archimedes' *The Sand-Reckoner* (mid-third century B.C.) might be fun; his problem-situation—how many grains of sand fill the universe?—is thought-provoking, and people can enjoy working out how to work out estimates of magnitude. This is now a more crucial skill than ever, with the advent of calculator manipulation; estimations and orders of magnitude are critical in learning to recognize absurd results. In this theorem, Archimedes proves the basic logarithm property, essentially, for discrete series:

> If there be any number of terms of a series in continued proportion, say
>
> $$A_1, \ A_2, \ A_3, \ldots A_m, \ldots A_n, \ldots A_{m+n-1}, \ldots$$
>
> of which $A_1 = 1$, $A_2 = 10$, and if any two terms as A_m, A_n be taken and multiplied, the product $A_m A_n$ will be a term in the same series and will be as many terms distant from A_n as A_m is distant from A_1: also it will be distant from A_1 by a number of terms less by one than the sum of the numbers of terms by which A_m and A_n respectively are distant from A_1. [1]

But this is put rather abstractly, and may be a little hard to motivate except in a hand-waving kind of way. For many students the assertion in a simpler form tied down to specific numbers might be easier to see, such as is found in the works of the French Renaissance writer Nicholas Chuquet.

In 1484, Chuquet extended the Babylonian progressions, as it were, to the twentieth term, and explicitly noted that addition in the arithmetic progression corresponded to multiplication in the geometric one.

Numbers	Denomination	Numbers	Denomination	Numbers	Denomination
1	0	128	7	16384	14
2	1	256	8	32768	15
4	2	512	9	65536	16
8	3	1024	10	131072	17
16	4	2048	11	262144	18
32	5	4096	12	524288	19
64	6	8192	13	1048576	20

[W]hoever multiplies 2^0 by 2^2, it comes to 4 which is a second number. Thus the multiplication amounts to 4^2. [Note that the raised 2 is not a power of 4 but registers the "denomination" of the number in that place; we would write the expression as $4x^2$.] For 2 multiplied by 2 makes 4 and adding the denominations, that is, it comes to second terms. Likewise whoever multiplies 2^1 by 4^2, it comes to 8^3. For 2 multiplied by 4 and 1 added with 2 makes 8^3. And thus whoever multiplies first terms by second terms, it comes to third terms. Also, whoever multiplies 4^2 by 4^2, it comes to 16 which is a fourth number, and for this reason whoever multiplies second terms by second terms, it comes to fourth terms. Likewise ... [8]

While this passage is useful for consolidating the idea, it is noticeable that Chuquet does not produce any reason for his interest—there is no situation that the student can explore beyond the world of number patterns—and the discussion is so leaden that teachers would do better to paraphrase the ideas and explore them away from Chuquet's words. Students could be asked to see whether it works for powers of 3, and other geometric progressions, generalize the observations, and so on. At any rate, by this stage (of the student's studies) the sense of observing operations in parallel progressions, and using them to some purpose, should be clear.

The question arising is—with hindsight—can this idea be made to work for *any* pair of numbers that you want to multiply? But, of course, just to ask this question shows a considerable depth of insight and lateral thinking; it involves seeing a need and also imagining an unprecedented way of satisfying it. This could be an opportunity for teachers to discuss, with their class, lateral thinking and flexibility of thought as qualities to be fostered in their education, and whether learning about approaches such as specializing and generalizing (see [13]) helps promote these qualities.

Interlude: Why Handle Big Numbers?

There is a paradox in trying to teach techniques devised solely for difficult problems. One tends to choose simple examples, for the laudable reason that they're easier to follow; but of course those are precisely non-examples of the real problems for which the technique was devised. While well-intentioned, choosing examples which are unrealistically simple can mislead students, through making it hard to see why anyone would have bothered to devise techniques for such problems; hard, indeed, to see what the problem was. There's a sense in which past achievements are diminished, if not rendered incomprehensible, through didactic parody.

Does this matter? Since the purpose here is not to teach past mathematics for its own sake, but to foster better understanding in students' minds, distortions matter only if they interfere with that understanding. Students will differ in the extent to which they value being motivated through an imaginative re-entering of the difficulties that called forth some mathematical advance. Here I think one just wants to stay alert to the possibility of subverting understanding by trivializing the past, and ensure that at some stage a sufficiently complicated calculation is shown or done.

A further difficulty today is that numbers having lots of digits are not thought of as presenting problems that a calculator or computer cannot deal with. My friend's perception is right, to the extent that the original motivation for logarithms has disappeared from everyday experience, which on the face of it presents motivational difficulties in teaching them now. It may be tempting for the teacher to leap straight in with a definition that cuts out all historical developments, in effect following the great eighteenth century mathematician Leonhard Euler, as in this example taken at random from a modern textbook:

> The logarithm of a number to a given base, assumed positive, is the index of the power to which the base must be raised in order to equal the number. [4, p. 316]

But this kind of thing may not bring enlightenment to all students, and at least two sorts of insight are lost by cutting the Gordian knot in this way: the more deeply rooted understanding of what is going on, which is encouraged by exploring progressions; and learning from the marvelous transition of logarithms from practical to theoretical tool, which I open up for discussion later.

Handling Astronomical Numbers

By the end of the sixteenth century, the most efficient way of multiplying two large numbers was to use the technique of prosthaphaeresis ([15], [16]): that is, to convert the problem into an addition or subtraction problem via one of the trigonometric identities

$$2\cos A\cos B = \cos(A-B) + \cos(A+B)$$

$$2\sin A\sin B = \cos(A-B) - \cos(A+B)$$

$$2\sin A\cos B = \sin(A+B) + \sin(A-B)$$

$$2\cos A\sin B = \sin(A+B) - \sin(A-B)$$

Exploring this process offers several interesting opportunities for enlarging students' trigonometrical understanding. For example, the relation between sine, cosine, and circle can be usefully reviewed. Reverting from ratios to a sixteenth-century consideration of sines and cosines as lengths (Figure 1) can help to demystify the subject: it certainly helps to make better sense of the *words* co[mplementary] sine, tangent and secant. Then the understanding can be imparted that with a large radius of one's choice, the trigonometric values can be both integers and as accurate as one wants (or can calculate—quite another problem!). And from this, the understanding is not far away that the prosthaphaeresis technique makes arbitrary multiplications possible—it's not just confined to numbers between −1 and 1 as students might at first guess.

The most off-putting aspect of prosthaphaeresis appears to be the name; but children enjoy large funny words, so perhaps naming the school cat, goldfish or gerbil "Prosthaphaeresis" would serve to familiarize the word and humanize the concept. At all events, the technique provides a good way of investigating the idea of doing multiplication by adding, as a prelude to that aspect of logarithms. An attempt to calculate using prosthaphaeresis would be a useful classroom introduction to the idea of arithmetical modelling of calculation, and exploring the problems—can you divide? can you take roots?—would make for an interesting project that would place Napier's achievement in perspective. A more general question arises out of this

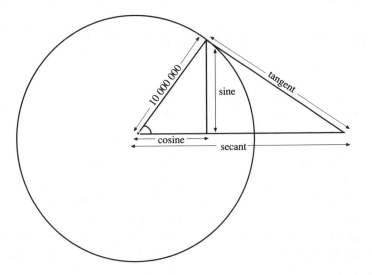

FIGURE 1

topic, suitable for a project for younger children: who needed big numbers anyway? Who needs them now? What kinds of numbers do people come into contact with in their everyday lives, and what do they do about them?

Motion in Geometry

Another example of a problem that has been perhaps over-trivialized, since the invention of a more sophisticated mathematics to deal with it, is how one uses or deals with movement mathematically. Its history since early Greek times was one of subtle concepts and sophisticated problems, understandable anxieties and prohibitions, paradox and avoidance. Have these problems been solved through the development of the calculus over the last three centuries, or just swept under the carpet?

There is an opportunity here to discuss motion in a pre-calculus way, which would have heuristic benefit in preparation for a later introduction to calculus, by considering how one might try to treat motion. Napier took small line segments, for putting his basic logarithm definition (which involved motion) into a mode one could calculate logarithms through. Napier's own handling may be too complicated to be useful to pursue here, but see the following paper by Victor Katz, showing the feasibility of a quasi-Napierian approach for an older age group.

At an earlier level, however, just getting students to see the difference in complexity between the measurement of distances and of speeds, say, is certainly a start. But another perspective which is increasingly coming back into junior education is of seeing motion not as a problem which needs re-expressing in static terms, but as an intrinsic quality which can be used to conceptualize things mathematically. Thinking, for example, of an angle in terms of a line being swept around from a fixed point [10], and of a line as the path of a moving point, are very old conceptions which, for reasons partly to do with computer displays, are coming back into mathematics education.

Napier's Logarithms

We can see Napier's logarithms as a confluence of the three conceptual streams mentioned:
- comparing arithmetic and geometric progressions,
- doing multiplication by means of addition,
- using the geometry of motion.

Students who have explored some or all of these ideas, as suggested, are then in a stronger position to take in quite rapidly, at the stage when it is appropriate, either an account of the Napierian approach (see the paper by Katz in this volume) or a fairly conventional definition of logarithms. The latter commonly follows Euler, as will be mentioned later. While Napier's own definition of logarithms, and his means for their calculation, have features which may be unhelpful or distracting if dwelt upon too much at this level, there are some jolly stories to be told for general educational purposes about Napier and the reception of logarithms—which indeed it is a schoolchild's right to know, as much as about the death of Socrates or the life of Michelangelo. Filling in the cultural background in this way should be recognized as one end of the spectrum of pedagogic use of the history of mathematics. The story of logarithms could and perhaps should be so enlivened, with a minimum of Napierian technicalities. A particular technicality which it may be wise to avoid at this level concerns the constructions of logarithms.

Although it is helpful to point out that logarithms themselves needed computation, and that this was by no means easy—in the absence of logarithm tables—practical demonstration of the difficulties might be unrewarding.

Today's students are perhaps advantaged over their predecessors, for the trees of log tables and their use can be discarded in favour of the wood of the significance of the concept of logarithms: what are logarithms for, nowadays? Why are they still on the syllabus? This is where ideas drawn from the historical development can fruitfully be employed.

From Practical to Theoretical Tool

Logarithms are a good and accessible example of something fundamentally changing its conceptual role within mathematics. If one is discussing such matters with students, then, the topic is useful for drawing attention to some characteristics of mathematics, to help students better understand the nature and potential of what they are studying. Indeed, there should be a role for discussions about mathematics in general studies classes, in which the case of logarithms could fruitfully figure.

In the 1650s it became clear that areas under a hyperbola have the logarithm property, or as Isaac Newton (1642–1727) wrote in his *Waste Book* in 1664–65:

In ye Hyperbola ye area of it bears ye same respect to its Asymptote wch a logarithme dot[h its] number. [18, p. 457]

This recognition had arisen from a massive treatise on the quadrature of the circle and other conic sections written by a Belgian Jesuit priest, Gregory St. Vincent. One of the readers of this work, a fellow Belgian Jesuit by the name of A. A. de Sarasa, noticed a simple property of areas under a rectangular hyperbola which appears to have escaped St. Vincent himself, that they relate to each other just as logarithms do. Put most simply, denote by $A(p)$ the area bounded by the hyperbola and the x-axis, between the values 1 and p; and similarly for another value q and also for their product pq. Then what follows immediately from a long and detailed argument by St. Vincent [19, pp. 221–2] is that $A(pq) = A(p) + A(q)$. (See Figure 2.)

Within a few years three essentially interconvertible definitions of logarithm were available:

- as an artificial number in the classical Napierian way,
- as an area measure of the hyperbola, and
- as an infinite series, such as Mercator's 1668 series

$$\log(1 + x) = x - \frac{1}{2}x^2 + \frac{1}{3}x^3 - \frac{1}{4}x^4 + \cdots$$

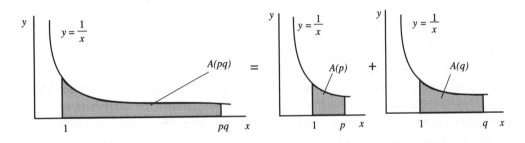

FIGURE 2

Attentive students today will want to ask, as some may have felt in the 1650s, whether the area redefinition is of any material help in calculating logarithms or whether the numerical computation of areas under a hyperbola is any simpler than a Napierian calculation. This is where the significance of the infinite series definition becomes clear, adding both a further conceptual tool to mathematics and a means of calculating logarithmic values. In short, logarithms were by the 1670s conceptually well on their way to their crucial role within modern mathematics, although it took a further seventy years or so for their presentation to reach the clarity and simplicity imposed by Euler.

Most modern textbook expositions follow Leonhard Euler (1707–1783) in coming straight to a student's first notion of logarithms through the use of exponents. Euler's definition of logarithms in his best-selling *Complete Introduction to Algebra* of 1770—a leading algebra textbook throughout the nineteenth century—goes as follows:

220 Resuming the equation $a^b = c$, we shall begin by remarking that, in the doctrine of Logarithms, we assume for the root a, a certain number taken at pleasure, and suppose this root to preserve invariably its assumed value. This being laid down, we take the exponent b such that the power a^b becomes equal to a given number c; in which case this exponent b is said to be the logarithm of the number c...

221 We see, then, that the value of the root a being once established, the logarithm of any number, c, is nothing more than the exponent of that power of a, which is equal to c; so that c being $= a^b$, b is the logarithm of the power a^b. [6, pp. 63–64]

The historical development of exponential perceptions is a fourth stream, to add to the three converging in Napier's thought, whose study will benefit students' all-around understanding of these matters, for it provides a good example of the power of good notation to clarify and advance thinking processes.

The Role of Symbols

The Chuquet passage of 1484 quoted earlier leaves the impression of someone at a turning point of mathematical history: making rather heavy weather of what appears to be a not-very-difficult perception, yet at the same time beginning to explore the role of symbols in encapsulating and conveying mathematical meaning. It was not until the time of René Descartes, some 150 years later, that the exponential notation for powers began to become widely used. His celebrated *La geométrie* (1637) [5] was the first text to make consistent use of the notations which we recognize today, a^3, x^4 and so on. (He and many of his successors used xx for the square of x—the general adoption of x^2 is rather recent.) The historian of mathematical notation, Florian Cajori, rightly pointed out that:

There is perhaps no symbolism in ordinary algebra which has been as well chosen and is as elastic as the Cartesian exponents. .... Our exponential notation has been an aid for the advancement of the science of algebra to a degree that could not have been possible under the old German or other early notations. Nowhere is the importance of a good notation for the rapid advancement of a mathematical science exhibited more forcibly than in the exponential symbolism of algebra. [3, p. 360]

The truth of this is well seen in the case of logarithms. Following his definition given above, Euler carried on to explain that $a^{\log c} = c$, that $a^{\log c + \log d} = cd$, and this elastic simplicity continues to pervade his discussion.

A useful exercise, both for reviewing notions of logarithms and for appreciating the power of notations, would be for students to examine how the logarithmic formulae would appear in some of the other symbolisms that at one time vied with Descartes' for acceptance by mathematicians (listed in [3]), and explore how easy it would be to manipulate them and understand them. Such a project would give a better understanding of the importance of good symbolism, through treating it somewhat as biologists treat competing life-forms in the Darwinian struggle for existence, appreciating what has survived better by analyzing what was lacking in those that fell by the wayside.

Conclusion

There is much which can be done, in modern teaching, with late seventeenth- and eighteenth-century developments in logarithms, especially with students who are beginning to explore the world of calculus. The intertwinings with other mathematical areas are by now very considerable, and it may be better to postpone their unfolding for another day. This is a subject that can easily start from exploring intuitions suitable for primary schools, and slide disconcertingly into higher levels.

From the perspective of modern pedagogy, an important lesson to draw from these developments—which may be accessible to different children at different ages—is precisely the variety of equivalent formulations of logarithmic ideas. What can seem rather confusing at one stage of a child's education, the idea that there is more than one "right" definition of something, can later be seen as a source of mathematics' richness and power. The example of logarithms in their historical development is a good way of letting this idea be understood, not as an arbitrary profusion but as a response to problems and inquiries within the practice of mathematics.

A final thought—in preparing this paper I looked at the treatment of logarithms and allied topics in quite a few textbooks written over the last century. This provided an interesting amount of prima facie evidence for what I had not set out to find here, a permeation of the discussions and examples with ideological concepts and value judgements about the structure of society. Evidently the spirit of Hammurabi lives on. Let those who believe that mathematics and its teaching is beyond such concerns and presuppositions look to their own textbooks!

Bibliography

1. Archimedes: *The Sand-reckoner*, in T. L. Heath (trans. & ed.): *The Works of Archimedes*. New York: Dover, 1953. Also in [7, pp. 150–152].
2. Buck, R. Creighton: "Sherlock Holmes in Babylon," *The American Mathematical Monthly* 87 (1980), 338–345; also in [7, pp. 32–40].
3. Cajori, Florian: *A History of Mathematical Notations*, Vol. 1. Chicago: Open Court, 1928.
4. Dakin, A., and R. I. Porter: *Elementary Analysis*. London: Bell, 1971.
5. Descartes, René: *La géométrie*, in David Eugene Smith and Marcia L. Latham (trans.): *The Geometry of René Descartes*. New York: Dover, 1954.
6. Euler, Leonhard: *Elements of Algebra* (John Hewlett, trans.). New York: Springer, 1984.
7. Fauvel, John, and Jeremy Gray: *The History of Mathematics: a Reader*. London: Macmillan, 1987.

8. Flegg, H. G., C. M. Hay, B. Moss (trans. & ed.): *Nicholas Chuquet, Renaissance mathematician*. Boston: Reidel, 1985; also in [7, pp. 247–249]
9. Gill, Dawn: "Politics of Percent," *Mathematics Teaching* 114 (1986), 12–14.
10. Griffiths, Peter: *Exploring Angle*. Milton Keynes: Open University Press, 1988.
11. Handcock, Percy: *The Code of Hammurabi*. SPCK, 1920.
12. Levi, Primo: *The Periodic Table*. New York: Schocken Books, 1984.
13. Mason, John: *Learning and Doing Mathematics*. London: Macmillan, 1988.
14. Neugebauer, Otto, and Abraham Sachs (eds.): *Mathematical Cuneiform Texts*. New Haven: American Oriental Society, 1945.
15. Pierce, R. C.: "Sixteenth-century astronomers had prosthaphaeresis," *Mathematics Teacher* 70 (1977), 613–14.
16. Resnikoff, H. L., and R. O. Wells: *Mathematics in Civilization*. New York: Holt, Rinehart and Winston, 1973.
17. van der Waerden, B. L.: *Science Awakening, I*. New York: Oxford University Press, 1961.
18. Whiteside, D. T.: *The Mathematical Papers of Isaac Newton*, vol. I. Cambridge: Cambridge University Press, 1967.
19. Whiteside, D. T.: "Patterns of mathematical thought in the later 17th century," *Archive for History of Exact Sciences* 1 (1961), 179–388.

Napier's Logarithms Adapted for Today's Classroom

Victor J. Katz

Students often ask what is "natural" about logarithms to the base e. John Napier (1550–1617), the Scottish laird who was the first to publish a table of logarithms along with rules for their use in 1614 and posthumously in 1619 [3], [4], did not deal with bases as such. But his definition of a logarithm and his method of constructing his table were, in some sense, "natural." Napier developed his logarithms for use in the extensive plane and spherical trigonometrical calculations necessary for astronomy. His table was therefore not a table of logarithms of numbers, but one of logarithms of values for sine functions. Since sines in his day were thought of as lengths of certain lines in a circle of large radius—Napier used 10,000,000—it was useful for him to set the logarithm of the "total sine," $(\log \sin 90° = \log(10,000,000))$ equal to 0 and have the values of the logarithm increase with decreasing sines. Since the modern logarithm is an increasing function with $\log 1 = 0$, one cannot use Napier's original method in a classroom today without an extensive, and probably confusing, digression. It is the purpose of this paper to present an adaptation of Napier's original method which, while still preserving his central ideas, can in fact be used in today's classrooms as a method of introducing natural logarithms to bright precalculus students.

The Use of Sequences

Beginning with the relationship of arithmetic and geometric sequences which was well known to mathematicians of the sixteenth century and earlier, we consider an arbitrary arithmetic sequence beginning with 0 and with constant difference $a > 0$. Place below it an arbitrary geometric sequence beginning with 1 and having a constant ratio r where $r > 1$.

$$0 \quad a \quad 2a \quad 3a \quad 4a \quad 5a \quad 6a \quad 7a \quad \cdots$$
$$1 \quad r \quad r^2 \quad r^3 \quad r^4 \quad r^5 \quad r^6 \quad r^7 \quad \cdots$$

Each number in the first sequence corresponds to the number directly below it in the second sequence; that is, $0 \leftrightarrow 1$, $a \leftrightarrow r$, $2a \leftrightarrow r^2$, $3a \leftrightarrow r^3$, and, in general, $na \leftrightarrow r^n$, for all $n > 0$. Use two notations to represent this correspondence, namely $f(na) = r^n$ and $g(r^n) = na$.

The basic reason for establishing this correspondence is that multiplication of elements in the second row corresponds to addition of elements in the first row. For example, to multiply r^3 by r^4, simply add the numbers directly above them, $3a$ and $4a$ to get $7a$. The number below $7a$, namely r^7, is then the product of the original pair. In terms of the notation, this property can be written as $g(r^n) + g(r^m) = g(r^n r^m)$, for arbitrary positive integers m, n. To write this

49

more concisely, let $x = r^n$ and $y = r^m$. Then

$$g(x) + g(y) = g(xy).$$

Of course, this "law" is only valid for numbers x and y which can be expressed as powers of r.

Similar laws hold for other operations. Namely, a division in the second row corresponds to a subtraction in the first whenever both operations are defined, or

$$g(x) - g(y) = g(x/y).$$

Next, raising an entry to a positive integral power in the second row corresponds to a multiplication in the first, or

$$g(x^n) = ng(x).$$

Finally, extracting a root in the second row, as long as it makes sense, corresponds to a division in the first row:

$$g(\sqrt[m]{x}) = g(x)/m.$$

These rules can also be written using the alternative rule of correspondence expressed in terms of f. Thus

$$f(u)f(v) = f(u + v)$$
$$f(u)/f(v) = f(u - v)$$
$$[f(u)]^n = f(nu)$$
$$\sqrt[m]{f(u)} = f(u/m)$$

It must be remembered, however, that the arguments of f are restricted to multiples of the fixed number a.

Using Napier's Approach

The laws above are very suggestive. They enable us, as they enabled Napier, to extend the correspondence. First, note, as did various mathematicians in the centuries before Napier, that both of the sequences can be extended to the left. Namely, one gets

$$\cdots \quad -4a \quad -3a \quad -2a \quad -a \quad 0$$
$$\cdots \quad \frac{1}{r^4} \quad \frac{1}{r^3} \quad \frac{1}{r^2} \quad \frac{1}{r} \quad 1$$

If the two functions are applied to these numbers as well, the laws above continue to hold. But to preserve the original definitions of these correspondences, it is necessary to write $1/r = r^{-1}$, $1/r^2 = r^{-2}$, and, in general, $1/r^n = r^{-n}$. That is, negative integral powers of the base r are now defined.

Next, take Napier's momentous step of converting from arithmetical reasoning to geometrical reasoning. Conceive of two number lines on which these sequences are represented (Figure 1) and points P and Q moving on the two lines as follows: P moves on the upper line with a constant velocity v. Therefore, P covers each marked interval, $[0, a], [a, 2a], [2a, 3a], \ldots$ in the

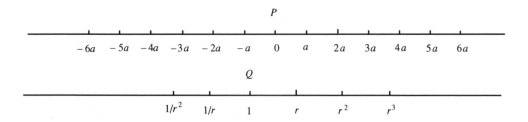

FIGURE 1

same time period. The second point Q moves along the lower line so that it too covers each marked interval, $[1, r], [r, r^2], [r^2, r^3], \ldots$ in the same time period. What does this imply about the velocity of Q? Consider this velocity as constant in each of the marked intervals. Therefore, Q travels from 1 to r in the same time as from r to r^2 and from r^2 to r^3. Since the distances traveled are $r - 1$, $r(r - 1)$, and $r^2(r - 1)$ respectively, these distances are in the same ratio as the coordinates of the points themselves, namely r. Thus the velocities in each interval must also be in the same ratio. It follows that the velocity at any of the marked points is proportional to that point's distance from 0.

It can also be seen that, given any two marked intervals $[\alpha, \beta]$, $[\gamma, \delta]$ of arbitrary length on the lower line with $\beta/\alpha = \delta/\gamma$, that is, $\alpha = r^j$, $\beta = r^{j+k}$, $\gamma = r^m$, and $\delta = r^{m+k}$, the time for Q to cover $[\alpha, \beta]$ is equal to its time to cover $[\gamma, \delta]$. For in each case, it takes k times as long as it takes to cover the interval from 1 to r.

If each of the points P, Q begins to move at the same time, P from 0 and Q from 1, with the initial velocity of Q equal to v, the correspondence $na \leftrightarrow r^n$ is exactly the correspondence between the distances of P and Q from the respective 0 points at given times, assuming that at those times P, and therefore Q, are at marked points. In effect, the original correspondence has been translated from one of numbers to one of points. The geometrical concepts, however, lead in a new direction.

With the above definitions, the velocity of point Q must jump at each marked point. Real objects usually don't abruptly change velocity when they move. To smooth out these jumps, introduce the points $s = \sqrt{r}$ between 1 and r, as well as all integral powers of s. Since $s^2 = r$, $s^4 = r^2$, $s^6 = r^3, \ldots, s^3$ must be between r and r^2, s^5 must be between r^2 and r^3, etc. Now assume Q changes velocity at each of the new points as well as keeps the same property as before. That is, Q continues to cover each new marked interval $[1, s], [s, s^2], [s^2, s^3], \ldots$ in the same time. Therefore, the velocity at each of the marked points is still proportional to the distance of that point from 0. The question then arises, where is P when Q is at s? Since it takes the same amount of time for Q to get from 1 to s as from s to $s^2 = r$, and since P travels at constant velocity, P must be halfway between 0 and a, that is, at $\frac{1}{2}a$, when Q is at s. Similarly, when Q is at s^3, P is at $\frac{3}{2}a$; when Q is at s^5, P is at $\frac{5}{2}a$, etc. (Figure 2). Rename $\frac{1}{2}a$ as b. Then the points on the upper line corresponding to $1, s, s^2, s^3, s^4, \ldots$ on the lower line are $0, b, 2b, 3b, 4b, \ldots$ In other words, the original correspondence has simply been extended to these new numbers, and the laws given above continue to hold. In particular, the property that given any two marked intervals $[\alpha, \beta]$, $[\gamma, \delta]$, of whatever length, such that $\beta/\alpha = \delta/\gamma$, the point Q covers each in equal times, continues to be valid.

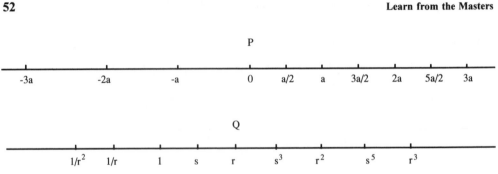

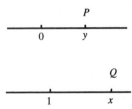

FIGURE 2

This idea can be extended. Take $\sqrt{s}$ as a new "first" point after 1, or perhaps $\sqrt[3]{s}$. Whichever number is taken, new points can be filled in on both lines and the correspondence continued. All of the properties will continue to hold. In particular, as before, if $\beta/\alpha = \delta/\gamma$, then the time for Q to cover the interval $[\alpha, \beta]$ is equal to that to cover $[\gamma, \delta]$.

Napier took the bold step of eliminating all velocity jumps by assuming that the velocity of Q increases smoothly so that at any point on the lower line that velocity is proportional to the moving point's distance from 0. Assume that P travels at constant velocity v as before and that Q starts at 1 with velocity v at the same time that P starts at 0. Suppose now that Q is at the number x at the same time as P is at the number y (Figure 3). Define as did Napier the **logarithm** of x, written $\log x$, to be y.

FIGURE 3

What are the properties of $\log x$? First of all, by definition, $\log 1 = 0$. Next, the basic property still holds, that if $\beta/\alpha = \delta/\gamma$, then Q covers $[\alpha, \beta]$ in the same time as it covers $[\gamma, \delta]$, since any four such numbers can be approximated to whatever degree of accuracy desired by "marked" points on the line for some marking by powers of a particular number. Because the property holds for these approximations, it must hold for the four points themselves. What does that imply about the logarithms? Since the logarithms are measured by the point P moving at constant velocity, the result is that $\log \beta - \log \alpha = \log \delta - \log \gamma$. Specialize this relationship by letting $\alpha = 1$. Then $\log \beta = \log \delta - \log \gamma$, or, since $\beta = \delta/\gamma$,

$$\log(\delta/\gamma) = \log \delta - \log \gamma.$$

This can be written differently. From $\log \beta = \log \delta - \log \gamma$, it follows that $\log \delta = \log \beta + \log \gamma$. Since $\delta = \beta\gamma$,

$$\log(\beta\gamma) = \log \beta + \log \gamma.$$

Repeated applications of this last property give, at least for integral n, m,

$$\log(\beta^n) = n \log \beta \qquad \text{and} \qquad \log(\sqrt[m]{\beta}) = (\log \beta)/m.$$

It follows that $\log(\sqrt[m]{\beta^n}) = \log((\sqrt[m]{\beta})^n) = (n/m) \log \beta$. Defining $\beta^{n/m}$ as usual as $\sqrt[m]{\beta^n} = (\sqrt[m]{\beta})^n$, the last two properties can be combined into one:

$$\log(\beta^r) = r \log \beta,$$

where r is any rational number. Since the two points are moving continuously, it can be assumed that this same rule holds even if r is not rational. After all, any irrational number can be approximated as closely as desired by a rational number.

The conclusion is that logarithms defined this way, that is by using Napier's basic ideas, have the same properties as the correspondence g given earlier. The difference is that now the logarithm is defined for any point to the right of 1 on the lower number line. One can also extend the definition to the points between 0 and 1 by having both P and Q move to the left. For those numbers, the logarithms will be negative.

What more can be said about logarithms given the definition and the basic properties? Napier was able to calculate a table of logarithms from the definition and the properties. It might be feasible to do the same on a limited scale in a class, but the current prevalence of calculators makes it very difficult to convince students to perform long arithmetic calculations when a simple push of one button gives them the answer instantaneously. A class would be better served, instead, if the students are shown how this Napierian definition leads to more modern ideas about logarithms.

Logarithms and Exponential Functions

The definition of logarithm above shows that as x increases, so does $y = \log x$. In fact, $\log x$ can get as large as desired provided x is large enough. Using continuity, it follows that for some number e, $\log e = 1$. Now consider e^y for any number y. What is its logarithm? The laws above show that $\log e^y = y \log e = y$. In other words, given any number y on the upper line, the number of which it is the logarithm is expressible as e^y for some fixed number e. That is, the coordinate x of the lower line can be expressed as $x = e^y$, where y is the coordinate of the upper line. In other words, the logarithm y of a number x is the exponent of x when x is written as a power of the fixed number e. The exponential properties are then easily derived from the logarithm properties. These are similar to the properties of the earlier correspondence f:

$$e^y e^z = e^{y+z}$$

$$e^y / e^z = e^{y-z}$$

$$(e^y)^z = e^{yz}$$

These properties hold for any real numbers y and z.

What does the velocity definition tell us about e? Since P moves with constant velocity v, the rate of change of y with respect to t is v, or $\Delta y/\Delta t = v$. [Here and in what follows, think of Δu for any variable u as denoting a "small" change in u.] On the lower line, the velocity of Q at x is proportional to the distance x itself. Since Q started with velocity v, that quantity is the constant of proportionality. In other words, the rate of change of x with respect to time t is vx, or $\Delta x/\Delta t = vx$. Combining the two relationships gives

$$\Delta x/\Delta y = (\Delta x/\Delta t)/(\Delta y/\Delta t) = vx/v = x,$$

or, the rate of change of x with respect to y is equal to x.

Since x depends on y, write Δx as $x(y+\Delta y) - x(y)$. The velocity relationship then gives

$$\frac{x(y+\Delta y) - x(y)}{\Delta y} = x(y)$$

or, recalling that $x(y) = e^y$,

$$\frac{e^{y+\Delta y} - e^y}{\Delta y} = e^y.$$

Using the properties of exponents, rewrite the numerator on the left as $e^y e^{\Delta y} - e^y$ or $e^y(e^{\Delta y} - 1)$. Dividing through by e^y and multiplying by Δy then gives the new equation

$$e^{\Delta y} - 1 = \Delta y \qquad \text{or} \qquad e^{\Delta y} = 1 + \Delta y.$$

Applying the properties of exponents, raise both sides of this equation to the power $1/\Delta y$. The result is

$$e = (1 + \Delta y)^{\frac{1}{\Delta y}}.$$

Since Δy is assumed "small," e can be "calculated" using calculators. For example, if $\Delta y = 0.001$, the right side becomes 2.717. Still smaller values of Δy give slightly different values for e; the calculator will show that the values one gets by taking Δy smaller and smaller form a convergent sequence. In fact, these values converge to a number whose first three decimal places are 2.718. This number was named e by Leonhard Euler (1707–1783) around 1730 [1, p. 258].

The logarithms defined here are called **natural logarithms**. Though the discussion has shown in what way they arise naturally out of Napier's definition through velocity and how the mysterious number e is determined as the base for which those logarithms are the exponent, Napier himself didn't get quite this far. He never thought of logarithms as exponents. And he was soon convinced by Henry Briggs (1561–1631) that instead of dealing with these natural logarithms, it would be more convenient to have logarithms whose value at 10 was equal to 1. A table of these logarithms, today called common logarithms, was computed by Briggs in the 1620s. [1, p. 257] It took, however, several developments over the next 120 years for the modern theory of logarithms to be fully developed.

Conclusion

Logarithms today are generally introduced to students as exponents to a given base. Before the 1970s, much class time was spent in using logarithms for intricate numerical calculations. The advent of calculators has virtually eliminated that part of the curriculum. Nevertheless, the concept of the logarithm function, and in particular of the natural logarithm function, remains an important one in calculus and more advanced subjects. Since the notion of logarithm as exponent is not primary in dealing with the logarithm function, and since the functional properties are, the students would be better served by beginning their study of logarithms with those properties. The method presented in this article, pioneered by Napier himself nearly 400 years ago, thus appears to be a reasonable approach to the teaching of the logarithm concept.

For Further Consideration

1. Explicitly derive the laws given for the correspondence f from the laws derived for g.

2. Show that when distances traveled at constant velocities in the same time are in a given ratio, the same is true of the velocities themselves.

3. Explicitly derive the logarithm properties for powers and roots from the properties for products and quotients.

4. Show how to define the logarithm for numbers between 0 and 1 in a way analogous to the methods developed in the article for numbers greater than 1.

5. Use a calculator to determine values of $(1 + \Delta y)^{1/\Delta y}$ for small values of Δy and show that the numbers calculated approach a limiting value as Δy approaches 0.

6. Show that the function $A(x)$ defined as the area under the curve $y = 1/x$ from 1 to x satisfies the basic logarithm property that $A(rs) = A(r) + A(s)$ by approximating A using appropriate sums of rectangles. Conclude that the definition of the natural logarithm as an integral, usually used today, gives the same function as the definition presented in this article.

Bibliography

1. Kline, Morris: *Mathematical Thought from Ancient to Modern Times*. New York: Oxford University Press, 1972.
2. Knott, C. G.: *Napier Tercentenary Memorial Volume*. London: Longmans, Green and Co., 1915.
3. Napier, John: *Description of the Wonderful Canon of logarithms*. 1614. Reprint edition, New York: Da Capo Press, 1969.
4. Napier, John: *Construction of the Wonderful Canon of logarithms*. 1619. Reprint edition, London: Dawsons, 1966.

René Descartes

Trigonometry Comes Out of the Shadows

Frank J. Swetz

It is significant that one of the few specific mathematical accomplishments ascribed to Thales of Miletus (625–547 B.C.) was that of "shadow reckoning." Thales, the first historical personage associated with mathematics in the West, used this ability to determine the heights of the pyramids in Egypt, as related by Plutarch in his *Banquet of the Seven Wise Men*:

> Although he [the King of Egypt] admired you [Thales] for other things, yet he par-
> ticularly liked the manner by which you measured the height of the pyramid without
> any trouble or instrument, for by merely placing your staff at the extremity of the
> shadow which the pyramid casts, you formed, by the impact of the sun's rays, two
> triangles and so showed that the height of the pyramid was to the length of the staff
> in the same ratio as their respective shadows. [3, p. 94]

With the simplest of instruments, a vertical staff of known length, and the natural phenomenon of sunlight casting shadows, Thales was able to utilize the principle of proportionality
between similar right triangles to determine the heights of the pyramids. It appears that many
ancient societies relied on shadow observations for agricultural and religious purposes. Shadow
lengths helped determine Summer and Winter solstices, that is, the times at which the sun is
farthest north and farthest south of the equator, thus fixing a year in time upon which planting
seasons could be scheduled. Egyptian and Hindu priests fixed religious rituals according to
the sun's position in the sky as determined by shadow lengths. In later Islamic societies, three
of the five prescribed times for daily prayer are based on shadow lengths. Existing evidence
indicates that the Babylonians, Egyptians, and Chinese developed rather precise celestial observation techniques using merely a vertical staff or pole and noting shadow positions. The
shadow-casting reference pole became standardized in various early societies. Such a pole is
usually referred to by its Greek name, *gnomon*. Shadow observations were tabulated and consulted for computational purposes. Such activities provide some of the first historical evidence
of human preoccupation with the right triangle and its adoption for scientific purposes. While
the initial concern of early shadow reckoners was to determine the relationship of the gnomon's
length to that of its shadow, vertical to horizontal, it is most probable that further exploration
and computational experimentation with measurements obtained resulted in the discovery of
the "Pythagorean Theorem" and a fuller mathematical appreciation of right triangle ratios and
properties.

The contents of the oldest known Chinese mathematical classic, the *Zhoubi suanjing (The
Arithmetical Classic of the Gnomon and the Circular Paths of Heaven)* (c. 300 B.C.) would seem
to support this theory. In this work, a conversation between the legendary Duke of Zhou and
his minister, Shang Gao, describes the foundation of early Chinese mathematics and astronomy

as being built around shadow reckoning and the use of the right triangle. As Shang Gao notes to his master:

> He who understands the earth is a wise man, and he who understands the heavens is a sage. Knowledge is derived from the shadow. The shadow is derived from the gnomon. And the combination of right angle with numbers is what guides and rules the ten thousand things. [11, p. 33]

It is in the *Zhoubi* that the first historically documented proof of the Pythagorean Theorem is found [15]. Thus, shadow reckoning was an important mathematical activity, one with a long history. It is from shadow reckoning that a primary intellectual appreciation of the right triangle and its properties emerged and from which trigonometry as a mathematical discipline eventually evolved. Truly, it can be said that "trigonometry came out of the shadows." This simple fact, the evolution of trigonometry from shadow reckoning techniques of old, should be revealed during its teaching and shared with students.

Shadow Functions

In using a vertical staff or gnomon and observing its shadow under either sunlight or moonlight, two lengths are involved, the length of the staff and that of its shadow. The length of the staff of course remains fixed for a fixed length of pole, but, the shadow's length varies according to the position of the light source producing the shadow. A particular point in the sky can be associated with a particular shadow length and, in modern terms, we may say that a functional relationship exists between shadow length and celestial point. Since, as in Figure 1, each point in the sky can similarly be associated with an angle of sight determined by the tip of the vertical staff, a functional relationship can be established between $\angle A$ and point P. In this situation two ratios, $BC : AC$ or $AC : BC$ can be used to express the relationship. Today, we designate these ratios as $\tan \angle A$ and $\cot \angle A$. These two shadow functions, the forerunners of our family of trigonometric functions, were intuitively appreciated and well used thousands of years ago.

The Egyptian *Rhind Mathematical Papyrus* of about 1650 B.C. contains three problems involving the determination of inclination of the sides of a pyramid. Egyptian computers used

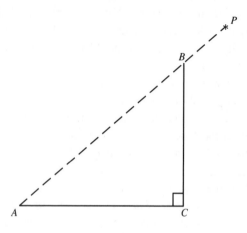

FIGURE 1

the concept of *seqt* to measure inclination. Today, we would recognize *seqt* as the cotangent ratio determined by an angle.

> If a square pyramid is $93\frac{1}{3}$ cubits high and the side of its base is 140 cubits long, what is its *seqt*? [4, problem 58]

To assist in such computational situations, tables of shadow lengths were compiled. The *Zhoubi* lists sun induced shadow lengths recorded over a prolonged period of time. Similar tables are found in Old Babylonian works (c. 1700 B.C.). (See [17] and [12, p. 74].) Such listings of shadow lengths imply the use of standardized gnomons and, indeed, the ancient Chinese used a gnomon (*biao*), of 10 or 8 units in length to obtain their shadows, while the Hindus, and later Arab shadow reckoners, preferred a gnomon of 12 divisions. To accurately record the length of a shadow, templates were developed and used. Records from the Zhou Dynasty speak of the Chinese use of such templates.

> The Ssu Thu (a high official), using a shadow template, determines the distance of the earth below the sun, fixes the exact (length of the) sun's shadow, and thus finds the center of the earth. [11, p. 286]

In official dynastic manuals, specific procedures were described for the leveling of the gnomon and the shadow surface. Quite early in Chinese history, the gnomon evolved into an "L-shaped" carpenter square (*ju*), with the horizontal leg serving as the shadow template, and the vertical leg the gnomon itself. This instrument was portable and accurate. Thus specific procedures and technology were developed around shadow reckoning and recording. [8] Frequently, as in the Chinese context, shadow reckoning was a state sanctioned activity incorporated into the functioning of the bureaucracy. Legendary Emperor Yu of the Xia Dynasty (c. 2000–1520 B.C.) is credited with using a gnomon to "survey the lands and bring the floods under control" [9, p. 2–4]. Data on numerous right triangle configurations were accumulated and stored for reference. In such an environment, a mathematical curiosity reinforced by computational experimentation could well have led to the discovery of the "Pythagorean Theorem." In societies in which shadow reckoning was established, for example, the Chinese and Babylonian, the Pythagorean relationship was known and used well before the time of Pythagoras.

Early Shadow Reckoning Theory and Computations

Little is recorded about the specific computational techniques of early Babylonian, Egyptian, and Greek shadow reckoners, other than that they utilized shadow observations for astronomical purposes. Gnomons in the form of obelisks were constructed in Babylonia, Egypt, and Greece to supply shadows for the reckoning of seasonal and daily time. Anaximander (c. 575 B.C.) is credited with erecting the first formal gnomon in Greece. It is known that Aristarchus, the Copernicus of antiquity (c. 260 B.C.), used principles of proportionality similar to those evident in shadow reckoning situations, to compute distances between the earth, the sun, and the moon. Thus, although we know of diverse instances in which ancient peoples in the West used shadow reckoning and shadow reckoning principles, their specific calculation methods remain obscure.

It is in the historical records of the Orient that the first technical descriptions of shadow reckoning procedures can be found. In the *Zhoubi*, the Duke of Zhou inquires "May I ask how to use the gnomon?" To which Minister Shang Gao replies, "Align the gnomon with the plumb line to determine the horizontal, lay down the gnomon to find the height, reverse the gnomon to

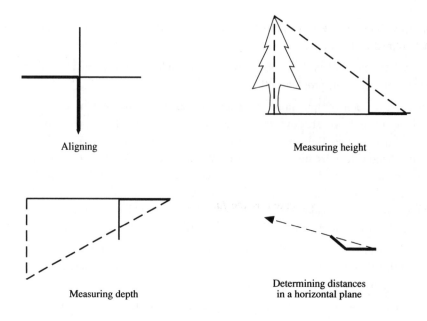

Aligning   Measuring height

Measuring depth   Determining distances
                                        in a horizontal plane

FIGURE 2

find the depth, lay the gnomon flat to determine the distance." [9, p. 31] The gnomon referred
to in these passages is of the carpenter square type. (See Figure 2.)

In developing its *Gai Tian* [covering heaven] theory, the *Zhoubi* notes that at the summer
solstice an eight foot high gnomon casts a shadow six feet long. Further, it states that if
the gnomon is moved from north to south, for every thousand miles of gnomon movement
the resulting shadow length changes by one inch. [9, p. 30] These facts were then used in
computing the height of the sun from the earth, which was erroneously found to be 80,000
miles. The assumption that "for every thousand miles the gnomon moves its shadow changes
by one inch" is wrong, as was later proven by Li Chunfeng of the Tang Dynasty (A.D. 618–907).
The *Zhoubi* calculations assumed a flat earth. [10] Although some of the results contained in
the *Zhoubi* are questionable, by the beliefs and standards of the times they represent a high
level of computational proficiency using right triangle proportions.

The first extensive extant discussion on the use of the right triangle, specifically in distance
reckoning problem situations, is found in the last chapter of *Jiuzhang suanshu* (*Nine Chapters
on the Mathematical Art*), a work summarizing the development of ancient Chinese mathematics
from the Zhou through the Han Dynasties (c. eleventh century B.C.–A.D. 200). This chapter is
called *Gougu*, the Chinese term for a right triangle, and contains 24 problems and their solution
schemes involving proportionality calculations based on the use of the right triangle [15]. It
is interesting that none of these problems depend on the use of shadows directly, although
shadow reckoning principles are inherent in the solution strategies of several problems. Indeed,
the problem situations and their solutions demonstrate complete confidence in working with the
mathematical properties of similar right triangles.

In A.D. 263, the mathematician Liu Hui published a commentary on *Jiu Zhang* in which
he supplied theoretical verifications of the given solutions, and enriched and expanded the text

with his own contributions. In particular, he considered the ninth chapter inadequate in its treatment of right triangle theory. In order to remedy this situation, he compiled a collection of 9 problems devoted to a technique of shadow reckoning called *Chong Cha* [double differences] as an addendum to *Gougu.* [2] Liu considered *Chong Cha* to be one of "the nine mathematical arts" of his time and was amazed that it had been omitted from the *Jiuzhang.* In prefacing this addendum, he illustrated the technique of *Chong Cha* by generalizing the "sun's height" problem from *Zhoubi* as follows:

> Erect two gnomons at the city of Loyang. Let the height [of each gnomon] be eight *chi.* [Both the gnomons erected] in the north-south direction are on the same level. Measure the shadows (of the gnomons) at noon on the same day. The difference in length of the shadows is taken as the *fa.* Multiply the difference in distance of the gnomons by their height and take the result as the *shi.* Divide the *shi* by the *fa* and add it to the height of the gnomon. The result is the height of the sun from the earth. Take the shadow length of the southern gnomon and multiply it by the distance between the gnomons to give the *shi.* Divide the *shi* by the *fa.* The result is the distance of the southern gnomon from the subsolar point in the south at noon. [13, p. 92]

Chi is a Chinese unit of measure equivalent to a "foot." *Fa* and *shi* are technical terms, indicating divisor and dividend, respectively, referring to positions on the Chinese computing board. An analysis of the problem situation using modern notation reveals that Liu's solutions are correct although the flat-earth assumption still invalidates the result obtained for the sun's height. (See Figure 3.)

In $\triangle PRA$ and $\triangle CNE$ and in $\triangle PAC$ and $\triangle CED$, the following proportions can be found:

$$\frac{RP}{NC} = \frac{RA}{NE} = \frac{PA}{CE} = \frac{AC}{ED}$$

Now if $ND = a_2$, $SB = a_1$, $PQ = y$, $QS = z$, $SN = x$, and $AS = CN = RQ = h$, then these relationships yield:

$$\frac{y - h}{h} = \frac{z}{a_1} = \frac{x}{a_2 - a_1}$$

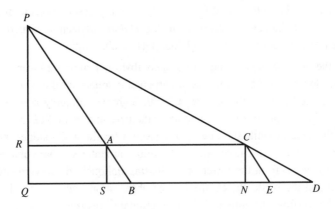

FIGURE 3

from which the desired distances can be found:

$$y = \frac{hx}{a_2 - a_1} + h \quad \text{and} \quad z = \frac{a_1 x}{a_2 - a_1}$$

The quantity $a_2 - a_1$ is known as the "shadow difference." Now if the concept of angle were imposed on the situation so that

$$m(\angle PBQ) = \alpha \quad \text{and} \quad m(\angle PDQ) = \beta,$$

then a solution expression could be obtained in terms of the trigonometric ratios tangent or cotangent. That is,

$$y = \frac{\tan \beta \tan \alpha (a_1 - a_2 - x)}{\tan \beta - \tan \alpha}$$

or

$$y = \frac{x + a_2 - a_1}{\cot \beta - \cot \alpha}.$$

Thus under a modern interpretation, the *Chong Cha* process of double differences could refer to differences of cotangent or tangent ratios.

Since Liu's nine problems all involved the use of shadow techniques in surveying situations, they in themselves were viewed as a manual on mathematical surveying techniques, and by the beginnings of the Tang Dynasty (A.D. 618–906) evolved into a separate mathematical work, the *Haidao suanjing* (*Sea Island Mathematical Manual*), so named in reference to the first problem presented, a variation of the "Loyang problem."

> Now for [the purpose of] looking at a sea island, erect two poles of the same height, 3 *zhang* [on the ground], the distance between the front and rear [poles] being a thousand *bu*. Assume that the rear and front poles are aligned [with the island]. By moving away 123 *bu* from the front pole and observing the peak of the island from ground level, it is noticed that the tip of the pole coincides with the peak. Then by moving backward 127 *bu* from the rear pole and observing the peak of the island from ground level again, the tip of the back pole also coincides with the peak. What is the height of the island and how far is it from the pole? [2, p. 105]

But then, in the remaining problems, Liu extended the technique to include more complex problem situations involving three and four sets of necessary observations to draw a conclusion. For example, consider the seventh problem in the *Haidao* collection, that of peering into an abyss, a problem which requires the use of four sets of observations.

> Now for (the purpose of) looking into a deep abyss containing a clear pool of water with white stones at the bottom, hold a carpenter's square on the edge of the abyss. Assume that the height of the square is 3 *chi*. Sighting obliquely downwards on the water level, the line of observation intersects the base at a point 4 *chi* 5 *cun* [from the corner of the square] while the line of observation for the white stones intersects the same base at 2 *chi* 4 *cun*. Erect another similar carpenter's square above [the first] such that it is 4 *chi* from the lower one. Sighting obliquely downwards again on the water level and the white stones, the line of observation intersects the base at 4 *chi* and 2 *chi* 2 *cun* [from the corner of the upper square], respectively. What is the depth of the water? [2, p. 107] (See Figure 4.)

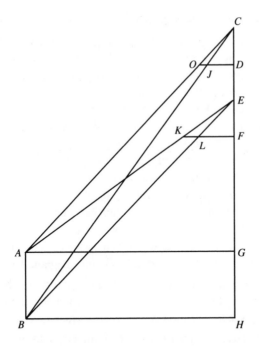

If one follows Liu's solution procedure based on the use of proportions where

$$GH = \frac{CE[JD(KF - OD) - OD(LF - JD)]}{(LF - JD)(KF - OD)},$$

the depth of the water is found to be 1 *chang* 2 *chi*. Once again, if the concept of angle is introduced into the situation where $m(\angle COD) = \alpha$, $m(\angle CJD) = \beta$, $m(\angle EKF) = \delta$, and $m(\angle ELF) = \gamma$, the solution becomes a function of cotangent ratios:

$$GH = \frac{CE[\cot \beta(\cot \delta - \cot \alpha) - \cot \alpha(\cot \gamma - \cot \beta)]}{(\cot \gamma - \cot \beta)(\cot \delta - \cot \alpha)}.$$

As is evidenced in the *Haidao*, the implicit use of tangent and cotangent ratios in the form of shadow lengths and shadow differences was perpetuated throughout early mathematical literature in China. The scholar Zhou Shung (A.D. third–fourth century), in preparing a commentary on the *Zhoubi*, expanded on its contents to develop a separate work *Ri gao tu shuo* (*Theory with Diagrams of the Sun's Altitude*) in which he illustrated the use of shadow measurement and techniques using a gnomon.

In the year 724, the State Astronomical Bureau of the Tang Dynasty initiated the most extensive meridian survey undertaken in the ancient world. Officials established 10 observation stations near the meridian 114° E and between latitudes 29° N and 52° N. From these stations, they determined the angular altitude of the north celestial pole above the horizon and measured the lengths of the shadows of a standard gnomon at noon on the days of the summer and winter solstices and equinoxes [6]. In analyzing the data obtained, the Buddhist monk I-hsing, the foremost astronomer of the time, devised and used formal tangent tables. This instance raised the art of shadow reckoning to a more formal trigonometric basis. Apparently I-hsing was influenced by the existence of primitive sine tables or more correctly tables of semi-chords of given arcs. Such tables found their way into China via Hindu astronomers employed at the Imperial Court and through the existence and use of Hindu reference works, particularly the *Surya-Siddhanta* (c. A.D. 300–400) [6, pp. 24–25].

Concerns with shadow reckoning can be found in Vedic literature of about 800 B.C. where the use of a gnomon and its shadow are recommended in determining East-West reference lines. In Sanskrit, a gnomon was designated as *canku* and its shadow, *chaya*. The comprehensive Jainian astronomical canon *Suryaprajapti* (c. A.D. 200–300) refers to shadow measurement in relation to time reckoning. The noted astronomer Aryabhata (A.D. 500) included a table of sine values in his classic *Aryabhatiya* and related shadow ratios, i.e., tangent and cotangent, to the ratios of sine and cosine. In making this relationship, Aryabhata used graphical techniques whereby a circle was scribed about a given gnomon and its shadow so that the tip of the gnomon lay on the circle and the foot of its shadow determined the center of the circle [1]. (See Figure 5.) This reference circle is called the *svavrtta*. Using this diagram, he then advised his readers:

> The distance between the gnomon and the *bhuja* is multiplied by the length of the
> gnomon and divided by the difference between the gnomon and the *bhuja*. The result
> is the length of the shadow of the gnomon from its foot.

Bhuja refers to the respective sine chord which in a graphic situation would appear as the side of a right triangle. In that same triangle, the base leg would represent the cosine chord and be known as the *koti*.

Thus, following Aryabhata's directions in analyzing the situation depicted in Figure 5, where point S can be considered a source of light and AB the gnomon, we get, because $BC/CD = AB/SD$ that

$$\frac{BC}{DC - BC} = \frac{AB}{SD - AB}$$

or, that the length of the shadow is

$$BC = \frac{AD(DC - BC)}{SD - AB} = \frac{(AB)(DB)}{SD - AB}.$$

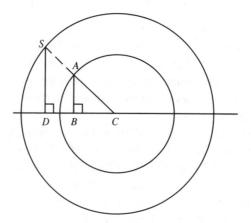

FIGURE 5

Further on in his discussions on shadow reckoning, Aryabhata poses and solves a problem strikingly similar to the first problem of the Chinese *Haidao*. [16] The problem situation is presented in Figure 6 where point S represents a source of light, AB and A_1B_1 are gnomons of equal length and their respective shadows are given by BC and B_1C_1.

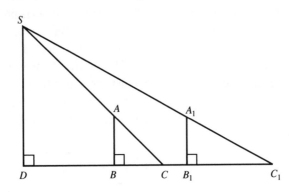

FIGURE **6**

In this figure, the *koti* [DC or DC_1] equals

$$\frac{(\text{distance between feet of shadows})(\text{shadow length})}{\text{difference between shadow lengths}}$$

and the *bhuja* equals

$$\frac{(koti)(\text{length of gnomon})}{\text{length of shadow}}.$$

By the sixth century, Hindu shadow reckoning techniques implicitly made use of all the basic trigonometric ratios. Special sections of mathematical works were devoted to shadow reckoning: Brahmagupta (c. 625) wrote on "measure by shadow," Mahavira (c. 850), "calculations relating to shadows" and Bhaskara (1150), "determination of shadows." In his *Lilavati,* Bhaskara challenges the reader's mathematical ability with a shadow problem:

> The ingenious man who tells the shadows of which the difference is measured by nineteen, and the difference of hypotenuses by thirteen, I take to be thoroughly acquainted with the whole of algebra as well as arithmetic. [5, p. 106]

Quite simply, given a gnomon with two opposing shadows whose difference is nineteen and if hypotenuses are constructed from the tip of the gnomon to the foot of its two shadows, the resulting hypotenuses differ by thirteen. It is required to find the length of the shadows.

Early Arab and Persian astronomers and mathematicians were experienced in shadow reckoning and formed a synthesis of their experiences with acquired Hindu and Greek theories to form a true discipline of mathematical trigonometry. The foremost Arab writer on astronomy was al-Battani (c. 920) who among his many accomplishments compiled tables of sines and tangent values and derived a trigonometric formula for determining the altitude of the sun. In the later middle ages, there appeared to be a genre of Arabic astronomical literature on shadow reckoning. Ibn al-Haytham (c. 965–1039) wrote a treatise on shadows as did the noted geometer Ibrahim bin Sinan (d. 946). But perhaps the most comprehensive work in this series was produced by the eleventh century astronomer al-Biruni. It was entitled the *Exhaustive Treatise on Shadows* [7].

Arab writers conceived of the cotangent, *al-zill al-mustawi*, as a shadow cast by a gnomon normal to the horizontal, and, for them, the hypotenuse of the direct shadow, *qutr al-zill*, became the cosecant. A horizontal gnomon casting a shadow on a vertical plane produced *al-zill al-ma kus*, the reversed shadow or tangent; the hypotenuse of the reversed shadow then became the secant. Al-Biruni, working with these relationships, derived several identities, whose modern rendering would appear as $\csc \alpha = 1/\sin \alpha$, $\cot \alpha = 1/\tan \alpha$. Further, he demonstrated that $\cot \alpha = \tan(\frac{\pi}{2} - \alpha)$. He also advanced shadow functions to arc functions. Al-Biruni defined four of the basic trigonometric ratios, tangent, cotangent, secant, and cosecant, in terms of shadows and then related them to the arc-derived functions of sine and cosine. His technique is illustrated in Figure 7.

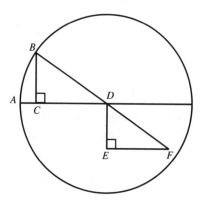

FIGURE 7

Let arc $AB = \alpha$, $BD = R$, DE = gnomon, g; then $BC = \mathrm{Sin}_R\alpha$, $CD = \mathrm{Cos}_R\alpha$, $DF = \mathrm{Csc}_g\alpha$, $EF = \mathrm{Cot}_g\alpha$. Note that $\mathrm{Sin}_R\alpha$ is the sine of arc α with respect to radius R, $\mathrm{Csc}_g\alpha$ is the cosecant of arc α with respect to gnomon g.

A successor of al-Biruni, Nasir al-Din al Tusi (c. 1250) refined these relationships further by establishing the identities

$$\frac{\tan \alpha}{R} = \frac{\sin \alpha}{\cos \alpha} \quad \text{and} \quad \frac{\tan \alpha}{\sec \alpha} = \frac{\sin \alpha}{R}.$$

Thus, from the thirteenth century onwards, Islamic astronomers and mathematicians utilized six trigonometric ratios.

Islamic shadow terminology became incorporated into the design of the astrolabe; the "shadow ladder," "shadow box," and "shadow square" all containing shadow scales were features of this instrument for hundreds of years. Arab writers referred to the shadow scales on the astrolabe as the "straight shadow" and the "turned shadow." These terms found their way into Latin as *umbra recta* and *umbra versa* respectively and were used in European mathematical literature to denote the legs of a right triangle well into the eighteenth century [14, p. 620].

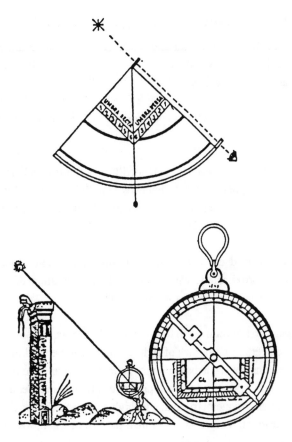

FIGURE 8

Conclusion

As trigonometry evolved from a descriptive, applied science to an analytic discipline, its shadow origins were forgotten. But the concepts and principles of shadow reckoning dominated calendrical computation and time keeping for over two thousand years. They shaped the sciences of land survey, cartography and navigation. Concern with shadow ratios led to the eventual derivation of the basic trigonometric functions that we know today, and probably provided a basis for the discovery of the Pythagorean theorem. Certainly, shadow reckoning constituted an important phase of early mathematical activity. Its history supplies an illuminating link on how early man used ingenuity; it harnessed a natural phenomenon and developed simple scientific and mathematical principles to quantify and, in a sense, control the environment. From these efforts a major mathematical discipline, trigonometry, evolved. These facts deserve to be acknowledged in the teaching of trigonometry.

For Further Consideration

1. Use a pole or staff of known height and replicate Thales' accomplishment. Determine the height of a distant object such as a flagpole or building. If possible, use direct measurements to check the accuracy of your results.

2. What are the historical and ethnological origins of the words sine, cosine, secant and tangent?

3. A colleague, in reading this article, contended that activities such as shadow reckoning do not fall in the realm of trigonometry. He believes that "trigonometry" begins with the use of clearly defined trigonometric functions in computational situations. What do you think—what is trigonometry and when did it begin?

4. In what historical period did recognizable trigonometric tables first appear in European mathematics books? Approximately, when was the first illustration of a sine curve used in a mathematics book? In what context was it employed?

5. When students are first introduced to trigonometric concepts, they usually begin by considering problems in which certain properties of right triangles can be used to determine inaccessible heights and distances. Their work is based on the ratio concept and the relationships of the sides of a right triangle. It allows practice in developing and solving simple equations. However, when these students encounter trigonometry again in higher grades, the teaching-learning approach is more formal and abstract. They now generalize the use of the ratios they previously learned to any triangles, not just right triangles, but even more confusing, the ratios are described as trigonometric functions and not ratios at all! How might this transition be used? What teaching strategies can be used to help bridge the ratio-function learning gap?

6. In the third century B.C., the Greek mathematician Eratosthenes used shadow reckoning principles to compute the circumference of the earth. He noted that during the summer solstice in the Egyptian city of Syene, the sun was directly overhead while at the same time in Alexandria, a city 5000 stadia (a Greek unit of linear measurement) distant, its rays were inclined $7°12'$ to the vertical. Because $7°12'$ is $\frac{1}{50}(360°)$, the theory of proportions shows that 5000 stadia $= (\frac{1}{50})$(circumference of the earth), or that the earth's circumference is 250,000 stadia. While we do not know the exact length of a stadium, we know it was approximately 183 meters, so, for his times, Eratosthenes obtained a good estimate for the earth's circumference. Eratosthenes technique can be used by modern day mathematics classes to obtain their own measurement for the earth's circumference. First, consider the situation depicted below in Figure 9 where two locations, C and B, are chosen on the earth's surface, so that one is due north of the other and the distance D between the locations is known. Two vertical staffs are erected, one at each location and at a specified time shadow measurements are taken. From these measurements the sizes of angles are determined. In order to obtain better accuracy, these measurements are retaken over a period of several days and average values for γ and β computed. Show that $\alpha = \beta - \gamma$. If this is true and we let the measure of the circumference of the earth equal E,

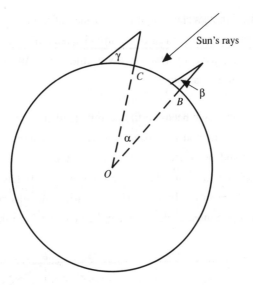

FIGURE 9

then

$$\frac{\alpha}{D} = \frac{360}{E} \qquad \text{or} \qquad E = \frac{360D}{\alpha}.$$

To undertake this project with a class, a cooperating partner class must be found in a school directly north or south of the first class's school.

7. If an observer is stationed at a point A above the surface of the earth, how far can he see to the horizon? Consider the situation as shown in Figure 10. Let $OC = OB = r$, the radius of the earth; let $AC = h$, the distance above the earth and $AB = d$, the distance to the horizon. Assume a value for r is known and thus treat it as a parameter. Develop a functional relationship between h and d, i.e., $d = f(h)$. If $r = 6374$ km, find the theoretical viewing distance for:

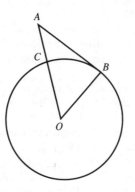

FIGURE 10

a. A man of height 2 meters standing on the surface of the earth.

b. A sailor in a look out on the mast of a ship 35 meters high.

c. An astronaut in a spacecraft 160 km above the earth. What percentage of the earth's surface is visible to the astronaut.

8. While Aryabhata may have been the first known mathematician to conceive of and use geometrical constructions to demonstrate trigonometric relationships, this instructional technique has remained popular and been used by many teachers over the centuries. Consider the geometrical situation depicted in Figure 11 where the given circle has a diameter $BE = 1$ unit. If $BD = \sin\theta$ and $BC = \tan\theta$, find the trigonometric functions corresponding to AB, AE, EC, and BF. What trigonometric identities can be deduced from this diagram?

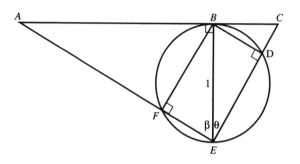

FIGURE 11

Bibliography

1. Amma, T. A. Sarasvati: *Geometry in Ancient and Medieval India.* Delhi: Motilal Banarsidass, 1979.
2. Ang Tian Se and Frank J. Swetz: "A Chinese Mathematical Classic of the Third Century: *The Sea Island Mathematical Manual of Liu Hui,*" *Historia Mathematica* 13 (1986), 99–117.
3. Burton, David: *The History of Mathematics: An Introduction.* New York: Allyn and Bacon, 1985.
4. Chace, A. B.: *The Rhind Mathematical Papyrus.* Reston, VA: The National Council of Teachers of Mathematics, 1978 (reprint of 1927, 1929 editions).
5. Colebrooke, Henry Thomas: *Algebra with Arithmetic and Mensuration from the Sanscrit of Brahmegupta and Bhascara.* London: John Marry, 1817 (available in reprint from University Microfilms).
6. Cullen, Christopher: "An Eighth Century Chinese Table of Tangents," *Chinese Science* 5 (1982), 1–33.
7. Kennedy, E. S.: *The Exhaustive Treatise on Shadows by Abu al-Rayhan Muhammad b. Ahmad al-Biruni: Translation and Commentary.* Aleppo, Syria: Institute for the History of Arabic Science, 1976.
8. Lam Lay-Yong and Shen Kangsheng: "Mathematical Problems on Surveying in Ancient China," *Archive for History of Exact Sciences* 7 (1986), 1–20.
9. Li Yan and Du Shiran: *Chinese Mathematics: A Concise History.* Oxford: Clarendon Press, 1987.
10. Lih Ko-Wei: "From One Gnomon to Two Gnomons: A Methodological Study of the Method of Double Differences." Paper presented at The Fifth International Conference on the History of Science in China, University of California, San Diego. August 5–10, 1988.

11. Needham, Joseph: *Science and Civilization in China*, Vol. 3. Cambridge: Cambridge University Press, 1959.
12. Pannekoek, A.: *A History of Astronomy*. London: George Allen and Unwin, 1961.
13. Qian Baocong (ed.): *Suanjing shi shu (The ten mathematical classics)*. Shanghai: Zhoaghua shuju, 1963.
14. Smith, D. E.: *History of Mathematics*, Vol. 2. New York: Dover Publications, 1958.
15. Swetz, Frank and Kao, T. I.: *Was Pythagoras Chinese? An Examination of Right Triangle Theory in Ancient China*. Reston, VA: National Council of Teachers of Mathematics, 1977.
16. Vogel, Kurt: "Ein Vermessungsproblem Reist von China noch Paris," *Historia Mathematica* 10 (1983), 360–367.
17. Weidner, E. F.: "Ein babylonishes Kompendium der Himmelskunde," *American Journal of Semitic Languages and Literature* 40 (1924), 186.

Johannes Kepler

Alluvial Deposits, Conic Sections, and Improper Glasses, or History of Mathematics Applied in the Classroom

Jan A. van Maanen

Introduction

Anyone wanting to promote the use of material from the history of mathematics in mathematics teaching is confronted with two types of questions. The first question is: "Why in teaching should one use historical material, or, more generally, refer to the history of mathematics?" The second question is posed by those who want to give history of mathematics a chance: "How should one apply history and what sort of material is available?"

The first question leads to a theoretical approach, in which the psychology of learning and views on mathematics and history must be considered. It is an important question, which needs thorough discussion. Unless there is a clear answer it will be impossible to convert the unbelievers, and it will be more difficult to set goals for the practical work. The following section of this paper is devoted to the first question.

The main emphasis in the paper, however, will be on the second question, for which I discuss themes from my research in the history of mathematics that eventually reached my mathematics classes. So many teachers say that they would like to use history in their teaching but they don't know how. Perhaps the concrete examples described below will help to provide them with ideas on this subject.

History in Mathematics Lessons: Why?

There are children who do mathematics because it appeals to them. They like every topic that the teacher offers, and their enthusiasm increases proportionally with the complexity of the task. Open problems are the best of all. Such pupils see mathematics as an art in its own right; they are convinced of its value and do not need much external motivation.

Usually classes do not contain very many pupils with such enthusiasm. Most pupils have to be stimulated to learn mathematics by an external source. The most direct source of motivation is, of course, the fact that mathematics occupies a central position in society. This central position is reflected by the rules set for admission to institutions of tertiary education, universities, for example. In the case of many academic disciplines one of the requirements for the admission of prospective students is that they have completed the mathematics curriculum of their school and have passed the corresponding part of the final examination. Pupils are conscious of the fact that unless they have studied mathematics to a certain standard they are cut off from many subsequent courses of study.

The history of mathematics can also serve as a stimulus for learning mathematics. What we nowadays know as an abstract formal theory, a set of eternal and unchangeable truths, usually originated as a reaction to a problem, and the truths that resulted were often found only after a tremendous amount of human endeavor. Our present-day theorems, absolute as they may be, remain the solutions to yesterday's problems. It is possible to learn a theory without knowing the problems that were behind it, but for many pupils a knowledge of this background can provide a sound and logical stimulus for learning.

Furthermore, the history of mathematics clarifies some of the problems that are inherent in certain areas of mathematics. The concept of "infinity," for example, has been discussed for centuries. It is my experience that modern pupils who are wrestling with the (psycho-)logical problems of "infinity" can profit from a flash-back to some of the arguments put forward in the past. Another argument for using history of mathematics in the classroom is that the study of alternative or possibly even forgotten topics and methods means that we can better evaluate our present techniques and recognize both their strengths and weaknesses. And of course history may reveal interesting starting points for mathematical research. Problems, for example, which would not arise in our times, but which are connected in a natural way with developments in the past, may serve as an unexpected source of inspiration, as a later section will show.

In the following sections of this paper I discuss three specific classroom applications of these instances.

Alluvial Deposition

The historical background. In 1355 an Italian professor of law, Bartolus of Saxoferrato (1313–1357), wrote a treatise on the division of an alluvial deposit. The problem that he discussed is shown in Figure 1: some landowners—Bartolus calls them Gaius, Lucius and Ticius—have neighboring properties beside the bank of a river. The river deposits silt so that new land is formed at the river bank. How is this new fertile soil to be divided up? [6]

A possible solution is to reconstruct the boundaries in the deposit by extending in a straight line the previous boundaries between the properties. This solution, however, could lead to conflict, (see Figure 1 and imagine that Lucius fervently sticks to the idea of extending the extant boundaries in a straight line!). In fact, Bartolus actually became involved in such a conflict while on holiday near the Tiber River. He quickly realized the real importance of the problem and decided to search for a solution. Of course this was not the first time such a problem existed. It had been treated in the oldest texts of Roman law, the *Institutes* of Gaius, c. A.D. 150, and in the work of Justinian, c. A.D. 530. These authors and their later commentators,

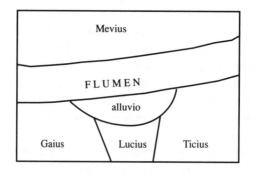

FIGURE 1

however, did not present a practical solution, but only formulated a general principle. In such texts alluvial deposit was considered as *accessio*, accretion. Since accretion is something new, a decision has to be made about who is the owner. The rule is obvious: accretion belongs to the owner of the good to which it has accreted; for example, a calf of my cow and the apples of my tree are my property.

Bartolus, and in this respect his work was new, pointed to the difficulty arising from this rule about the ownership of accretion. The rule can be applied in the case of alluvial deposit only if it is clear to whose land the new land has been added. Bartolus proposed a mathematical criterion:

New land will be the property of the owner of the nearest old land.

According to this rule, points which are equidistant from two existing properties form the boundaries in the new land. Bartolus applies his rule to numerous geographical situations, which are shown in a series of maps, and—a good teacher —he increases the difficulty step by step. He starts with the case in which the bank of the river originally consisted of segments of straight lines. In this case, the new boundary will be the perpendicular to the old bank if the latter was a straight line. See Figure 2. Or it will be the bisector of the angle in the old bank as shown in Figure 3.

Returning for a moment to Lucius, who was in favour of extending the old boundaries in a straight line, we must conclude that he fares badly with Bartolus' rule, for in all three cases that are shown in the previous figures, he has to reduce his claims considerably.

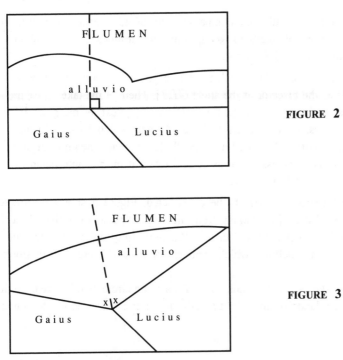

FIGURE 2

FIGURE 3

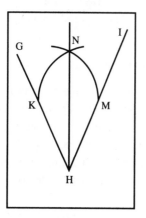

FIGURE 4

FIGURE 5
The construction of the bisector by Bartolus.
Bibl. Vat.: *Barb.Lat.* 1398 f.164^r

The Euclidean constructions with ruler and compass are discussed extensively, both for the perpendicular and for the bisector; this leads to passages and illustrations—rather uncommon in a law treatise—like the quotation and Figure 4:

> Draw two straight lines, GH and HI, which meet in a point H, above which you want to draw the line [i.e., the bisector of the angle GHI]. Then let me take on each of these lines a point at equal distance [i.e., from H]: K, M. Next let me place in K the point of the compass, which I open as far as M and which I rotate in the direction in which I want to draw the line. Subsequently let me place the point of the compass in M and open the compass as far as K and rotate it in the same manner. Then these two circles intersect in point N.

Line HN will be the bisector that was to be constructed. Figure 5 shows the quoted passage in the manuscript that I used [8], namely from the words "ponas duas lineas" at the end of line 12 to "se perfindunt in puncto .n." in line 18; interesting too is the reference in line 11: "ut probatur per .x. primi Euclidis" which explicitly refers to Theorem 10 of Book 1 of Euclid's *Elements*.

The next case studied by Bartolus is the bank which is part of a circle. See Figure 6. In this case the boundaries in the new land are given by the radii of the circle, so one has to connect

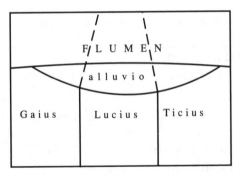

FIGURE 6

the end points of the boundaries in the old land to the center of the circle. Indeed, Bartolus does present the Euclidean construction of the center of a given circular arc, as is clear from Figure 7, which stems from one of the first printed editions of the treatise. [1, p. 101]

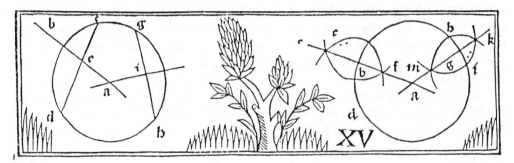

FIGURE 7
Construction of the center of a circular arc

More and more complex situations are considered, in which the old bank is made up of combinations of straight lines and circular arcs. Arbitrary curves are not found in Bartolus, but he is not far in his work from a general discussion of the normal to a curve at a given point.

Much more could be said about the history of mathematics as reflected in this treatise. One could study the mathematical style of Bartolus and consider to what extent the treatise gives information about the mathematical knowledge of an educated layman in the Middle Ages, but it seems better now to concentrate on the classroom application of the described material.

The classroom application.   The goals set for my work in the classes (three first-year classes of a Dutch grammar school: pupils about 11 years old, who follow a curriculum which includes 3 hours of Latin and 4 hours of mathematics per week) were:

- to demonstrate the importance of mathematics in society;
- to show how profitable it can be to work together in solving open problems;
- to integrate disciplines, i.e., Latin and mathematics;

and more specifically related to the mathematics program:

- to let pupils "invent" a number of constructions with ruler and compass (in The Netherlands this subject no longer belongs to the basic program, but it is a very natural extension of work on symmetry and reflection);
- to let the pupils solve some legal problems using the constructions that they had "invented" earlier.

The project was split into a series of three lessons concerning ruler and compass constructions and a month later a series of five lessons: 2 of Latin, 3 of mathematics, on the division problem. In both series the work was done in small groups averaging 3 pupils each.

The first series started with the following exercise:

Draw a line segment. There are two equilateral triangles which have the drawn line segment as a side. Construct these triangles with ruler and compass. You now have a quadrangle with one diagonal. Calculate the size of the angles in this figure. Draw the second diagonal of the quadrangle. Calculate the size of the angles that you have constructed by the last step.

In this exercise the possibility of bisecting an angle (limited to the special case of 60°) and constructing a right angle were hinted at. In the next exercises this construction was generalized step by step so that the pupils became acquainted with most of the elementary constructions (bisector, mid-point of a line segment, lines through a given point and perpendicular or parallel to a given line, the center of a given circular arc, etc.). The pupils worked with enthusiasm. They were seen to be discussing intensely, gesticulating, drawing, and making one discovery after another.

Then came the Bartolus part of the project. After some preliminary work, in which Bartolus was introduced and the division problem was posed (when you have not yet had to work with the distance criterion the problem leads to heated discussions: to some pupils it was not at all obvious that the new land should be added to the properties of those who already owned so much land!). The distance criterion was then formulated:

New land will belong to the owner of the nearest old land.

So the clump of grass A is added to the land of Gaius, for it is nearer the old land of Gaius than the old land of Lucius. The clump in B is added to the property of Lucius. The clump in C lies on the new boundary, for it is just as far from the old land of Gaius as it is from the old land of Lucius.

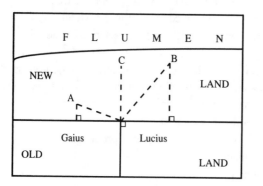

FIGURE 8

Having been given this criterion the groups of pupils were asked to apply it with the use of ruler and compass constructions to a number of division problems, which were based on Bartolus or directly borrowed from him. Figures 9, 10 and 11 present a cross-section of the proposed division problems.

In the two Latin lessons, the pupils were confronted with the linguistic aspects of the text. A fragment of the text, like the one given in Figure 5, had to be deciphered, and an edited fragment, about the bisector, was translated. For a class that had just started Latin, the text was, of course, much too difficult to read without help, but that was not the aim. The aim was to lead the pupils to realize that it is impossible to interpret the sources of Western culture without knowledge of classical languages. By and large that aim was achieved.

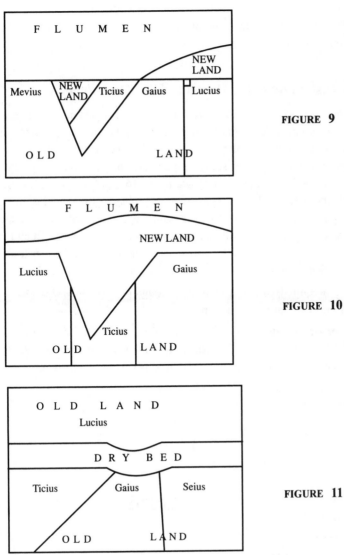

FIGURE 9

FIGURE 10

FIGURE 11

Looking back on the project.   The historical context is not a necessary condition for using the division problem in the classroom but history did provide an important extra dimension. The classes saw someone struggling with a problem in real life, someone who realized that he could solve it with the help of mathematics. His solution, the distance criterion, is still in use today. In the course of time it became canonized and reached present-day international law via the Napoleonic Codes. On 25 April 1988 there was a news item about Japanese efforts to keep the coral island Okino Torishima above sea level by means of concrete dikes. The reason for their efforts was the distance criterion—the Japanese will lose their rights to the surrounding fishing waters and to the exploitation of the ocean bottom when the island disappears. The Falklands War, which was related to the claims of Great Britain to Antarctica, is yet another example of the application of Bartolus's solution to a division problem—by the distance criterion, occupying the Falkland Islands gave the British a claim to part of Antarctica.

Making contact with Bartolus was only possible via deciphering and translating, but that was simply an extra attraction to most of the pupils. They learned to work with point-sets in plane geometry, and simultaneously their knowledge of general history increased. Last but not least, they were greatly stimulated to learn Latin.

Seventeenth Century Instruments for Drawing Conic Sections

Background.   As an introduction to conic section drawers, which is the name I shall give to instruments for drawing conic sections in a continuous movement, I shall first present some background information about geometrical constructions (for a detailed account, see [2]).

Conic section drawers became a subject of study in the course of the seventeenth century in connection with the new method, introduced by Descartes in 1637, of solving geometrical construction problems with the help of algebra. Construction problems were fundamental to mathematics in Greek antiquity. The execution of constructions was governed by strict rules: one should use only ruler and compass and should perform only the actions which Euclid had declared practicable in postulates 1 to 3 of his *Elements*. An example of such a construction problem, belonging to the ancient tradition of the "division of figures," is shown in Figure 12.

Given: a triangle ABC. Required: to construct a line through point P on AB so that
the triangle is divided into two parts of equal area.

A possible solution along pre-Cartesian lines goes like this. Draw PC and draw a line through A, parallel to PC. The latter meets BC extended at a point which is called D. Draw PD. Since PC and AD are parallel, area $(\triangle APC)$ = area $(\triangle DPC)$, and therefore, area

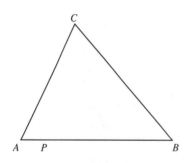

FIGURE **12**

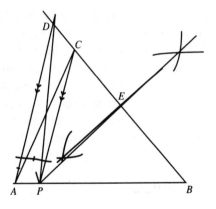

FIGURE 13

$(\triangle DPB)$ = area $(\triangle ABC)$. Let E be the mid-point of line-segment BD. Since the medians of a triangle divide the triangle into two triangles of equal area, PE divides $\triangle DPB$ into two parts of equal area, so area $(\triangle PBE) = \frac{1}{2}$ area $(\triangle DPB) = \frac{1}{2}$ area $(\triangle ABC)$. The conclusion is that PE is the line that had to be constructed. In Figure 13 the construction has been executed by ruler and compass.

Clearly the crucial point is to find out which construction steps are needed to solve the problem (the analysis stage). When these steps have been discovered the proof of the correctness of the construction (the synthesis stage) is generally easy. Descartes showed that algebra could be used to analyse such geometrical construction problems. His analysis would have run like this. (The following more or less paraphrases a passage from Descartes' *Géométrie*).

Start by making a sketch in which you imagine that the problem has already been solved. Represent the lengths of the line-segments which appear in the figure by letters, the known lengths by letters from the beginning of the alphabet, the unknown ones by letters from the end of the alphabet. Try to express algebraically in two different ways one of the quantities that appears in the figure, and set these two expressions equal to each other. This gives you an equation, and the solution of this equation relates the length of the unknown line-segment which has to be constructed to the given lengths. The geometrical interpretation of this algebraic solution yields the construction.

In our example, we make a drawing in which we suppose that line PU solves the problem. See Figure 14.

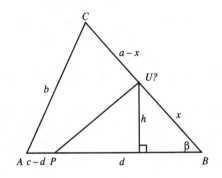

FIGURE 14

In the figure the lengths of AB, AC, and BC (c, b, and a respectively) and the distances AP and PB (here called $c - d$ and d) are known. Unknown are the lengths of BU and CU, here called x and $a - x$, and the height h of $\triangle PBU$, which, however, can be expressed in x directly: $h = x \sin \beta$. The subsequent steps speak for themselves: area $(\triangle ABC) = \frac{1}{2}ac \sin \beta$ and area $(\triangle PBU) = \frac{1}{2}dx \sin \beta$; and since the latter area must be half of the first area, one has $dx = \frac{1}{2}ac$, which is the equation relating the unknown x to the given lengths a, c and d. On the basis of this equation x has to be constructed; this can be done using the equivalent proportionality $x : c = \frac{1}{2}a : d$, which can be constructed via two similar triangles.

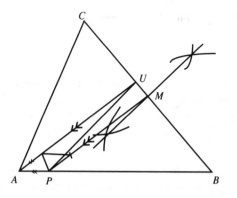

FIGURE 15

First construct (to produce a length $\frac{1}{2}a$) the mid-point M of BC, and draw PM. In $\triangle PBM$ the lengths $\frac{1}{2}a$ and d occur; c is the length of AB. So one has to construct a triangle which has a side AB and which is similar to $\triangle PBM$. This is easily done by drawing through A a line parallel to PM, which intersects BC in U. Then triangles ABU and PBM are similar, so $BU : AB = BM : PB$, that is $BU : c = \frac{1}{2}a : d$. We conclude that BU satisfies the proportionality $x : c = \frac{1}{2}a : d$, which implies that PU is the required line through P which divides $\triangle ABC$ into two parts of equal area.

For Descartes this procedure was a new and powerful tool for investigating whether a construction could be executed by ruler and compass. For, if the analysis leads to an equation of first or second degree and if a solution exists, then a ruler and compass construction is possible, a fact proved explicitly by Descartes. If the resulting equation is of third or fourth degree, there might be a ruler and compass construction, but it will exist only in special cases. In general, said Descartes, one will have to use other means to construct a solution. Referring to the ancient Greek notion of "solid" problem which could be constructed with the help of conic sections, he proved that, if x satisfies an equation of third or fourth degree, x can be constructed geometrically if one allows not only circles and straight lines as construction curves but also conic sections.

This approach, however, caused a serious problem. Circles and straight lines can be drawn in a continuous movement (at least: that was postulated by Euclid; the straight ruler of course exists only as a mental concept), but can conic sections also be drawn continuously? They can be constructed point by point, but if one wants to intersect a conic section with other construction curves, a point by point construction does not suffice. So the admission of conic sections as construction curves is practicable only if there are instruments to draw them in a continuous movement. This explains the growing interest in conic section drawers around the middle of the seventeenth century.

The conic section drawers of van Schooten.   In the *Géométrie*, Descartes did not say anything about conic section drawers. Frans van Schooten (1615/6–1660), the Leiden professor of mathematics, who filled in many gaps in the *Géométrie* and who was the great propagator of Cartesian mathematics, studied conic section drawers to make up for this deficiency of Descartes. The magnum opus of van Schooten was the Latin edition of Descartes' *Géométrie* (1649, 2nd enlarged ed. 1659-1661), which contained, in addition to the Latin text, commentaries by van Schooten and results of investigations by his students into the power of the new analytic geometry. Van Schooten's first publication of his own creative work was his treatise on conic section drawers, *De organica conicarum sectionum in plano descriptione, tractatus*, (Leiden, 1646), reprinted in *Exercitationum Mathematicarum libri quinque*, (Leiden, 1656/7). Figures 16, 17 and 18 are taken from the treatise. They show three conic section drawers.

The instruments are depicted as seen from above, lying on a horizontal plane on a sheet of paper. The instrument in Figure 16 is operated by moving the ruler GI along the ruler EG, perpendicular to it. The rhombus $BFGH$ ($BF = FG = GH = HB$) has hinges in B, F, G, and H and is fixed to the plane by a pin in B, and to ruler GI by a pin in G. At F a slotted ruler is attached to the rhombus. A pin in H goes through the slot, so that the slotted ruler always forms the diagonal of $BFGH$. In D, where the rulers FH and GI (which is also slotted) intersect, a style, passing through the two slots, draws a curve. Triangles BHD and GHD are congruent, for they have HD as a common side and furthermore $BH = HG$ (since $BFGH$ is a rhombus) and $\angle BHD = \angle GHD$ as in the rhombus the diagonal FH bisects the

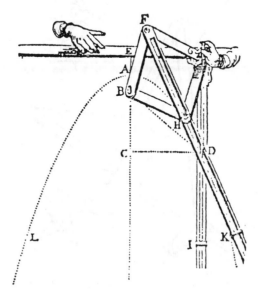

FIGURE 16
Tractatus p. 74

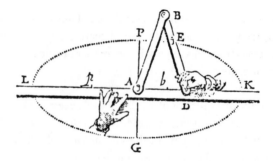

FIGURE 17
Tractatus p. 26

angle H both internally and externally. From the congruence it follows that $BD = GD$, so D is at the same distance from B as from the ruler EG, and therefore the curve drawn by the style in D is a parabola.

In Figure 17, A is a point on the ruler KL. AB and BD are of equal length. AB rotates around A, and BD is hinged to AB at B. The instrument is operated by moving D along the ruler KL. In a fixed point E of BD a stylus is placed, which draws an ellipse.

The functioning of the third instrument as shown in Figure 18 may now be clear. One should know that $CD = GF$ and $DG = CF$. The style, held by the hand, draws a hyperbola, of which the vertices (E and K; CD is made equal to EK) and the distance between the foci (CF) are given.

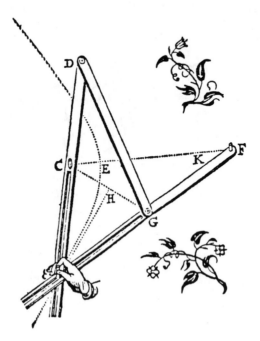

FIGURE 18
Tractatus p. 57

The proof that the second and the third instruments produce conic sections is left to the reader (a well-known Cartesian phrase). A possible approach may be derived from the problem that is quoted below. It was set in a leaving certificate examination of the grammar school where I teach. (In the Dutch school system, half of the final mark is determined by a government-set examination, which is the same for all schools of the same type; the other half of the mark depends on tests that are set by the schools themselves. The problem quoted comes from such a school test).

The subject was "conic sections," but at an earlier stage the pupils had already studied plane curves, given by a one-parameter representation, and subjects from the history of mathematics. The problem is as follows:

> For a long time mathematicians treated conic sections with a certain suspicion, since there were no instruments by which you could draw them as accurately as, for example, a straight line with the use of a ruler or a circle with the use of a pair of compasses. In the seventeenth century the Leiden professor of mathematics Frans van Schooten (1615–1660) studied this problem. As a result he describes in his book *Exercitationum Mathematicarum libri quinque*, among other things, some methods for drawing an ellipse mechanically. One such method is the well-known one with the string (see Figure 19).

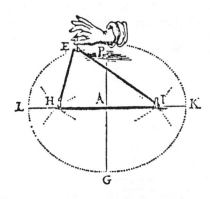

FIGURE 19
From Frans van Schooten,
Exercitationum Mathematicarum

a. Explain why the hand with the pencil in E draws an ellipse.

A second instrument works as follows (see Figure 20; compare Figure 17). A bar AB of fixed length a rotates around A; at B a second bar BE is hinged to it, on which there is a point D which moves along the x-axis such that $BD = BA = a$.

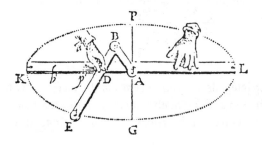

FIGURE 20
From Frans van Schooten,
Exercitationum Mathematicarum

The length DE is called b. Van Schooten claims that the pencil in E describes an ellipse if D moves along the x-axis.

b. Let A be the origin. If we take it for granted that the set of points E is indeed an ellipse, what is the equation of that ellipse?

Finally we shall prove that the set of points E really is an ellipse. Again let A be the origin, and let E be (x,y) as in Figure 21, so for convenience take $x > 0$ and $y > 0$. As above, $AB = BD = a$ and $DE = b$. Every direction of line BE with angle φ determines a point $E(x,y)$.

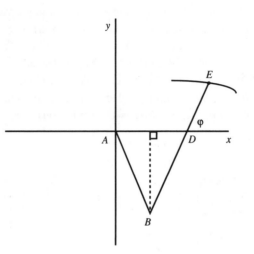

FIGURE 21

c. Express the coordinates of B, D, and E in terms of a, b, and φ.
d. Derive from c. an equation that is satisfied by the coordinates of the points E; formulate a conclusion.
e. Would this instrument of van Schooten's be as suitable for drawing conic sections as the compass for drawing circles?

Not an easy problem, the candidates said, but a nice one.

In this case, history is a source of inspiration to the teacher, but it is also important to the pupils. They were already familiar with the technique of studying a curve from its parametrization, and here history provided a curve to which this technique could be applied in a very natural way. They had learned the technique since it is prescribed in the curriculum, but here they saw its usefulness.

Improper Glasses

Every teacher who has discussed improper integrals in class remembers the incredulous faces of pupils when they realize for the first time that an infinitely extended part of the plane, for example, the part between the graph of a function and the x-axis, can have a finite area. The same disbelief, or at least surprise, arises when they are confronted with an infinitely extended solid, the volume of which is finite.

While teaching improper integrals some years ago, I told the pupils that even top ranking mathematicians could not believe their eyes when they discovered this phenomenon, in about the middle of the seventeenth century, and I added that the phenomenon had actually led to a crisis. Two girls sitting in the front row challenged me to investigate this matter thoroughly, and fortunately they suggested that I should make a written report of my findings. Their suggestion enables me now to finish this paper with the report that I read to the class and then submitted to the girls who felt very proud. A much more detailed version of this report is found in [7].

In 1641 the Italian mathematician Torricelli (1608–1647) discovered that the volume of the infinitely long solid that you get by rotating an infinite part of an orthogonal hyperbola about its asymptote is finite. This is easy to see now that we have integral calculus at our disposal. Consider, for example, $y = \frac{1}{x}$ for $x \geq 1$, and rotate this part of the hyperbola about its asymptote, the x-axis. See Figure 22. The volume of the infinitely long solid that is formed is:

$$\int_1^\infty \pi \left(\frac{1}{x}\right)^2 dx = \lim_{t \to \infty} \int_1^t \pi x^{-2} \, dx = \lim_{t \to \infty} [-\pi x^{-1}]_1^t = \pi,$$

so the solid of revolution does indeed have finite volume. Torricelli's proof, of course, went quite differently, since he did not have integral calculus at his disposal; that was invented only about 1670.

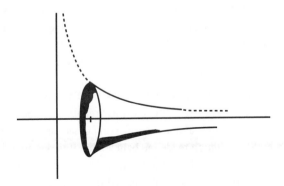

FIGURE 22

It is not completely clear whether Torricelli was the first to discover that a figure with infinite dimensions can have a finite volume. The French philosopher Oresme (c. 1320–1380) has made similar statements. What is clear, however, is that Torricelli's discovery was a sensation of the first order in the scientific world of his time. We can follow the reactions in the correspondence of Marin Mersenne (1588–1648) [9]. This French monk and scientist exchanged innumerable letters with scientists all over Europe, and in this way he greatly facilitated the diffusion of new scientific discoveries. Nowadays, scientists exchange new ideas by publishing them in scientific journals, but the first of these was founded only in the second half of the 17th century;

so Mersenne was an extremely important pivot in the scientific community of his time. Mersenne thought Torricelli's new theorem so important that he communicated it in many of the letters that he wrote between 1643 and 1646. And via his correspondence the news spread like wildfire. It led to a number of other studies on the same subject. I shall give one other example.

In 1658, Huygens (1629–1695) and Sluse (1622–1685) studied methods for determining the surface area and volume of infinitely extended solids of revolution. Sluse, in a letter to Huygens, referred to Torricelli's discovery as an example. Then they discovered something similar: a solid that one gets by rotating a certain curve about its asymptote, the surface area of which is finite (so that it could be made from a finite amount of material), has infinite volume.

Sluse, in his letter to Huygens, expressed the fundamental property of the solid in a very lively way by making the proud remark that, using the discovered curve, he now could give the measurements of a drinking glass (or vase), that has a small weight, but that even the hardest drinker could not empty ("levi opera dedicator meansura vasculi, pondere non magni, quod interim helluo nullus ebibat.") [5, vol. 2, p. 168].

[WARNING: When I originally wrote this, I believed it to be the proper interpretation. At present I know that I misread Sluse, who did NOT consider the surface area, and who did NOT claim that a solid exists with finite surface area and infinite volume. In fact, such a solid does not exist, as is proved in an appendix. I discuss my error after the next paragraph.]

To summarize: the discovery that objects with infinite extension can have a finite area or volume was made (or made anew) in the first half of the seventeenth century. The discovery spread quickly and was widely studied, so it certainly had an impact on scientists. The use of the word "crisis" to characterize the reaction of the mathematical community seems rather too strong, considering the information presented here, but it is possible that further research will reveal sources to support this characterization. (See [3], [5, vol. 2, pp. 164, 212; vol. 14, pp. 199, 200, 306, 312], [9, vol. 12–14].)

There is a lesson to be learned from my mistake. I have opted for a lesson in public, although it could have been private if I had chosen to "repair" the above quotation.

What happened? When reading about improper integrals for my pupils, I made a wrong interpretation of a text by Sluse, which made me believe that it is possible to construct a solid of revolution with finite surface area and infinite volume. I inserted that wrong idea of mine into the text that I wrote for my pupils, and literally quoted this text at the Kristiansand conference. Then it was pointed out to me that a solid of revolution with finite surface area must also have a finite volume.

So the question arose what did Sluse and Huygens really claim? The answer is that Sluse found an infinitely extended solid, not with finite surface area but with finite volume so that it could be made from a finite amount of material. This solid, however, enclosed a cavity of infinite volume. So Sluse indeed designed a "glass" that, when empty, has a small weight, but that even the hardest drinker could not empty. To understand what he and Huygens were doing, we must turn to the sources, published in [5].

In 1658 Huygens and Sluse were studying the cissoid (for its definition, see [5, vol. 1, pp. 264–266]). Let ABC be an arc of the cissoid, and GD its asymptote. See Figure 23. Sluse lets C and G tend to infinity and he rotates ABC about the asymptote GD. He proves that the solid produced in this manner has finite volume. Huygens adds in his answer that the area

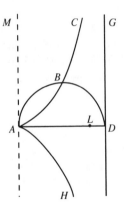

FIGURE 23

of the infinitely long region bounded by arc ABC and the straight lines AD and DG is the triple of the area of the cissoid's generating half circle ABD. Next Huygens reflects arc ABC in the axis AD so as to have arc AH as its image. Then he is able, with the help of Pappus's theorem, to locate the center of gravity of the region bounded by the two arcs and the asymptote in point L of AD (L is on AD for reasons of symmetry) so that $AD = 6LD$.

Then, says Sluse in his next letter, let the region between $CBAH$ and its asymptote rotate about AM, the parallel to the asymptote which passes through A. The volume of the produced solid as shown in Figure 24 is finite as well (again an implication of Pappus's theorem, which gives for the volume of the solid the product of the length of the orbit of L under one rotation with the area of the rotated region; both factors are finite). Therefore it can be made from a finite amount of material. The cavity that is left open around the axis of rotation AM, however, has infinite volume, which proves Sluse's claim.

I missed the track at the introduction of the second rotation, about AM. Preparing the story for the two girls I read too superficially and thought that the solid produced by the rotation about the asymptote already produced the "glass that even the hardest drinker could not empty." It will remain a structural problem for teachers, who want to integrate history in their lessons,

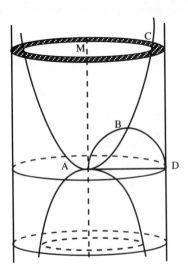

FIGURE 24

that studying the sources takes much more time than is available for preparing lessons. In such circumstances a mistake may perhaps be excused, but in a paper on the history of mathematics it should not appear, and therefore I am very grateful to Tony Gardiner and Man-Keung Siu for pointing out that there was something wrong, which enabled me to correct my error.

Conclusion

What we have learned, or rather, what I have learned in the last few years, is that this kind of application of the history of mathematics in the classroom is very fruitful. The drawback is that one has to do quite a lot of research before one has the material at hand which one can really use in the classroom.

With respect to this conclusion I would make two comments. In the first place, historians of mathematics should realize that there is a challenge here (just like the challenge made by the two girls in the front row): a challenge to transfer historical sources into material that can be used in schools, to furnish ideas, to suggest projects. And secondly, since most things in life are governed by conservation laws, a conservation law will be involved: history of mathematics in the classroom requires much energy, but the profits are high as well (interested pupils and, as a consequence, pleasant lessons). "Ordinary" lessons require less energy, but the profits diminish in proportion.

Appendix

In this appendix it is proven that a solid of revolution that has finite surface area has finite volume as well.

Take the graph of a function $f \in C^1(D)$ where D is an arbitrary, connected subset of the real numbers, rotate it about the x-axis, and assume its surface area to be finite. See Figure 25. First remark that f must be bounded. For suppose that f is not bounded, then $\lim_{x \to a} |f(x)| = \infty$ for some real number a. There is a b near a such that $(a, b] \subset D$ or $[b, a) \subset D$. The surface area that one gets by rotating the graph of f on $[a + \varepsilon, b]$ about the axis

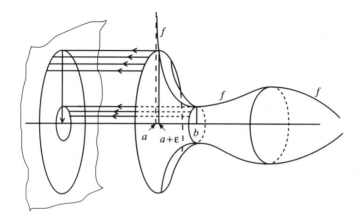

FIGURE 25

is greater than the area of its projection on a plane perpendicular to the axis. The latter area, $\pi(f^2(a+\varepsilon) - f^2(b))$, becomes infinite if $\varepsilon \downarrow 0$. The same argument can be used if $|f(x)| \to \infty$ for $x \to \pm\infty$. So, if the surface area is finite, then there is a positive constant M such that $|f(x)| < M$ for every $x \in D$.

Now look at the volume V of the solid, constructed in the beginning, the surface area S of which was assumed to be finite.

$$V = \int_D \pi f^2(x)\, dx \leq \int_D \pi M |f(x)|\, dx = \frac{M}{2} \int_D 2\pi |f(x)|\, dx$$

$$\leq \frac{M}{2} \int_D 2\pi |f(x)| \sqrt{1 + (f'(x))^2}\, dx = \frac{M}{2} S$$

And therefore, if the surface area of the solid is finite, the volume of the solid must be finite as well.

Bibliography

1. Bartolus de Saxoferrato: *Opera Omnia,* Lyon, 1510–1511.
2. Bos, H. J. M.: "Arguments on motivation in the rise and decline of a mathematical theory, the 'Construction of equations'," *Archive for the History of Exact Sciences* 30 (1984), 331–380.
3. Boyer, C. B.: *The History of the Calculus.* New York: Dover, 1959.
4. Heath, T. *A History of Greek Mathematics*, 2 vols. New York: Dover, 1981.
5. Huygens, Christiaan. *Oeuvres Complétes*, 22 vols. The Hague, 1888–1950.
6. van Maanen, Jan. "Over het verdelen van aangeslibid land. Een Brugklas projekt," *Euclides* 60 (1984/85), 161–168; "Teaching Geometry to 11 year old 'medieval lawyers'," *The Mathematical Gazette* 76 (March, 1992), 37–45.
7. van Maanen, Jan. "From quadrature to integration: thirteen years in the life of the cissoid," *The Mathematical Gazette* 74 (March, 1991), 1–15.
8. Vatican Library, Barberiniani Collection, Lat. 1398, f 164.
9. de Waard, C. *et al.* (eds.) *Correspondance du P. Marin Mersenne.* Paris: G. Beauchesne, 1932–.

Blaise Pascal

An Historical Example of Mathematical Modeling:
The Trajectory of a Cannonball

Frank J. Swetz

Cannon first appeared on the European scene in the fourteenth century. An extant illustration of these "infernal devices" can be found in Edward III's *De Officus Regnum* (*The Duties of Kings*) compiled in 1327. Early cannon were extremely primitive. As a weapon of war, their impact was primarily psychological—the noise they produced frightened horses, thus disrupting cavalry charges. Constructed from metal staves which were banded together into barrels, the power, accuracy and safety of early cannon were quite limited. Due to the improvement of technology, staved barrels were replaced by solid castings, inhibiting the likelihood of barrel fatigue, and the consistency of a shot's trajectory was improved by the machine-boring of gun barrels. Cannon became safer (for the user) and more accurate, and by the middle of the fifteenth century, cannon had progressed into a dreaded tool of warfare. Primarily established as siege weapons, cannon were effective in breaching ramparts and destroying castle walls. The privilege and protection of fortified isolation associated with the Age of Feudalism had ended. Artillery pieces obtained personalities and names as well as reputations—the famous "Mons Meg" of Edinburgh was 13 feet long, held a charge of 105 lbs. of powder and cast a cannon ball of approximately the same weight 4/5 of a mile. Powerful and temperamental, each artillery piece was an individual with peculiarities known only to its master gunner who alone, relying on experience, could gauge its range and effectiveness.

Spurred by the technological advances of the Early Renaissance, an empirically based science of warfare evolved. Military engineers puzzled over the logistics of supplying and transporting artillery pieces, of establishing effective firing patterns and of concentrating the impact of shot to breach enemy walls. Geometry and mathematics now became the tools of military strategy. (See Figure 1.) As the range of cannon increased beyond the scope of a gunner's vision, the questions of predicting and controlling that range became paramount. What was the path or trajectory of a cannon ball? Could it be described and analyzed by the use of geometry, the most powerful mathematics of the time? These questions were readily taken up by contemporary mathematicians, many of whom themselves served as military engineers. Further, such concerns fit into the scientific and intellectual controversies of the era as to the path of a projectile moving through space.

Theoretical Considerations

The intellectual milieu of the Late Middle Ages and Early Renaissance was dominated by the traditions of scholasticism which held that the writings and theories of Aristotle accurately

93

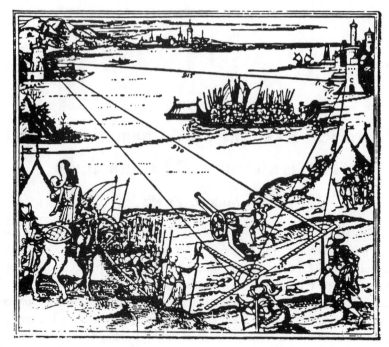

FIGURE 1
From Niccolo Tartaglia, *Nova Scientia*

described all natural phenomena. For Aristotle an inanimate object could not move without the existence of an ever present motive force. Unnatural motion had to be caused by external agents. But what about the flight of an arrow? Scholastics would answer that the arrow's flight is due to the medium in which it travels—the bow as it imparts a force to the arrow also disturbs the air around the arrow and this disruption causes the air to suck or pull the arrow along. Accordingly, motion in a vacuum could not exist. Of course, under the scientific conditions of the thirteenth and fourteenth centuries such theories could not be physically tested, but they could be intellectually challenged. Such a challenge was issued by the Parisian scholar, Nicole Oresme (1320–1382), who rejected Aristotle's concept of motion and replaced it with a theory of "impetus." According to Oresme, impetus was a quality imparted to a moving body by its initiating force. An inanimate body moved through space because it had impetus. Thus, for Oresme, it was possible to throw a heavy body farther than a feather since the heavy body had more mass than the feather and it absorbed more impetus. A projectile in the air is supported by its impetus, and as the impetus relaxes, it falls to the ground. Oresme theoretically partitioned the flight of a projectile into three segments: the first, in which it moves in contact with its propellant and is accelerated; the second, in which it leaves the propellant and in which its velocity continues to increase; and the last, where with impetus exhausted, natural motion towards the earth takes place.

This theory was expanded by the natural scientist, Albert of Saxony (c. 1365) who regarded the path of a projectile as influenced by three mutually opposing forces: the impetus, the resistance of movement through the air, and gravity. (See Figure 2.) During the first part of flight, impetus is greater than both gravity and resistance and the projectile travels in a straight

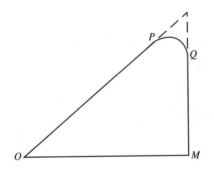

FIGURE 2

line, OP; as the excess of gravity over resistance increases, the projectile is still driven by the impetus but its path curves in a circular arc, PQ and finally, when gravity overcomes both resistance and impetus, the projectile falls to the ground in a straight line, QM. Thus a sketch of a curved path could be made to model the flight of projectiles including cannon balls.

Leonardo da Vinci (1452–1519) adopted this analysis with some slight modifications in his studies of ballistics. Other theorists ignored the curved section of the trajectory as being insignificant and modelled the path of a projectile by a right triangle. This right triangle model had several attractions. Experience with cannon had convinced mathematicians that the height and range of a trajectory were directly proportional to the strength of the powder charge employed for the shot. With a right triangle model of trajectory, range and height could be represented by the legs of the triangle. Further, range could be computed as a function of the angle of elevation.

The first comprehensive, published thesis on the trajectory of cannonballs, *Nova Scientia* (*The New Science*) appeared in 1537. It was the work of the Italian mathematician, Niccolo Tartaglia, who was initially inspired to the task by an inquiry on a practical problem of gunnery, namely, how to position a cannon in order to obtain the maximum range for a shot. Through observations and calculations, Tartaglia determined that the maximum range could be achieved if the cannon barrel was elevated 45° to the horizon. He concluded that range was a function of the angle of barrel elevation and devised a gunner's quadrant or square calibrated into 12 equal parts (Figure 3a). When placed in the barrel of a cannon this quadrant permitted a determination of the angle of elevation and thus an adjustment for range (Figure 3b).

Now, if a gunner used standard powder charges, a table of ranges could be compiled for a given cannon and consulted in combat situations. Tartaglia determined that in order to be effective, an artilleryman must be able to calculate the range to his target and must know the extent of his shots according to various cannon elevations. He set out to devise mathematical methods for both purposes. In doing so, he established the 45° maximum range rule and determined that any intermediate range can be achieved by firing at two different elevations whose angles were complementary. Tartaglia noted that air resistance was a real factor of concern in the study of ballistics and that the trajectory of a projectile, due to the effects of gravity, is curved at every point of flight. But despite this latter assumption, in his calculations and drawings he still approximated the path of a cannon ball by straight lines and circular arcs (Figure 4).

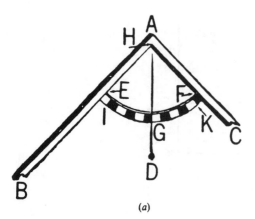

(a)

(b)

FIGURE 3
From Niccolo Tartaglia, *Nova Scientia*

FIGURE 4
From Niccolo Tartaglia, *Nova Scientia*

In many respects, Tartaglia's ballistics satisfied the empirical data known about artillery fire at that time and, although some of his theories were incorrect, they became popular and were incorporated into many manuals on the use and disposition of cannon. A more scientific and exacting theory of ballistics would appear in the next century.

Mathematical Considerations

By the middle of the seventeenth century, cannon could fire several miles and a more exacting theory of trajectory and range was required. Galileo Galilei (1564–1642) and his disciples were among those who undertook the task. Through careful experimentation and observation, Galileo noted that if a cannon, elevated above a plane, was fired horizontally across the plane using different powder charges, the ranges achieved would be different; however, the shots always hit the ground at the same elapsed time interval. He concluded that although the horizontal velocities changed due to the use of different powder charges, the vertical velocities remained the same. In essence, the velocity of a moving body could be analyzed into two components: a horizontal velocity and a vertical velocity.

In experimenting with this trajectory, Galileo also noted that if a body is projected along a surface horizontally and allowed to fall into free space its distance of descent obtained in successive time intervals: $t_1, t_2, t_3, \ldots, t_k$, would be in the ratios given by the sequence of numbers $1, 4, 9, \ldots, k^2$. It followed that the motion described by the falling body and the horizontally projected cannon ball is a parabola. While he discovered that the path of a cannon ball was parabolic in nature and used this fact to compile range tables, he did not publish or widely discuss his theory.

The credit for being the first person to demonstrate that in the absence of air resistance a projectile follows a parabolic path went to the mathematician Bonaventura Cavalieri (1598–1647). Cavalieri, a student of Galileo's, amplified upon his master's theories and published a text on trajectories in 1632. In *Lo Specchio Ustorio* (*The Burning Glass*), he noted that a projectile was acted upon by two distinct forces, that of the propellant and that of gravity. If each acted alone, the projectile would follow a rectilinear path; however, combined they produce a curved motion. Cavalieri went on to plot the curve using a system of rectangular coordinates, calculated corresponding horizontal and vertical distances and demonstrated that it was a parabola. Galileo was infuriated at being upstaged by one of his students. He countered by publishing a more detailed theory of parabolic trajectories in *Discourses and Mathematical Demonstrations Concerning Two New Sciences* in 1638. While theoretically appealing and correct, both Cavalieri's and Galileo's revelations remained remote from the pragmatic needs of gunners. It would remain for another pupil of Galileo, Evangelista Torricelli (1608–1674), to clarify further the parabolic nature of trajectories and to supply mathematical theories readily acceptable and usable by gunners.

Torricelli offered geometrical demonstrations of his theories. The demonstrations are concise and supply insights into the mathematical methods of this time. His demonstration that the path of a projectile is truly parabolic depends on the constructions shown in Figure 5 and is explained as follows:

> Let AF be the semi-trajectory of a projectile with F the vertex of the trajectory. Let HF be drawn tangent to AF at point F, and construct $AG \parallel HF$. Let AB be the tangent line drawn to the curve at point A. Construct $BG \perp HF$ intersecting AG at

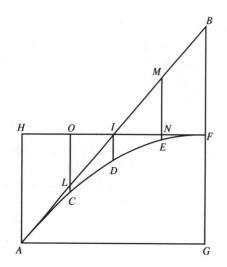

FIGURE 5

G. Because the greatest height of an ascending body is half the height which it would have reached if it had ascended with a uniform rather than a decelerating vertical velocity, we have $BF = FG$. Partition AB into several equal sections: AL, LI, IM, MB. Vertical lines through the points of partition intersect HF and AF in points O, I, N and C, D, E respectively. Clearly, $HO = OI = IN = NF$. For a moving projectile these increments represent equal horizontal distances traversed in equal time periods t_1, t_2, t_3, t_4. However for a projectile moving along AB in interval t_1, it will drop a vertical distance LC; for t_2 the distance is ID; t_3, ME; and t_4, BF. By Galileo's principle, these vertical distances are in the ratio $1 : 4 : 9 : 16$ respectively. Now since $OL = MN = \frac{1}{2}BF$ and $BF = 16LC$, $MN = 8LC$, but $ME = 9LC$. Therefore, $NE = (ME - MN) = (9LC - 8LC) = LC$. By similar arguments, we find that the distances NE, ID, OC, and HA are in the ratios $1 : 4 : 9 : 16$ and curve $ACDEF$ is a parabola. [4, p. 92]

Torricelli used rather ingenious techniques of geometric analysis in determining properties of a given trajectory. For example, given the angle of projection α and the initial velocity v_0 for a projectile, a diagram similar to Figure 6 would be constructed. The diagram would be

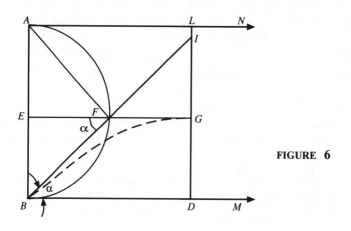

FIGURE 6

drawn to a predetermined scale so that AB would represent the distance a body would fall under the effects of gravity to obtain the velocity v_0. Now a semicircle is constructed with AB as a diameter, and BM and AN are drawn tangent to the semicircle with $BM \parallel AN$. With B as the initial point, a ray is constructed at an angle α to BM cutting the semicircle in point F; EG is drawn through F where $EG \parallel BM$ and $EF = FG$ and AF is drawn. Through point G a line is constructed perpendicular to EG intersecting BM at D and AN at L so that $BALD$ is a rectangle. BF extended intersects DL at point I. Employing concepts from algebra and trigonometry, the following relationships emerge from this diagram:

Because AB is the distance a body must fall to obtain velocity v_0, we get from elementary physics that $AB = \frac{1}{2}gt^2$ and $v_0 = gt$. Eliminating t from these two equations gives

$$AB = \frac{v_0^2}{2g}.$$

Now

$$BI = 2BF = 2AB\cos(90 - \alpha) = 2AB\sin\alpha.$$

But then

$$GD = \frac{1}{2}ID = \frac{1}{2}BI\sin\alpha = AB\sin^2\alpha.$$

It follows that

$$GD = \frac{v_0^2}{2g}\sin^2\alpha,$$

and this is the maximum height reached by the projectile.

Because the complete trajectory will be symmetric about GD, the range will equal $4EF = 4FG$. Now,

$$FG = \frac{IG}{\tan\alpha} = \frac{DG}{\tan\alpha} = \frac{v_0^2}{2g}\sin\alpha\cos\alpha.$$

Thus, using modern values and setting $g = 32$, the range = $4FG = (v_0^2/16)\sin\alpha\cos\alpha$. With the aid of Figure 7, we also show that the range will be a maximum when $\alpha = 45°$. Let $\alpha = \angle CAB$; then $\sin\alpha = h/b$, $\cos\alpha = b/c$, and $\sin\alpha\cos\alpha = h/c$. Now suppose that the length c remains fixed as the diameter of the circle. If h varies, the question becomes in what position would its length be a maximum, thus maximizing the quantity h/c? The answer is that h must be a radius of the circle; thus $\alpha = 45°$ for maximum range.

Torricelli also demonstrated that any range r could be expressed as a function of the maximum range R by the formula $r = 2R\sin\alpha\cos\alpha$ or $r = R\sin 2\alpha$. With this formula, a

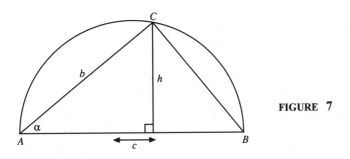

FIGURE 7

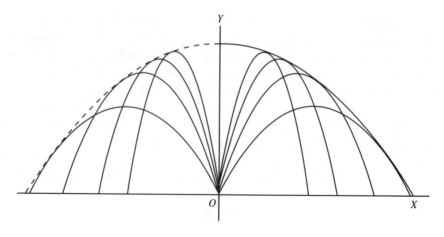

FIGURE 8

cannon's range could be determined by the use of a sine table. Further, he noted that if a gun is fired from a point with a fixed initial velocity but with varying angles of elevation, a series of parabolic trajectories are formed, the envelope of which is itself a parabola whose focus is the gun. (See Figure 8.) This all encompassing parabola is called the parabola of surety as under its cover, a gunner finds safety.

Using modern techniques based on Galileo's findings, a student of today if given $\mathbf{v_0}$ as the initial velocity and θ as the angle of projection, would treat $\mathbf{v_0}$ as a vector quantity and resolve its effect into two components, a vertical component v_y and a horizontal component v_x. Since $v_x = v_0 \cos\theta$ and $v_y = v_0 \sin\theta - 32t$, where $v_0 = \|\mathbf{v_0}\|$, it follows that the two components of the position vector are given by $x = v_0 t \cos\theta$ and $y = v_0 t \sin\theta - 16t^2$. Setting $y = 0$, we get $-16t^2 + v_0 t \sin\theta = 0$, or $t(-16t + v_0 \sin\theta) = 0$. The two solutions are then $t_1 = 0$ and $t_2 = (v_0/16)\sin\theta$, where t_1 represents the time at the start of the trajectory and t_2 represents the time to impact of the projectile. Substituting t_2 into the expression for x, we find a formula for the range:

$$x = \frac{v_0^2}{16}\sin\theta\cos\theta.$$

In considering this expression further, since $\sin(90 - \theta) = \cos\theta$ and $\cos(90 - \theta) = \sin\theta$, we have

$$x = \frac{v_0^2}{16}\sin\theta\cos\theta = \frac{v_0^2}{16}\sin(90 - \theta)\cos(90 - \theta).$$

As Tartaglia hypothesized, the same range can be obtained for two different angles of a gun elevation as long as the angles are complements of each other.

Classroom Teaching Implications

The brief historical-mathematical scenario just considered has several implications for teaching. As an example of a modeling process that results in the development of formulas and equations, the sequence of described events demonstrates a conflict that frequently arises in modeling situations, namely, the need to reconcile pragmatic concerns with theoretical considerations. Modeling a trajectory by the use of a right triangle may seem absurd to a knowledgeable reader, but is it? If I, as a Renaissance gunner, am primarily interested in finding a range as a function of cannon barrel elevation, this model may work fine. In admitting this fact, it should be appreciated that until the technological advances of the nineteenth century, mainly the rifling of barrels and the streamlining of projectiles, the firing of artillery was not an accurate process. Thus what a Renaissance gunner sought was an approximate answer to his needs. The triangular model supplied that approximation as did, with some justification, the line-arc-line model. Galileo, in using experimental techniques, provided the parabolic model, one soundly based on theoretical principles. His method in ascertaining the parabolic nature of trajectories, if demonstrated before a mathematics class, provides a classic example of an instance where mathematical formulation evolved from empirically based observation. Further, Galileo's conclusion that planar motion could be resolved into two mutually perpendicular components initiated the development of vector arithmetic. The involvement of historically significant personages in the unfolding story of trajectories, Oresme, Tartaglia, Galileo, Cavalieri and Torricelli, is noteworthy as are the algebraic and geometric methods they employed. In the three hundred years from the deliberations of Oresme to those of Cavalieri, one sees the trend of mathematical thinking moving from the purely geometric to the algebraic-trigonometric.

Of course what has been examined in the above discussion is but one historical phase in the mathematical modeling of trajectories. A little later, Newton was to employ the analytic techniques of his calculus to investigate questions of air resistance on a moving body, a negligible concern to gunners but one of importance to the scientific philosophers of the period. In more modern times, partial differential equations have provided a valuable computational means to understand the properties of trajectories and ballistics. Contemporary secondary school students can explore a variety of problem-solving situations involving trajectories and use their knowledge of geometry, algebra, trigonometry, vectors, calculus and even computer programming. The appropriateness of the use of problem-solving situations involving guns and artillery in a classroom may be open to question. However, it must be remembered that such examples are historically correct and while they may reflect on mankind's involvement with war, they also demonstrate the ingenuity and persistence of the human intellect. Harnessing of gunpowder through the use of military ordinance has served as a great impetus in the development of various sciences: chemistry, metallurgy, physics, and most especially, mathematics.

Bibliography

1. Anderson, Robert: *The Genuine Use and Effects of the Gunne*. London: J. Darby, 1674.
2. Drake, Stillman and Drabkin, I. E.: *Mechanics in Sixteenth Century Italy*. Madison: The University of Wisconsin Press, 1969.
3. Egg, Jobe *et al. Guns: An Illustrated History of Artillery*. Greenwich, Ct.: New York Graphic Society, 1971.
4. Hall, A. R.: *Ballistics in the Seventeenth Century*. New York: J. J. Harper, 1969.
5. Hogg, O. F. G.: *Artillery: Its Origin, Heyday and Decline*. London: Archon Books, 1970.
6. Swetz, Frank.: "An Example of Mathematical Modeling: The Trajectory of a Cannonball," *International Journal of Mathematical Education in Science and Technology* 20 (1989), 731–741.

Gottfried Wilhelm Leibniz

PART II

History in Higher Mathematics

Mathematics, as it is taught at most universities and institutions of higher education, is a rather abstract science. Today, an undergraduate studies much more mathematics and at a more condensed level than his predecessor did even a decade ago. Topics formerly encountered at the periphery of graduate research, such as time-series analysis, chaos theory or dynamic modeling, have found their way into undergraduate studies. More mathematics is being taught, but is it being taught better? Many instructors of higher mathematics feel that the pedagogy of the subject lies within the logic of its formal structure. Thus to teach mathematics "well," one only needs to enunciate the "definition-theorem-proof" sequence smoothly and student understanding will follow. Unfortunately, far too often, this is not the case particularly when teaching non-mathematics majors. Students who fail to grasp the logical structure of mathematics can become overwhelmed by its abstractness and frustrated in their studies. In such cases, rigor becomes self defeating.

Some observers of this educational scene will contend that we are teaching too much mathematics and not enough about it, i.e., its origins, evolution and applications, in brief, its human roots. It is exactly at this juncture of pedagogical concerns that the history of mathematics can serve as a teaching and learning resource. History can supply the often needed "human connection" with the mathematics in question, shed light on its origins, trace its evolution and provide insights into the process of mathematical conceptualization. Thus, the history of mathematics can serve as a major pedagogical tool for the teaching of higher mathematics. A sharing of thoughts and experiences on this matter are provided by the following authors.

Man-Keung Siu considers the very basic and unifying concept of a mathematical function and reviews its various historical definitions and applications. In this survey, he clarifies those historical aspects of the function concept which readily lend themselves to the teaching of higher mathematics. In a second contribution, Siu discusses heuristic reasoning in the work of Leonhard Euler and shows how contemporary teachers can use his work in the classroom.

V. Frederick Rickey is noted for his ability to enliven the teaching of calculus through relevant historical references. He shares some of his experiences in the calculus classroom and in doing so, illustrates just how historical awareness can enrich calculus teaching.

Michel Helfgott has actually designed a calculus teaching sequence based on historical considerations. He relates his experiences in providing a cultural background for calculus.

Joel Lehmann proposes a model for the classroom development of mathematical concepts and applies it to the particular case of the study of series. Lehmann's model, while based on an historical understanding of mathematics, also reflects an awareness of the psychology of concept formation.

103

Lars Mejlbo takes us to infinity by sharing some of his experiences in teaching transfinite arithmetic. He historically examines the difficulty in working with the concept of infinity.

Victor Katz provides a background for linear algebra by showing that the ideas from the subject possess a long history. Mathematical problems involving linear expressions have been studied for thousands of years, and this history can easily be made use of in today's classroom.

Otto Bekken reviews the work of the Norwegian, Caspar Wessel, who provided a graphical meaning to the concept of vectors in two dimensions and attempted to extend it to three dimensions. Bekken's historical comments on vectors reflect on both mathematics and physics teaching.

Karen Reich points out the confusion surrounding the use and designation of vectors. She traces this situation to the historical development of the subject and explains how the confusion can be alleviated through an understanding of the history of mathematics.

Israel Kleiner supplies an historical perspective to the teaching of abstract algebra. He cites some basic mathematical issues and problems facing students and demonstrates how these difficulties can be resolved by using the history of mathematics.

Adding another dimension to the teaching of abstract algebra, David Burton and Donovan Van Osdol trace the development of ring theory. Their discussion delinates the evolution of a particular mathematical theory and notes the many human contributions to its realization.

Anthony Gardiner uses historical insights in discussing a pedagogy for the teaching of group theory. He sounds a note of caution on the use of historical based pedagogy.

Eric Aiton reviews some of the work of Newton and Leibniz as it pertains to celestial mechanics. In his discussion, Aiton develops connections between the early development of calculus, mathematical modeling and celestial mechanics.

In the concluding essays of this section, Man-Keung Siu and Abe Shenitzer discuss their ideas on developing and teaching a "topics course" in the history of mathematics. Such courses demonstrate how an instructor can incorporate a humanistic approach to mathematics learning and teaching—one that emphasizes mathematics as a human accomplishment.

Concept of Function—Its History and Teaching

Man-Keung Siu

Objective

In the last century, Felix Klein strongly advocated an emphasis on the function concept in teaching as a unifying idea permeating all mathematics. How basic is the function concept? We shall try to trace its development and attempt to incorporate this mathematical-historical vein into the teaching of mathematics at various levels, from secondary school to university. It is hoped that, by doing so, not only can the understanding of this important concept be enhanced but a sense of history can be imparted to a wider audience. With this in mind, instead of giving a comprehensive historical account/analysis, we discuss its implications in learning/teaching. To render this article more usable as a teaching-aid we adopt a format by which questions for discussion are woven into the text and some illustrative examples are compiled as exercises at the end. There is no dearth of relevant scholarly material on the topic in general. I shall list only those on which I have drawn in the Bibliography at the end. Some of them are not specifically referred to in the text, but they have been so helpful that I record them if only to acknowledge my appreciation. (I also wish to express my gratitude to my colleagues Israel Kleiner and Abe Shenitzer for providing me with translations of [29] and [31].)

What Is a Function?

It is of interest to start with three definitions as a "warm-up." The first one by Johann Bernoulli (1718) [41, p. 72] is classical, vague but more or less concrete. The third one by Patrick Suppes (1960) [44, pp. 59, 86] is modern, precise but formidable. The second one by Édouard Goursat (1923) [41, p. 77] lies somewhere in between.

(1) One calls here a function of a variable a quantity composed in any manner whatever of this variable and of constants.

(2) The modern definition of the word function is due to Cauchy and Riemann. One says that y is a function of x if to a value of x corresponds a value of y. One indicates this correspondence by the equation $y = f(x)$.

(3) A is a relation $\leftrightarrow (\forall x)(x \in A \rightarrow (\exists y)(\exists z)(x = < y, z >))$.
    f is a function $\leftrightarrow f$ is a relation & $(\forall x)(\forall y)(\forall z)(xfy \ \& \ xfz \rightarrow y = z)$.

What did Bernoulli mean by "in any manner whatever" in (1)? In what way do functions depicted in (1) and (2) differ? What advantages does (3) have over the other two definitions?

How well do these definitions fit with your intuitive idea of a functional dependence? How did the notion of a function evolve from that depicted in (1) to that depicted in (3)?

Browse through books such as:

- M. Abramowitz, I. A. Stegun, *Handbook of Mathematical Functions — With Formulas, Graphs, and Mathematical Tables*, National Bureau of Standards, 1964 (and subsequent revisions),
- J. Spanier, K. B. Oldham, *An Atlas of Functions*, Hemisphere Publishing Corporation, 1987,
- American Institute of Physics, *American Institute of Physics Handbook*, McGraw Hill, 1957 (and subsequent editions),

and think about the following questions: Is a function a formula (analytical expression)? a graph (curve)? a table of values (correspondence)? a law of dependence? How well or how inadequately do the descriptions above apply to a function? What else does a function signify? Nikolai Nikolaievich Luzin said that no single formal definition can include the full content of the function concept [31]. Comment on this. How did such vague but useful intuitive ideas of a function contribute to the emergence and evolution of its concept in the history of mathematics? Read pp. 62–65 of J. E. Littlewood's *A Mathematician's Miscellany*, Methuen, 1953 and comment.

When Did the Function Concept Originate?

Authors differ on the origin of the function concept. Some samples of opinion are shown below.

E. T. Bell: It may not be too generous to credit them [ancient Babylonians] with an instinct for functionality; for a function has been succinctly defined as a table or a correspondence. [1, p. 32]

O. Pedersen: But if we conceive a function, not as formula, but as a more general relation associating the elements of one set of numbers (viz, points of time t_1, t_2, t_3, ...) with the elements of another set (for example, some angular variable in a planetary system), it is obvious that functions in this sense abound throughout the *Almagest*. Only the word is missing: the thing itself is there and clearly represented by the many tables of corresponding elements of such sets. [37, p. 36]

W. Hartner, M. Schramm: The question of [the] origin and development [of the concept of function] is usually treated with striking one-sidedness: it is considered almost exclusively in relation to Cartesian analysis, which in turn is claimed (erroneously, we believe) to be a late offspring of the scholastic *latitudines formarum*. [20, p. 215]

A. F. Monna: The notion of a function has no place in Greek mathematics. [35, p. 58]

A. P. Youschkevitch: The mathematical thought of antiquity created no general notion of either a variable quantity or of a function. ... Occurring some time after the downfall of antique society, the new flowering of science in countries of Arabic culture did not, as far as is known, bring about essentially new developments in functionality. ... The notion of function first occurred in a more general form three centuries later [in the fourteenth century], in the schools of natural philosophy at Oxford and Paris. ... still I do not maintain that this role [played by ideas of both the Oxford and the Paris schools of natural philosophy] was dominant, the more so as a new interpretation of functionality came to the fore in the 17th century. ...As a consequence of all this, a new method of introducing functions was brought into being, to become for a long time the principal method in mathematics and, especially in its applications. [46, pp. 44, 45, 50, 51]

C. Boyer: The function concept and the idea of symbols as representing variables does not seem to enter into the work of any mathematician of the time [of Descartes and Fermat]. [4, p. 156]

D. E. Smith: After all, the real idea of functionality, as shown by the use of co-ordinates was first clearly and publicly expressed by Descartes. [42, p. 376]

F. Cajori: Some of the mathematicians of the Middle Ages possessed some idea of a function. ...But of a numeric dependence of one quantity upon another, as found in Descartes, there is no trace among them. [5, p. 127]

M. Kline: From the study of motion [by Galileo] mathematics derived a fundamental concept that was central to practically all of the work for the next two hundred years—the concept of a function or a relation between variables. [25, p. 338]

In view of the remarks above the following questions may be instructive. In primary/junior secondary school a student will encounter mathematical tables. How can these help to instill the notion of functionality? How far is the 'instinct for functionality' embodied in tables from the notion of functionality? What is missing? (Would one suspect a formula like $\sin(x + y) = \sin x \cos y + \cos x \sin y$ by staring at a sine/cosine table?) Is there similarity between tables and the notion of correspondence which is stressed in the modern definition of function? How much does (and actually did, in the past) the study of nature influence and benefit the development of mathematics?

From Fourteenth Century to Eighteenth Century

Quantitative description of (physical) change, e. g., velocity and acceleration, was embodied in the doctrine of *latitude of forms* developed by Nicole Oresme (*De configurationibus,* c. 1350). Although this dim idea of a functional dependence exerted minor influence later, it indicated: (i) quantitative laws of nature as laws of functional dependence, (ii) conscious use of general ideas about independent/dependent variables, (iii) graphic representation of a functional dependence. For more detail, read chapter 6 of [7].

According to Alistair C. Crombie [9, vol. II, section 1.4], this idea of functional relationships was developed without actual measurement and only in principle. Youschkevitch ascribed this to a lack of computational technique: "An obvious disproportion [developed] between the high level of abstract theoretical speculations and the weakness of mathematical apparatus" [45, p. 49]. We can ask: What moral do we gain from this historical incident concerning the balance of concepts and technical skills in our teaching?

Further impetus came from the study of motion by Johannes Kepler and Galileo Galilei in the early seventeenth century. By that time, arithmetic (extension of the concept of numbers) and algebra (symbolic algebra) had also developed to a stage which made possible the wedding of algebra and geometry by René Descartes and Pierre de Fermat, with the invention of calculus by Isaac Newton and Gottfried Wilhelm Leibniz to follow. As a consequence of all these events, time was ripe for the introduction of the notion of function.

Let us look at the notions of function by Descartes and Fermat.

P. Fermat (*Ad Locos Planos et Solidos Isagoge*, 1629; published in 1679): As soon as two unknown quantities appear in a final equation, there is a locus, and the end point of one of the two quantities describes a straight or a curved line. [46, p. 52]

R. Descartes (*La Géométrie*, 1637): If then we should take successively an infinite number of different values for the line y, we should obtain an infinite number of values for the line x, and therefore an infinity of different points, such as C, by means of which the required curve could be drawn. [11, p. 34]

Hermann Hankel commented: "Modern mathematics dates from the moment when Descartes went beyond the purely algebraic treatment of equations to study the variation of magnitudes that an algebraic expression undergoes when one of its generally denoted magnitudes passes through a continuous series of values" [2, p. 8]. Friedrich Engels also said: "The turning point in mathematics was Descartes' variable magnitude. With that came motion and hence dialectics in mathematics, and at once also of necessity the differential and integral calculus, which moreover immediately begins" [14, p. 199].

Infinitesimal calculus arose from geometric and kinematic problems. Although in the seventeenth century it was geometric rather than analytic in nature, and was not yet a calculus of *functions* as we know it today, it did however induce further study into the notion of function by providing more examples of functions, cloaked in various forms such as (i) fluent by Newton, (ii) abscissa, ordinate, subtangent, subnormal, etc. by Leibniz, (iii) expansion of function into infinite power series by Nicholas Mercator, James Gregory and Newton.

The most explicit definition of the function concept in the seventeenth century, although not by that name, was given by Gregory (*Vera Circuli et Hyperbolae Quadratura*, 1667): "We call a quantity composed (*compositum*) of other quantities if that quantity results from those other quantities by addition, subtraction, multiplication, division, extracting of roots or by any other imaginable operations" [46, p. 58]. Gregory referred to a quantity obtained through the first five operations as "composed analytically," where the word "analytic" was used in the sense of François Viète in his *In Artem Analyticem Isagoge* of 1591 [43, pp. 75–76]. The sixth operation meant some rather general infinite process.

The word "*function*" first appeared in a manuscript of Leibniz ("Methodus tangentium inversa, seu de functionibus," 1673): "Other kinds of lines which, in a given figure, perform some function" [46, p. 56]. Further on in the same manuscript, the term "*function*" took on a new meaning as a general term for various geometric quantities associated to a variable point on the curve. (This also appeared in Leibniz's later articles in 1692 and 1694. The word was also used in the same sense by Jakob Bernoulli in 1694.) In a letter dated September 2, 1694 of Johann Bernoulli to Leibniz, in which Bernoulli expanded the integral $\int n\,dz$ in an infinite series

$$nz - \frac{1}{1 \cdot 2} z^2 \frac{dn}{dz} + \frac{1}{1 \cdot 2 \cdot 3} z^3 \frac{d^2 n}{dz^2} - \cdots,$$

he said that "by n I understand a quantity somehow formed from indeterminate and constant [quantities]" [46, p. 57]. In a letter dated July 29, 1698 of Leibniz to Johann Bernoulli, he said that "I am pleased that you use the term function in my sense." The first explicit definition of a function (as an analytical expression) was by Johann Bernoulli (see the definition in the section "What Is a Function?"). Bernoulli used the notation φx, without brackets. Brackets, as well as the sign f for function were due to Euler in his article of 1734 [46, p. 60].

In the preface to Book I of his *Introductio in analysis infinitorum* (1748), Leonhard Euler claimed that mathematical analysis is the general science of variables and their functions [15, Book I, p. vi], thereby endowing the function concept a central prominence in analysis. His entire approach was algebraic and no longer geometric. Concerning function he defined:

1. A constant quantity is a determinate quantity keeping the same value perma-
nently. ... 2. A variable quantity is an indeterminate or universal quantity which comprises in itself all determinate values. ... 4. A function of a variable quantity is an analytical expression composed in any manner from that variable quantity and numbers or constant quantities. [15, Book I, pp. 2–3]

Note the use of: (i) analytical expression (with power series as a universal form), (ii) generality of the variable. A consequence was a tenet in eighteenth century mathematics on "analytical continuity": If two functions agree on an interval, they agree everywhere [29, section 1.9].

In Book II of the *Introductio* Euler extended his notion of function to include the so-called "*discontinuous*" functions. Care must be taken not to confuse Euler's use of the term "continuous" with what we know today (due to Bernard Bolzano and Augustin-Louis Cauchy). (See the section "Fourier Series and the Function Concept"). A function (curve) is *E-continuous* if it is given by a single analytical expression throughout. A function (curve) is *E-discontinuous* (also called *mixed* or *irregular* by Euler) if it is given by two or more analytical expressions on different intervals [15, Book II, p. 6]. (Later, Euler included also curves which were drawn freehand, i.e., the analytic expression changed from point to point, so to speak.) [30, p. 301; 46, p. 68]

Towards the end of the 18th century, Joseph Louis Lagrange and Sylvestre-François Lacroix defined the concept of a function in a seemingly more general way.

J. L. Lagrange (*Théorie des fonctions analytique*, 1797): One calls function of one or several quantities any expression for calculation in which these quantities enter in any manner whatever, mingled or not with some other quantities which are regarded as being given and invariable values, whereas the quantities of the function can take all possible values. ... We denote, in general, by the letter f or F placed before a variable any function of this variable, that is to say any quantity depending on this variable and which varies with it together according to a given law. [41, p. 73]

S. F. Lacroix (*Traité du calcul différentiel et du calcul intégral*, 1797): Every quantity whose value depends on one or more other quantities is called a function of these latter, whether one knows or is ignorant of what operations it is necessary to use to arrive from the latter to the first. [2, p. 36]

But Lagrange's and Lacroix's apparently general definitions of a function are in fact still "Eulerian." Lacroix had the implicitly given functions in mind when he said "whether one knows ... to the first." Lagrange even showed that any given function can be expanded as a power series. "Algebraification" of analysis was at its height!

It may be of interest to look at the Chinese terminology for function. In 1859 Li Shanlan (李 善 蘭), together with Alexander Wylie, translated *Elements of Analytical Geometry and of Differential and Integral Calculus* by Elias Loomis (1850). The word "function" was translated as 函 數 (literally meaning "quantity that contains") with the explanation that "if the variable quantity contains another variable quantity, then the former is a function (函 數) of the latter" [28, pp. 207–208]. Subsequent illustration indicates that the term is to be understood in the Eulerian sense of an analytical expression.

What do we learn from this piece of historical development about the understanding of the function concept? Are we retracing the steps in learning the function concept in school?

Controversy About the Vibrating String

The main impulse for further development of the function concept in the eighteenth century came from a controversy over a problem in mathematical physics, viz. the motion of a tense string fixed at two ends when it is made to vibrate. (This problem actually turned out to play a central role in the development of the whole of analysis.) In a nutshell, the dispute concerned the type of functions which could be allowed in analysis from the standpoint of a mathematician, a physicist, and a then emerging type of scholar: a mathematical physicist. For additional reading the following are recommended: [10; 39; 43, pp. 351–368].

The standpoint of a mathematician was represented by the work of Jean le Rond d'Alembert in 1747. He said, "I propose to show in this paper that there exist an infinity of curves different from the elongated cycloid [sine curve] which satisfy the problem under consideration" [43, p. 352]. D'Alembert deduced from the equation describing the motion of the string

$$\frac{\partial^2 y}{\partial t^2} = \frac{\partial^2 y}{\partial x^2}, \quad y(0,t) = y(L,t) = 0$$

the solution $y(x,t) = f(x+t) + f(x-t)$. The only restrictions he imposed on the function f were that it be periodic, odd, and everywhere (twice) differentiable.

The standpoint of a mathematical physicist was represented by the work of Euler in 1748 and later the work of Lagrange in 1759. Euler rederived the wave equation

$$\frac{\partial^2 y}{\partial t^2} = c^2 \frac{\partial^2 y}{\partial x^2}, \quad y(0,t) = y(L,t) = 0$$

and the functional solution, formally identical with d'Alembert's. He claimed that f can be deduced solely from initial conditions. If $Y(x)$, $V(x)$ are the initial position and velocity of the string, then

$$y(x,t) = \frac{1}{2}\left(Y(x+ct) + Y(x-ct) + \frac{1}{c}\int_{x-ct}^{x+ct} V(s)\,ds\right).$$

Further, Euler proclaimed that $Y(x)$, $V(x)$ need not be functions in the ordinary sense, but may be any curve drawn freehand (e.g., the "plucked" string, the "snapped" string). He said (*Institutiones calculi differentialis,* 1755): "If, therefore, x denotes a variable quantity then all the quantities which depend on x in any manner whatever or are determined by it are called its functions, ..." [41, p. 73]. However, throughout the book, only E-continuous functions were considered! Lagrange discretized the problem as that of a loaded string and found

$$y(x,t) = \frac{2}{L} \int_0^L dX Y(X) \times$$

$$\left[\sin\left(\frac{\pi X}{L}\right) \sin\left(\frac{\pi x}{L}\right) \cos\left(\frac{\pi ct}{L}\right) + \sin\left(\frac{2\pi X}{L}\right) \sin\left(\frac{2\pi x}{L}\right) \cos\left(\frac{2\pi ct}{L}\right) + \cdots \right]$$

$$+ \frac{2}{\pi c} \int_0^L dX V(X) \times$$

$$\left[\sin\left(\frac{\pi X}{L}\right) \sin\left(\frac{\pi x}{L}\right) \cos\left(\frac{\pi ct}{L}\right) + \frac{1}{2} \sin\left(\frac{2\pi X}{L}\right) \sin\left(\frac{2\pi x}{L}\right) \cos\left(\frac{2\pi ct}{L}\right) + \cdots \right]$$

One can ask: How close was Lagrange to the Fourier series? Why did he miss it? [2, pp. 30–33; 10, p. 36]

The standpoint of a physicist was represented by the work of Daniel Bernoulli in 1753. He said: "I do not the less esteem the calculations of Messrs. d'Alembert and Euler, which certainly contain all that analysis can have at its deepest and most sublime, but which show at the same time that an abstract analysis which is accepted without any synthetic examination of the question under discussion is liable to surprise rather than enlighten us. It seems to me that we have only to pay attention to the nature of the simple vibrations of the strings to foresee without any calculation all that these two great geometers have found by the most thorny and abstract calculations that the analytical mind can perform" [43, p. 361]. Bernoulli argued that the solution must be a sum of the fundamental and higher harmonics (principle of superposition):

$$y(x,t) = A_1 \sin\left(\frac{\pi x}{L}\right) \cos\left(\frac{\pi ct}{L}\right) + A_2 \sin\left(\frac{2\pi x}{L}\right) \cos\left(\frac{2\pi ct}{L}\right) + \cdots$$

Fourier Series and the Function Concept

Jean Baptiste Joseph Fourier developed in his *Sur la propagation de la chaleur* of 1807 the theory of the series which today bears his name in his investigation of heat conduction that won him a prize from the Institut de France in 1812. His theory was made more widely known in a later treatise (*Théorie Analytique de la Chaleur*, 1822). For more details on the subject, read [27]. The main idea is as follows.

The solution to $\frac{\partial^2 u}{\partial x^2} + \frac{\partial^2 u}{\partial y^2} = 0$ with conditions $u(0,y) = u(\pi,y) = 0$, $u(x,0) = \varphi(x)$, is the function $u(x,y) = \sum_{n=1}^{\infty} b_n e^{-ny} \sin nx$, when $b_1, b_2, b_3, \ldots$ are so chosen that $\varphi(x) = \sum_{n=1}^{\infty} b_n \sin nx$ for x lying between 0 and π.

Fourier's first heuristic approach was to assume $\varphi(x)$ to be an odd function expanded into its Taylor series and to compare it with the original series with $\sin nx$ expanded also into an infinite series, thus obtaining an infinite system of linear equations. He solved the truncated system, passed to the limit, and obtained $b_n = \frac{2}{\pi} \int_0^{\pi} \phi(x) \sin nx \, dx$. He then pointed out that this formula can be "verified" by the now standard procedure for evaluating Fourier coefficients making use of the orthogonality of the sine function on the interval $[0,\pi]$ [13, pp. 304–306].

Physical constraints in the vibrating string problem and in the heat conduction problem differ in nature. The shape of the string (geometry) is visible, while temperature distribution (algebra) is not. This may explain the freeing from geometric perception of a function and the emergence of a general notion of function in the nineteenth century. Concerning function, Fourier defined:

In general, the function $f(x)$ represents a succession of values or ordinates each of which is arbitrary. An infinity of values being given to the abscissa x, there are an equal number of ordinates $f(x)$. ... We do not suppose these ordinates to be subject to a common law; they succeed each other in any manner whatever, and each of them is given as if it were a single quantity. [41, p. 73]

Subsequent investigation into the Fourier series representation of a function led to the breakdown of the "Eulerian" notion of function. Although Fourier's definition sounds like our modern notion, it appears that he only had "discontiguous" functions (see Exercise 4 for its definition) in mind. Still, Fourier's work was instrumental in the following aspects:

- representation of an "arbitrary" function by an analytical expression (recall Daniel Bernoulli's claim),
- renewed emphasis on analytical expression,
- reexamination of the function concept,
- elimination of the tenet on 'analytical continuity' held by eighteenth century mathematicians.

In his *Cours d'analyse* of 1821 Cauchy defined:

When the variable quantities are linked together in such a way that, when the value of one of them is given, we can infer the values of all the others, we ordinarily conceive that these various quantities are expressed by means of one of them which then takes the name of independent variable; and the remaining quantities, expressed by means of the independent variable, are those which one calls functions of this variable. [2, p. 104]

Again, Cauchy's definition is in practice more limited than it sounds. Immediately after the definition, he classified functions into "simple functions" and "compound functions." The first group consists of eleven functions, viz.

$$a + x, \ a - x, \ ax, \ \frac{a}{x}, \ x^a, \ A^x, \ \log x, \ \sin x, \ \cos x, \ \arcsin x, \ \arccos x \ ,$$

where A is a nonnegative number and $a = \pm A$, while the second group consists of functions made up of the "simple functions" by composition. In chapter 8 when he came to complex function, he said: "when the constants or the variables included in a given function are assumed imaginary after first having been considered real, ..." [29, section 4.2]. Cauchy still had the "Eulerian" notion of function in mind. But he did mention: "As for methods, I have sought to give them all the rigor which exists in geometry, so as never to refer to reasons drawn from the generalness of algebra" [2, p. 102]. In the same text, Cauchy also defined the notion of continuity as we know it today: "The function $f(x)$ will be a continuous function of the variable x between two assigned limits if, for each value of x between those limits, the numerical value of the difference $f(x + a) - f(x)$ decreases indefinitely with a." (Bolzano gave the same definition in slightly different and more precise language in 1817 [18, p. 87; 35, p. 62].)

It is interesting to ask the following questions: Why was the "Eulerian" concept of function maintained so long after the realization that it was inadequate? What lesson do we learn from this experience? (If only a particular form is used, students unconsciously accept that particular form as the definition. We witness the same psychological effect in mathematicians of the seventeenth/eighteenth centuries. A new concept receives recognition only when it is relevant to current usage. This is as true in research as in teaching. What would a student think of a function if all he needs to work with are algebraic expressions?)

Function Concept in the Nineteenth and Twentieth Centuries

In a letter to his teacher Christoff Hansteen dated March 29, 1826, Niels Henrik Abel complained:

> It [analysis] lacks at this point such plan and unity that it is really amazing that it can be studied by so many people. The worst is that it has not at all been treated with rigor. There are only a few propositions in higher analysis that have been demonstrated with complete rigor. Everywhere one finds the unfortunate manner of reasoning from the particular to the general, and it is very unusual that with such a method one finds, in spite of everything, only a few of what may be called paradoxes. It is really very interesting to seek the reason. In my opinion that arises from the fact that the functions with which analysis has until now been occupied can, for the most part, be expressed by means of powers. As soon as others appear, something that, it is true, does not often happen, this no longer works and from false conclusions there flow a mass of incorrect propositions that link together. [2, pp. 86–87]

Mathematicians in the nineteenth century sought to provide new rigor for analysis. This began with the work of Carl Friedrich Gauss, Abel, Bolzano, Cauchy, and Peter Gustav Lejeune Dirichlet and was furthered by Karl Weierstrass, Richard Dedekind, Georg Friedrich Bernhard Riemann, and Georg Cantor. Many factors came together to bring about this new attitude. (See, for example, [17].) This trend brought with it a new conception of function as the following sampling shows.

N. I. Lobachevsky ("On the convergence of trigonometric series," 1838): General conception demands that a function of x be called a number which is given for each x and which changes gradually together with x. The value of the function could be given either by an analytical expression, or by a condition which offers a means for testing all numbers and selecting one of them; or lastly, the dependence may exist but remain unknown. [46, p. 77]

P. G. L. Dirichlet ("Über die Darstellung ganz wilkürlicher Funktionen durch Sinus- und Cosinusreihen," 1837): One thinks of a and b as two fixed values and of x as a variable quantity that can progressively take all values lying between a and b. Now if to every x there corresponds a single, finite y in such a way that, as x continuously passes through the interval from a to b, $y = f(x)$ also gradually changes, then y is called a continuous function of x in this interval. It is here not at all necessary that y depends on x according to the same law throughout the entire interval; indeed one does not even need to think of a dependence expressible by mathematical operations. Presented geometrically, that is with x and y thought of as the abscissa and ordinate, a continuous function appears as a connected curve which for every value of the abscissa contained between a and b has only one point. ... As long as one has determined the function for only a part of the interval, the manner of its extension to the rest of the interval remains completely arbitrary. [2, p. 197]

G. F. B. Riemann ("Grundlagen für eine allgemeine Theorie der Funktionen einer veränderlichen complexen Grösse," 1851): Let us suppose that z is a variable quantity which can assume, gradually, all possible real values then, if to each of its values there corresponds a unique value of the indeterminate quantity w, w is called a function of z; and if, as z continuously passes through all the values lying between two fixed values, w also continuously changes, then this function is said to be continuous within this interval. ... Obviously, this definition establishes, entirely, no law between the single values of the function as, if this function has been defined for a certain interval, the manner of its continuation outside of the interval is completely arbitrary. [2, p. 215; 41, p. 75]

H. Hankel ("Untersuchungen über die unendlich oft oszillierenden und unstetigen Funktionen," 1870): y is called a function of x when to every value of the variable quantity x within a certain interval there corresponds a definite value of y, no matter whether y depends on x according to the same law in the entire interval or not, or whether the dependence can be expressed by a mathematical operation or not. ... This purely nominal definition, which in the following I will associate with the name of Dirichlet because it reverts fundamentally to his works on Fourier series which clearly demonstrated the indefensibility of all the older concepts, is however no longer sufficient for the needs of analysis, in that functions of this kind do not possess general properties, and with this all relationships between the values of the function for various values of the argument fall to the wayside. [2, pp. 197–198]

Dirichlet ("Sur la convergence des séries trigonométriques qui servent à représenter une fonction arbitraire entre les limites données," 1829) proved that if a function f has only finitely many discontinuities and finitely many maxima and minima in $(-L, L)$, then f is represented by its Fourier series on $(-L, L)$. In proving this one has to have a clear understanding of the function concept. Dirichlet was the first to take seriously the notion of a function as an *arbitrary* correspondence. As an example of a function that does not satisfy his conditions, he gave the celebrated "Dirichlet function": $f(x) = c$ if x is rational and $f(x) = d$ if x is irrational, $c \neq d$.

Riemann ("Über die Darstellbarkeit einer Funktion durch eine trigonometrische Reihe," written in 1854) further investigated the problem on representation by use of Fourier series and in the course of this investigation developed his theory of integration. As an important example he gave an integrable function which is not continuous, indeed, with infinitely many points of discontinuity in any (small) interval [19, pp. 157–158]:

$$f(x) = \phi(x) + \phi(2x)/2^2 + \phi(3x)/3^2 + \cdots$$

where $\phi(x)$ is the difference between x and its nearest integer (zero if x is half-way). Riemann's work marked the beginning of a theory of the mathematically discontinuous. According to Thomas Hawkins, "the history of integration theory after Cauchy is essentially a history of attempts to extend the integral concept to as many discontinuous functions as possible; such attempts could become meaningful only after the existence of highly discontinuous functions was recognized and taken seriously" [22, p. 3].

In 1872 Weierstrass startled the mathematical community with his famous example of a continuous nowhere-differentiable function:

$$f(x) = \sum_{n=1}^{\infty} b^n \cos(a^n \pi x)$$

where a is an odd integer, b a real number in $(0, 1)$ and $ab > 1 + 3\pi/2$ [25, p. 956].

A host of such "pathological" examples of function brought about a change of emphasis in the late nineteenth century. Luzin described this change as: "the main difference between methods of studying functions within the framework of mathematical analysis and the theory of functions is that classical analysis deduces properties of any function starting from the properties of those analytical expressions and formulae by which this function is defined, while the theory of functions determines the properties of function starting from that property which a priori distinguishes the class of functions considered" [46, p. 81].

But not every mathematician was happy about this change. Henri Poincaré had said: "Formerly, when a new function was invented, it was in view of some practical end. To-day they are invented on purpose to show our ancestors' reasonings at fault, and we shall never get anything more than that out of them. If logic were the teacher's only guide, he would have to begin with the most general, that is to say, with the most weird, functions. He would have to set the beginner to wrestle with this collection of monstrosities" [38, pp. 125–126]. This prompts us to ask: What role is played by examples/counterexamples in the development of mathematics? in the teaching and learning of mathematics? In light of Poincaré's saying, are "pathological" examples good or bad in pedagogy? (But certainly, history can provide the motivation and a sense of history will help.)

Although Euler declared in Book I of the *Introductio* that the most general form of an E-continuous function is a power series, later he expressed his confidence in the fact that his E-discontinuous functions are not generally analytic. Dirichlet proved in 1829 that certain continuous functions can be expanded as Fourier series. It was then believed that all continuous functions can be so expanded until Paul du Bois-Reymond proved in 1876 that there exists a continuous function whose Fourier series diverges at a point. However, in 1885, Weierstrass proved his celebrated theorem that every continuous function is the limit of a uniformly convergent sequence of polynomials. Ulisse Dini posed ("Fondamenti per la teorica della funzioni di variabili reali," 1878) the question "if every function can be expressed analytically, for all values of the variable in the interval, by a finite or infinite series of operations on the variable." René Louis Baire (*Sur les fonctions de variables réelles*, 1899) called the class of continuous functions class 0; and for any countable ordinal α, the class of functions not in any of the preceding classes, but which are representable as limits of sequences of functions in preceding classes, class α. For example, the Dirichlet function is of class 2, viz.

$$\chi(x) = \lim_{m \to \infty} \lim_{n \to \infty} (\cos m! \, \pi x)^{2n} = \begin{cases} 1 & \text{if } x \text{ is rational} \\ 0 & \text{if } x \text{ is irrational.} \end{cases}$$

Henri Lebesgue (*Sur les fonctions représentables analytiquement*, 1905) showed further that: (i) a function is analytically representable if and only if it is of Baire class α for some countable α, (ii) for every countable α, there exists a function of Baire class α; a function is of Baire class α for some countable α if and only if it is Borel-measurable, (iii) there exists a measurable function that is not of any Baire class, i.e., not analytically representable.

This investigation led to the discovery of logical/philosophical difficulties inherent in the universal, hence nonalgorithmic, definition of a function. Hermann Weyl, in his *Philosophie der Mathematik und Naturwissenschaft* of 1927 said:

Nobody can explain what a function is, but that is what really matters in mathematics: A function f is given whenever with every real number a there is associated a number b (as for example, by the formula $b = 2a + 1$). b is said to be the value of the function f for the argument value a. Consequently, two functions, though defined differently, are considered the same if, for every possible argument value a, the two corresponding function values coincide. [45, p. 8]

For more detail, read [12; 35; 36, section 2.3].

Function As a Correspondence

With the impact of Cantor's set theory and development in algebra, the notion of a function as a mapping became dominant towards the end of the 19th century. Let us sample a few definitions in this light.

R. Dedekind (*Was sind und was sollen die Zahlen*, 1887): By a mapping of a system S a law is understood, in accordance with which to each determinate element s of S there is associated a determinate object, which is called the image of s and is denoted by $\phi(s)$. [41, p. 75]

G. Peano ("Sulla definizione di funzione," 1911): The function is a special relation, by which to each value of the variable there corresponds a unique value. One can define in symbols:

$$\text{Functio} = \text{Relatio} \cap u \ni [y; x\varepsilon u.\ z; x\varepsilon u\ \mathbf{\supset}_{x,y,z} \cdot y = x],$$

[which translates as] a function is a relation u such that, if two pairs $y; x$ and $z; x$, having the same second element, satisfy the relation u, it necessarily follows that $y = x$ whatever x, y, z may be. [41, p. 76]

C. Carathéodory (*Vorlesungen über reelle Funktionen*, 1917): The modern concept of function coincides with that of a correspondence. [6, p. 71]

F. Hausdorff (*Grundzüge der Mengenlehre*, 1914; *Mengenlehre*, 1937): Ordered pairs make possible the introduction of the concept of function. [21, p. 16]

K. Kuratowski (*Topologie*, 1933; *Introduction to Set Theory and Topology*, 1961): Let X and Y be two given sets. By a function whose arguments run over the set X (domain) and whose values belong to the set Y (range) we understand the subset f of the cartesian product $X \times Y$ with the property that for every $x \in X$ there exists one and only one y such that $< x, y >\in f$. The set of all these functions f is denoted by Y^X. We usually write $y = f(x)$ instead of $< x, y >\in f$. [26, p. 47]

N. Bourbaki (*Théorie des ensembles (fascicule de résultats)*, 1939): Let E and F be two sets, which may or may not be distinct. A relation between a variable element x of E and a variable element y of F is called a functional relation in y if, for all $x \in E$, there exists a unique $y \in F$ which is in the given relation with x. We give the name of function to the operation which in this way associates with every element $x \in E$ the element $y \in F$ which is in the given relation with x; y is said to be the value of the function at the element x, and the function is said to be determined by the given relation. Two equivalent functional relations determine the same function. [3, p. 351]

The aforementioned definitions all have their basis in set theory. Since the 1960s there has been considerable discussion of a foundation for category theory (and for all of mathematics) not based on set theory. The notion of function, in terms of composition of functions, is axiomatized into a primitive term. It is interesting to note that this is one example where a notation (representing a function by an *arrow* in topology by William Hurewicz in about 1940) led to a concept (category theory, by Samuel Eilenberg and Saunders Mac Lane in 1942) [32, p. 29]. For more detail, read [32, pp. 398–402].

In view of the development discussed above, it is instructive to ask: How can we motivate the (modern) abstract definition of a function in teaching mathematics (or even, how much should we teach) when most students feel that the classical definition is good enough? Frederick Rickey cites this page of history on a formal definition of function as an "example of how a knowledge of the history of mathematics indicates what we should not teach" [40]. Comment on this.

Generalized Functions

Euler had introduced his E-discontinuous functions for physical reasons. Later he stressed that these inevitably emerged in solving partial differential equations. He had the vision of the development of a calculus of E-discontinuous functions ("Eclaircissemens sur le mouvement des cordes vibrantes," 1766)[30, p. 303]: "But if the theory [of the vibrating string] leads us to a solution so general that it extends to all discontinuous as well as continuous figures, one must admit that this research opens to us a new road in analysis by enabling us to apply the calculus to curves which are not subject to any law of continuity, and if that has appeared impossible until now the discovery is so much more important." In a second memoir ("Sur le mouvement d'une corde qui au commencement n'a été ébranlée que dans une partie," 1767) [30, p. 304] he also urged others to work on these problems: "This part, of which we so far know barely the first elements, certainly deserves the united efforts of all geometers for its investigation and development." According to Jesper Lützen, this project was completed a little less than two centuries later: "All the *ad hoc* definitions of generalized solutions from the first half of this century were incorporated in the theory of distributions created by L. Schwartz during the period 1945–1950 as a result of his work with generalized solutions to the polyharmonic equation. The theory of distributions probably constitutes the closest approximation to Euler's vision of a general calculus one can obtain, for in that theory any generalized function is infinitely often differentiable" [30, p. 305].

Near the end of the last century Oliver Heaviside ("On operations in physical mathematics," 1892/93) had the creative imagination to differentiate the function

$$f(x) = \begin{cases} 1 & \text{if } x \geq 0 \\ 1/2 & \text{if } x = 0 \\ 0 & \text{if } x < 0 \end{cases}$$

to yield the impulse "function"

$$\delta(x) = \begin{cases} 0 & \text{if } x \neq 0 \\ \infty & \text{if } x = 0. \end{cases}$$

The latter was made famous when Paul Adrien Maurice Dirac (*The Principles of Quantum Mechanics,* 1930) introduced it as a convenient notation in the mathematical formulation of quantum theory. What is important are not the values assumed by δ at x, but rather the way δ and its derivatives operate on functions. It took another 15 to 20 years for mathematicians to discover the mathematical foundations of a correct formulation of the definition and properties of such "functions" as the "Dirac delta-function." Laurent Schwartz began publishing his researches on generalized functions in 1944, subsequently developed fully in his treatise

Théorie des Distributions (1950/51). A distribution is a continuous linear functional on a space
$\mathcal{D}$ of infinitely differentiable functions (called "test functions") that vanish outside some closed
interval. For more detail, read [23].

"Eadem Mutata Resurgo"

Anthony Gardiner likens the evolution of the function concept to a "creative tug-of-war" between
two mental images: the *geometric* and the *algebraic* [16, p. 256]. Israel Kleiner adds a third—
the *logic* (correspondence)—coming in subsequently [24, p. 282]. What are the highlights of
this tug-of-war in the evolution of the function concept?

What implications in teaching can we learn from this tug-of-war, in view of the following
saying of Richard Courant: "The presentation of analysis as a closed system of truths without
reference to their origin and purpose has, it is true, an aesthetic charm and satisfies a deep
philosophical need. But the attitude of those who consider analysis solely as an abstractly
logical, introverted science is not only highly unsuitable for beginners but endangers the future
of the subject" [8, vol. I p. vi].

At the end of his paper [24, p. 300] Kleiner says that the function concept has been
modified, generalized and finally "generalized out of existence" (category theory). He then
asked: Have we come full circle? I tend to think that history does go in circles, but in a
modified sense. (See the figure for a schematic summary.) Perhaps we can better describe the
evolution by borrowing the motto alongside a logarithmic spiral engraved on the tombstone of
Jakob Bernoulli:

        EADEM MUTATA RESURGO (I shall arise the same though changed.)

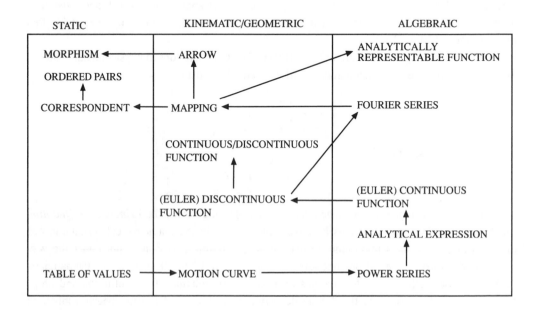

Exercises

1. A commentary by Liu Hui (c. 3rd century) on Problem 1 of Chapter 7 of *Jiuzhang Suan-shu* (*Nine Chapters on the Mathematical Art*) suggested an explanation of the solution via a viewpoint of functional dependence (rather than solving simultaneous equations). By making up a table of x (number of persons), S (total sum according to the first rule), S' (total sum according to the second rule) and $S - S'$, explain how the solution can be obtained. Problem 1 of Chapter 7 is as follows: "A certain number of persons want to buy an article. If each contributes 8 dollars (m), there will be an excess of 3 dollars (n). If each contributes 7 dollars (m'), there will be a deficiency of 4 dollars (n'). How many persons (x) are there? How much is the cost of that article (y)?" The solution is given as $x = (n + n')/(m - m')$ and $y = (mn' + m'n)/(m - m')$.

2. State a theorem on power series which can be regarded as the modern version of "analytical continuity" mentioned in the section "From Fourteenth Century to Eighteenth Century."

3. Fill in the detail of Euler's expansion of the logarithmic function as a power series outlined below [15, Book I, pp. 94–95]:

For the base a, $\log_a x$ is the exponent y such that $a^y = x$. Write $a^\varepsilon = 1 + k\varepsilon$ where ε is infinitely small. (In hindsight, what is k?) Let $N = y/\varepsilon$, then

$$a^y = (a^\varepsilon)^N = (1 + k\varepsilon)^N = 1 + N\left(\frac{ky}{N}\right) + \frac{N(N-1)}{1 \cdot 2}\left(\frac{ky}{N}\right)^2 + \cdots$$

$$= 1 + ky + \frac{1}{1 \cdot 2}\frac{N(N-1)}{N^2}k^2y^2 + \cdots.$$

N is infinitely large, so

$$a^y = 1 + \frac{ky}{1!} + \frac{k^2y^2}{2!} + \frac{k^3y^3}{3!} + \cdots.$$

Euler introduced the famous number e as the value of a for which $k = 1$, i. e.,

$$e = 1 + \frac{1}{2!} + \frac{1}{2!} + \frac{1}{3!} + \cdots.$$

Write $1 + x = a^y = a^{N\varepsilon} = (1 + k\varepsilon)^N$ so that $\log_a(1 + x) = N\varepsilon$. Then $1 + k\varepsilon = (1 + x)^{1/N}$, so $\varepsilon = [(1 + x)^{1/N} - 1]/k$ and $\log_a(1 + x) = N[(1 + x)^{1/N} - 1]/k$. Put $k = 1$ so that $a = e$, write $\log_e$ as simply $\log$, then $\log(1 + x) = N[(1 + x)^{1/N} - 1]$. By the expansion into binomial series, Euler obtained

$$\log(1 + x) = x - \frac{1}{2}x^2 + \frac{1}{3}x^3 - \frac{1}{4}x^4 + \cdots.$$

4. Find an E-continuous function which is not continuous. Find an E-discontinuous function which is continuous. What do we call an E-continuous function today? Is the definition of E-continuous function ambiguous? Discuss Cauchy's example given in 1844, viz. $f(x) = \sqrt{x^2}$ [46, p. 73]. In 1787 Louis Arbogast wrote a paper which won a prize offered by the Academy of St. Petersburg concerning "arbitrary functions." He called a curve *discontiguous* if the different parts of the curve do not join with each other [46, p. 71]. What do we call the function of such a curve today?

5. Work out the following examples given by Fourier in 1807: Extend $f(x) = x/2$ defined on $[0, \pi]$ into an odd function on $[-\pi, \pi]$ and compute its Fourier series. Extend $f(x) = x/2$ defined on $[0, \pi]$ into an even function on $[-\pi, \pi]$ and compute its Fourier series. (Note that two different expressions represent the same thing on the domain $[0, \pi]$.)

6. Discuss the mathematics in Weierstrass's example of a continuous nowhere-differentiable function mentioned in the section "Function Concept in the Nineteenth and Twentieth Centuries." (See pp. 351–352 of E. C. Titchmarsh's *The Theory of Functions* (2nd edition, 1939).)

7. Explain why the Dirichlet function is of (Baire) class 2.

8. Discuss the following example of Cauchy (1823) which shows that even a function infinitely differentiable at a given point can fail to be analytic at that point [46, p. 74]:

$$f(x) = \begin{cases} \exp(-1/x^2), & \text{if } x \neq 0; \\ 0, & \text{if } x = 0. \end{cases}$$

Bibliography

1. Bell, E. T.: *The Development of Mathematics*, (2nd Edition). New York: McGraw Hill, 1945.
2. Bottazzini, U.: *The Higher Calculus: A History of Real and Complex Analysis From Euler to Weierstrass*. New York: Springer-Verlag, 1986.
3. Bourbaki, N.: *Elements of Mathematics: Theory of Sets*. Paris: Hermann, 1968 (original edition published in 1939).
4. Boyer, C.: *The History of the Calculus and Its Conceptual Development*. New York: Dover, 1959.
5. Cajori, F.: *A History of Mathematics*. New York: Chelsea, 1985 (original edition published in 1893).
6. Carathéodory, C.: *Vorlesungen Über Reelle Funktionen*. Leipzig: Teubner, 1917.
7. Clagett, M.: *The Sciences of Mechanics in the Middle Ages*. Madison: University of Wisconsin Press, 1959.
8. Courant, R.: *Differential and Integral Calculus*, Vol. I and II, 2nd English Edition, (E. J. McShane, trans.). London: Blackie, 1934–36.
9. Crombie, A. C.: *Augustine to Galileo*. London: Heinemann, 1959.
10. Crummett, W. P. and Wheeler, G. F.: "The Vibrating String Controversy," *American Journal of Physics* 55 (1987), 33–37.
11. Descartes, R.: *The Geometry of René Descartes* (D. E. Smith and M. L. Latham, trans.). New York: Dover, 1954.
12. Dugac, P.: "Des Fonctions Comme Expressions Analytiques Aux Fonctions Représentables Analytiquement," in J. W. Dauben, ed.: *Mathematical Perspectives: Essays On Mathematics and its Historical Development*. New York: Academic Press, 1981, 13–36.
13. Edwards, Jr. , C. H.: *The Historical Development of the Calculus*. New York: Springer-Verlag, 1979.
14. Engels, F.: *Dialectics of Nature* (C. Dutt, trans.). New York: International Publishers, 1940 (original edition published in 1925).
15. Euler, L.: *Introduction to Analysis of the Infinite*, Books I and II, (J. D. Blanton, trans.). New York: Springer-Verlag, 1990.
16. Gardiner, A.: *Infinite Processes: Background to Analysis*. New York: Springer-Verlag, 1982.
17. Grabiner, J. V.: "Is Mathematical Truth Time-Dependent?" *American Mathematical Monthly* 81 (1974), 354–365.

18. Grabiner, J. V.: *The Origin of Cauchy's Rigorous Calculus*. Cambridge: MIT Press, 1981.
19. Grattan-Guinness, I. *et al.*: *From the Calculus to Set Theory: 1630–1910*. London: Duckworth, 1980.
20. Hartner, W. and Schramm, M.: "Al-Biruni and the Theory of the Solar Apogee: An Example of Originality in Arabic Science," in A. C. Crombie, ed.: *Scientific Change*. London 1963, 206–218.
21. Hausdorff, F.: *Set Theory*, 3rd Edition, (J. R. Aumann, trans.). New York: Chelsea, 1957 (original edition published in 1937).
22. Hawkins, T.: *Lebesgue's Theory of Integration: Its Origins and Development*, 2nd Edition. New York: Chelsea, 1975.
23. Horváth, J.: "An Introduction to Distributions," *American Mathematical Monthly* 77 (1970), 227–240.
24. Kleiner, I.: "Evolution of the Function Concept: A Brief Survey," *College Mathematics Journal* 20 (1989), 282–300.
25. Kline, M.: *Mathematical Thought From Ancient to Modern Times*. New York: Oxford University Press, 1972.
26. Kuratowski, K.: *Introduction to Set Theory and Topology* (L. F. Boron, trans.). Oxford: Pergamon Press, 1961.
27. Langer, E.: "Fourier's Series: The Genesis and Evolution of a Theory," *American Mathematical Monthly* 54, Slaught Memorial Paper 1 (1947).
28. Liang, Z. J.: *A Brief History of Mathematics*. (In Chinese) Shenyang: Liaoning People's Press, 1980.
29. Lützen, J.: "Funktionsbegrebets Udvikling Fra Euler Til Dirichlet" (The Development of the Concept of Functions From Euler to Dirichlet), *Nordisk Mat. Tiddskr.* 25–26 (1978), 5–32.
30. Lützen, J.: "Euler's Vision of a General Partial Differential Calculus for a Generalized Kind of Function," *Mathematics Magazine* 56 (1983), 299–306.
31. Luzin, N. N.: "Function" in *The Great Soviet Encyclopedia*. (c. 1930), 313–333.
32. MacLane, S.: *Categories For the Working Mathematician*. New York: Springer-Verlag, 1971.
33. MacLane, S.: *Mathematics: Form and Function*. New York: Springer-Verlag, 1986.
34. Malik, M. A.: "Historical and Pedagogical Aspects of the Definition of Function," *International Journal of Mathematics Education in Science and Technology* 11 (1980), 489–492.
35. Monna, A. F.: "The Concept of Function in the 19th and 20th Centuries, in Particular With Regard to the Discussions Between Baire, Borel and Lebesgue," *Archive for History of Exact Sciences* 9 (1972), 57–84.
36. Moore, G. H.: *Zermelo's Axiom of Choice: Its Origin, Development, and Influence*. New York: Springer-Verlag, 1982.
37. Pedersen, O.: "Logistics and the Theory of Functions", *Archive International d'Histoire de Sciences* 24 (1974), 29–50.
38. Poincaré, H.: *Science and Method* (F. Maitland, trans.). New York: Dover, 1952.
39. Ravetz, J. R.: "Vibrating Strings and Arbitrary Functions" in *The Logic of Personal Knowledge*. London: Routledge & Kegan Paul Ltd., 1961, 71–88.
40. Rickey, F.: "A Function Is Not a Set of Ordered Pairs." Preprint, Bowling Green State University, 1987.
41. Rüthing, D.: "Some Definitions of the Concept of Function From Joh. Bernoulli to N. Bourbaki," *Mathematical Intelligencer* 6 (4) (1984), 72–77.
42. Smith, D. E.: *History of Mathematics*. New York: Dover, 1958 (original edition published in 1923).
43. Struik, D.: *A Source Book in Mathematics, 1200–1800*. Cambridge: Harvard University Press, 1969.
44. Suppes, P.: *Axiomatic Set Theory*. Princeton: Princeton University Press, 1960.
45. Weyl, H.: *Philosophy of Mathematics and Natural Science* (O. Helmer, trans.) Princeton: Princeton University Press, 1949 (original edition published in 1927).
46. Youschkevitch, A. P.: "The Concept of Function Up to the Middle of the 19th Century," *Archive for History of Exact Sciences* 16 (1976/77), 37–85.

C. F. Gauß.
Thou, nature, art my goddess, to thy laws
My services are bound.

Carl Friedrich Gauss

My Favorite Ways of Using
History in Teaching Calculus

V. Frederick Rickey

No one ever forgets their first love. So it is with teaching. My first success in using history in the classroom will always be remembered fondly. In 1975, I was teaching integration techniques and, in a search for some good examples, I opened Michael Spivak's *Calculus* [19], which has always been one of my favorites. One problem (pp. 325–26) asked for the evaluation of the integral of the secant using "the world's sneakiest substitution," $t = \tan(x/2)$. Spivak suggested that when this approach is used the answer is more naturally expressed as

$$\int \sec x \, dx = \ln \left| \tan \left(\frac{x}{2} + \frac{\pi}{4} \right) \right|,$$

rather than in the usual form $\ln | \sec x + \tan x |$. Then he added a historical note that got my attention:

> This last expression was actually the first one discovered, and was due, not to any mathematician's cleverness, but to a curious historical accident: In 1599 Wright computed nautical tables that amounted to definite integrals of sec. When the first tables for the logarithms of tangents were produced, the correspondence between the two tables was immediately noticed (but remained unexplained until the invention of the calculus). [19, p. 326]

When I presented this example in class the next day and related Spivak's historical note, all of us realized there must be more to this story. I told my students that I would see if I could learn more of the history, and then headed for the library. The next day, I told them what little I had discovered about our problem.

When they requested I find out more, I realized something fascinating was happening. Of course, I was hooked on the problem, but so was my class. For the next several weeks, every class began with some curious tidbit that I had discovered about the problem. If I did not begin class this way, someone would ask, with a touch of both hopeful anticipation and disappointed resignation in their voice, what more had I learned.

This example explains why I like to use history in teaching mathematics. History provides a wonderful way to get the students interested in mathematics. Telling students that mathematics is an exciting field that draws much of its strength from real world problems does not suffice. We must show them that this is the case and history helps us do it.

Unfortunately, now that I know the history, this exciting teaching experience cannot be replicated. I am not that good of an actor. This episode was something that just happened. It was good for the students to see me interested in a problem, to see me doing historical research. They learned about the nature of mathematics, the process of discovery, and the role that mathematics plays in the world. These are the kinds of things that we must show our students. I keep hoping that this same type of experience will be repeated with another class and another problem, but thus far it has not happened. Perhaps if we freely admit what we do not know, it will.

There is no reason to give the historical details of this problem here, for Phil Tuchinsky and I published them as "An application of geography to mathematics: History of the integral of the secant" [18]. I am also pleased to say that this approach has been adopted by textbook writers, for example, in the popular calculus book of Thomas and Finney [21, pp. 458, 462–464].

The Uses of History

Many advocate using history in teaching mathematics. The question is, how should history be used? My answer is that it needs to be tied very closely to the material being discussed in class. General historical comments are nice, but not sufficient. For several years, I have been gathering historical information about individual topics in the calculus, to enrich my own teaching, and, eventually, so that I can prepare a little book of historical ideas for other teachers. I have been amazed at how difficult it is to find this information. The reason is that the kind of detail that I think is necessary for use in class is much greater than the generalities that are presented in most history books. For example, it is easy to find a detailed history of the brachistochrone problem, but other topics are problematic. I have learned the history of the Folium of Descartes, and of the hyperboloid of one sheet. Many others, such as why we use m for slope, or the history of the world's trickiest substitution, have thus far eluded me. What I intend to do below is to present a few examples of how I have successfully used history in teaching calculus. This accounts for the autobiographical tone of this paper. My hope is that this will inspire others to try my ideas in the classroom and also to share their own successes in using history.

Perrault and the Tractrix

When Leibniz was inventing the calculus in Paris in 1676, Claude Perrault (1613–1688) placed his watch in the middle of a table and pulled the end of its watchchain along the edge of the table. He asked: What is the shape of the curve traced by the watch?

If the name Perrault sounds familiar, it is probably because you have seen it on the title pages of such *Tales of Mother Goose* (*Contes de ma Mère l'Oye*, 1697) as "Cinderella" and "Puss in Boots." However, these stories were written by Claude Perrault's younger brother Charles (1628–1703). Claude Perrault was trained as a physician and practiced unnoticed for twenty years before he was invited, probably through the intervention of his brother, to become a founding member of the Académie des Sciences in 1666. There he took an interest in many of the scientific problems of the day, earning a reputation for the careful and detailed anatomical descriptions which he published based on dissections performed on deceased animals from the King's menagerie. In 1667 he joined a committee that was responsible for the design of the entrance façade of the Louvre. Perrault also designed the Paris Observatoire which Jean Baptiste Colbert (1619–1683), finance minister to Louis XIV, hoped would be the center of the Academy's activities. In 1688, Perrault died of an infection incurred as a result of dissecting a camel. He is remembered for his annotated translations of the *De architectura* of Vitruvius (d. c. 25 B.C.), and for a work on the design of columns which was influential throughout the eighteenth century. Thus he was well known as anatomist and architect. However, Perrault's only contribution to mathematics, as far as I am aware, was to pose the problem that concerns us here. For additional biographical information see [10] and [15].

At this point I would like to advocate the use of biography, for it fleshes out history. For some years, I tended to downplay the importance of the use of biography in the classroom, but I have come to appreciate the role that it plays in helping us—and especially our students—to understand mathematics. For example, the simple fact that René Descartes (1596–1650) traveled to The Netherlands where he met Frans van Schooten (1615–1660), explains how the Dutch became involved in the development of analytic geometry. What interests students most is the anecdotes we tell about mathematicians, but that is only a small part of what they can learn if—while we have their attention—we continue and present the larger picture. I do not mean to downplay anecdotes, for a catchy line will be remembered by students, and if they remember the mathematician, they have a better chance of remembering the mathematics. The image of Perrault and his camel is too vivid to forget. Another line that I have used with great effect when discussing falling bodies comes from Jacob Bronowski: "Galileo was a short, square, active man with red hair, and rather more children than a bachelor should have." [4, p. 200]. For a most interesting discussion of using biography in the classroom, see [17].

Perrault's question about his watch was one of the earliest inverse tangent problems. These are problems where some characterization of the tangent to a curve is given and the goal is to find the curve itself; today we call them differential equations. The first inverse tangent problem was posed by Florimond Debeaune (1601–1652) in 1638; it asks for the solution of $dy/dx = a/(y - x)$. While Debeaune's problem is usually part of a differential equations course, it is a little too complicated for the usual calculus course. However, Perrault's problem is an ideal example.

I begin my classroom presentation of Perrault's problem with a delightful picture from the first English edition of *Mathematical Snapshots* by Hugo Steinhaus [20]. (See Figure 1.)

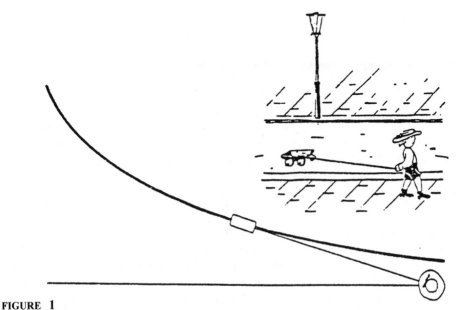

FIGURE 1
From Hugo Steinhaus, *Mathematical Snapshots*, p. 270.

It shows a young boy pulling a little wagon on a string. Note that the top view even shows the ribbon on his hat. He is walking along the sidewalk and the wagon is in the street. This picturesque way of viewing the problem gives the curve its name. From the Latin *trahere*, which means "to drag or pull," comes our word *tractrix*. (When we extract roots of equations we go back to the same Latin root.)

The important mathematical constraint on this problem is that the length of the string does not change. In the seventeenth century this was expressed by saying that the length of the tangent is constant, so we have an "equitangential curve." This way of speaking sounds strange to us, since we think of the tangent line as the whole line, and hence not having length, but in those days, before they had a good grasp of such elementary parts of analytic geometry as the equations of lines, mathematicians viewed the tangent as the line segment on our tangent line between the point of tangency to the curve and the x-axis.

Gottfried Wilhelm Leibniz (1646–1716) claimed to have solved the problem at once [14], but no trace of his original work survives. Isaac Newton (1642–1727) also solved the problem, but again no worksheets survive [16, vol. 3, p. 26, note 31]. Priority in publication goes to Christiaan Huygens (1629–1695), in a letter to Henri Basnage de Beauval (1656–1710) of 1693 [9]. For a detailed treatment of these solutions and the tractrix problem in general, see [6]. I do not know when the problem was first included in a textbook. In fact, I know of no research whatsoever on the issue of when certain problems entered our calculus textbooks. This would be a most interesting research project.

One aspect of this problem that I particularly like is that students are forced to grapple with the geometry of the situation in setting up the differential equation to solve. Students need to be given many opportunities to model real world problems, for this skill is much more important that the formal calculus techniques which dominate our books. If they are to master this skill then they need to be given many simple examples like this to set up for themselves.

Suppose the wagon is placed at the point $(a, 0)$, and the young boy starts at the origin, and then walks up along the y-axis. As he proceeds, the wagon, which is now at the point (x, y) in the diagram, steadily approaches the y-axis as asymptote. Consequently any value of x (with $0 < x \le a$) determines a unique value of y—so y is a function of x. Now sketch the curve.

Draw the tangent at an arbitrary point (x, y) on the tractrix and complete the triangle as in Figure 2. From the picture the slope of the tangent line is

$$-\frac{\sqrt{a^2 - x^2}}{x}.$$

(Reflecting the picture in the x-axis would eliminate both the minus sign and a little lesson for the student.) By definition the slope of the curve is dy/dx, so we obtain the equation

$$\frac{dy}{dx} = -\frac{\sqrt{a^2 - x^2}}{x}.$$

The hard part is now finished (this always surprises students when I say it, but it is true). Now all we have to do is integrate (Exercise 1). Any student who is moderately adept with basic integration techniques will be able to find the equation of the tractrix:

$$y = -\sqrt{a^2 - x^2} + a \ln\left(\frac{a + \sqrt{a^2 - x^2}}{x}\right).$$

I have used this as a classroom example and also as a problem on a take home examination. When it was an exam question a little of its history was included in the statement of the problem,

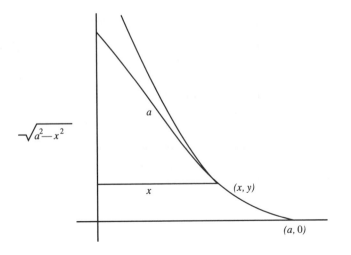

FIGURE 2

and I presented the rest later in class. In both instances, Perrault's tractrix problem was well received.

Suspension Bridges

Several years ago, I was teaching an "honors" calculus class, where the emphasis was on concepts, not computation. When we came to the simple differential equations that are discussed in calculus, I asked my students the shape of the cables on a suspension bridge. The answer to this question is not as widely known as it should be. Many mathematics teachers have told me that the shape of the cable on a suspension bridge is a catenary. Let us see why that is wrong.

To build a suspension bridge, we erect two tall towers, and hang a cable between them. From this cable we suspend a large number of small vertical cables called suspenders, which are used to support the roadway of the bridge. In an actual bridge, the roadway is almost horizontal, and its weight is very large compared to the total weight of the various cables. Thus it is reasonable to ignore the weight of the cables in our mathematical model. This is the crucial idea: only the weight of the roadway matters. It is also important that the weight of the roadway is uniformly distributed in the horizontal direction.

Our goal is to find the shape of the main cable on a suspension bridge. Since the cable is obviously symmetric with respect to its low point, let the y-axis pass through this point, and only consider the right portion of the bridge above the interval $[0, x]$. First consider just the cable above this interval (Figure 3). All of the forces acting on this segment of the cable must be in equilibrium or it would be in motion like "Galloping Gertie," the bridge over Puget Sound in the state of Washington that was torn apart on November 7, 1940 by high winds. Let $T(0)$ be the tension on the left end of the cable. Since this is the low point of the cable, the force acts horizontally. Let $T(x)$ denote the tension on the right end of the segment we are considering. This tension vector pulls the cable up and to the right and acts along the tangent line, which is at an angle α with the horizontal. When this vector is resolved into vertical and horizontal components we obtain the situation illustrated in Figure 3.

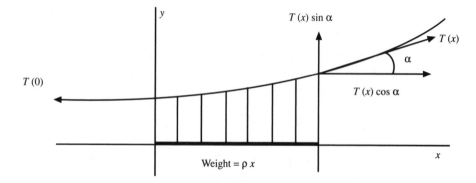

FIGURE 3

The assumption that the roadway is uniformly distributed tells us that over the interval $[0, x]$ the roadway has weight ρx, where ρ is its constant density. This weight acts vertically; it has no horizontal component. Consequently if we equate the horizontal forces we obtain the first equation below. The second comes from the vertical forces:

$$T(0) = T(x)\cos(\alpha)$$

$$\rho x = T(x)\sin(\alpha).$$

What we now desire is a differential equation. All we need do is to remember the basic idea that $dy/dx = \tan\alpha$. Then, from the above equations we obtain:

$$\frac{dy}{dx} = \tan(\alpha) = \frac{\sin(\alpha)}{\cos(\alpha)} = \frac{\rho x/T(x)}{T(0)/T(x)} = \frac{\rho}{T(0)}x.$$

Integrating we obtain,

$$y = \frac{\rho}{2T(0)}\,x^2 + h_0.$$

Thus we see that the shape of the cable on a suspension bridge is a parabola. Unfortunately, I do not know the history of this problem.

After I did this problem in class, I related that this was just a warm up for a more difficult and very famous problem, the catenary problem. This time we have only the main cable. What is its shape?

Galileo (1564–1642) had suggested that a heavy rope suspended from both ends would hang in the shape of a parabola, a conjecture which was disproved by Joachim Jungius (1587–1657) and published posthumously in 1669, although Huygens had an unpublished refutation in 1646. If you read the geometric proof of Huygens [5, p. 601], you will see what a great accomplishment the new calculus of Leibniz and Newton was. The true shape of the curve was not known until 1690/91 when Huygens, Leibniz, and Johann Bernoulli (1667–1748) replied to a challenge of Jakob Bernoulli (1654–1705). The name "catenary" was introduced by Huygens in a letter to Leibniz in 1690; it derives from the Latin *catena*, which means "chain." This was the first independent work of Johann Bernoulli, who was immensely proud that he had solved the catenary problem [2, vol. 1, 48–51] and that his brother Jakob, who had posed it, had not.

Writing to Pierre Remond de Montmort (1678–1719) years later, on 29 September 1718, Johann boasted:

> The efforts of my brother were without success; for my part, I was more fortunate, for I found the skill (I say it without boasting, why should I conceal the truth?) to solve it in full and to reduce it to the rectification of the parabola. It is true that it cost me study that robbed me of rest for an entire night. It was much for those days and for the slight age and practice I then had, but the next morning, filled with joy, I ran to my brother, who was still struggling miserably with this Gordian knot without getting anywhere, always thinking like Galileo that the catenary was a parabola. Stop! Stop! I say to him, don't torture yourself any more to try to prove the identity of the catenary with the parabola, since it is entirely false. The parabola indeed serves in the construction of the catenary, but the two curves are so different that one is algebraic, the other is transcendental. [3, pp. 97–98; translation from 11, p. 473]

After giving all of this history to my class, time had run out, so I announced that we would omit the derivation of the equation of the catenary. To my great surprise, the students howled in protest and insisted that we do the proof next time. Naturally, I was happy to oblige, but this event was so unique that I have ever since attributed it to the fact that I had presented the problem in its historical setting. I have no stronger example of history as a motivating force.

So now let us derive the equation of the catenary. The notation in Figure 3, with a few changes, will suffice. Of course, the roadway is no longer there. This time the weight, which consists only of the cable, is distributed uniformly along the cable, not uniformly in the horizontal direction. This is the main difference from the suspension bridge. The derivation becomes more complicated since we must introduce the parameter s, which denotes arc length. Consequently, the weight of the cable is ρs. The horizontal forces are the same as before, so we have

$$T(0) = T(x)\cos(\alpha),$$

but the downward vertical force is ρs, so this time we have

$$\rho s = T(x)\sin(\alpha).$$

Eliminating $T(x)$ between these two equations and then solving for s we have $s = k\tan(\alpha)$, where $k = T(0)/\rho$. If we differentiate x and y with respect to arc length s we have:

$$\frac{dx}{ds} = \cos(\alpha) \quad \text{and} \quad \frac{dy}{ds} = \sin(\alpha).$$

By the chain rule (what could be more fitting to use in this problem),

$$\frac{dx}{d\alpha} = \frac{dx}{ds} \cdot \frac{ds}{d\alpha} = \cos(\alpha) \cdot k\sec^2(\alpha) = k\sec(\alpha)$$

$$\frac{dy}{d\alpha} = \frac{dy}{ds} \cdot \frac{ds}{d\alpha} = \sin(\alpha) \cdot k\sec^2(\alpha) = k\sec(\alpha) \cdot \tan(\alpha).$$

Integrating each of these we have

$$x = k\ln|\sec(\alpha) + \tan(\alpha)| \quad \text{and} \quad y = k\sec(\alpha).$$

Finally, if we expend a little effort and eliminate α from these two equations we obtain $y = k \cosh(x/k)$ (Exercise 4), which is the equation of a catenary. For a somewhat different solution see [8, pp. 371–373].

Thus we have seen that when a suspension bridge is being erected, and only the cable is up, then it assumes the shape of a catenary. However, when the roadway is installed below, then the cable changes shape to a parabola.

Perhaps this is an opportune point to mention the issue of historical accuracy in the classroom. Contrary to the professional historian of mathematics, the classroom teacher need not be a slave to historical details and methods. The teacher must be accurate in what is said— we should not lie—but it is not necessary to tell the whole story. To provide an overabundance of detail will bore the students and will not advance our goal of using history to motivate and instruct the students. In particular, it is not necessary, and seldom desirable, to use the same methods to derive results that their inventors did. The above derivation for the catenary is stated in modern language, and I would certainly not apologize for doing so in class.

Cauchy's Famous Wrong Theorem

My final example again stems from my own teaching experience. Some years ago, when I came to the topic of sequences and series of functions while teaching an undergraduate analysis class, I realized that the book had done a particularly poor job on this topic (to protect the guilty, no reference will be given). Thus something had to be done. Since I had been rereading Imre Lakatos's delightful little book, *Proofs and Refutations,* I decided to see if his analysis of Cauchy's famous wrong proof could be adapted to the classroom [12, pp. 123–141].

In presenting the topic of sequences and series of functions, I began, as always, with a goodly supply of carefully chosen examples, drew pictures of some of them, and left others for homework. After noting how nicely the examples behaved, I coaxed the following observation out of the students.

Theorem. *A convergent series of continuous functions converges to a continuous function.*

After congratulating my students for this brilliant conjecture and attributing it to Augustin Cauchy (1789–1857), I pulled his *Cours d'Analyse* [7] from my briefcase, and then presented his proof. This dusty tome lent authority to the argument to be given below, and the students were pleased that I had taken the trouble to go back to original sources. I like to take relevant books to class, and after explaining how they bear on the material, pass them around for the students to look at. If that is impossible, I try to have an overhead transparency to show the relevant passages. The students appreciate this.

Before giving the proof, some notation needs to be introduced. Cauchy is dealing with a series of functions whose sum is the function s. The nth partial sum of this series is denoted s_n, and the remainder by r_n. Today all of this is usually presented in terms of sequences, but I wanted to follow Cauchy fairly closely. Next, I read Cauchy's proof to them and wrote it on the board in translation:

> Assume that the terms of the series contain the same variable x, and this series is convergent, and its different terms are continuous functions of x, in the neighborhood of a particular value assigned to this variable; and s_n, r_n and s are again three functions of the variable x, of which the first is evidently continuous with respect to

x in the neighborhood of the particular value in question. This assumed, let us consider the increases that these three functions have when x increases by an infinitely small quantity; the increase of s_n will be, for all possible values of n, an infinitely small quantity; and that of r_n will become insensible at the same time as r_n [sic], if one assigns to n a very considerable value. Hence, the increase of the function s can only be an infinitely small quantity. From this remark, one deduces immediately the following [read preceding] proposition. [7, p. 120]

Since the students were familiar with Weierstrassian ϵ-δ techniques, I took the time to carefully formulate Cauchy's proof in the modern language that they were learning to understand.

Proof. Let $\epsilon > 0$ be given. Then
(1) Since each function is continuous, their partial sum, $s_n(x)$, is continuous,

$$\exists \delta \, \forall \alpha \quad |\alpha| < \delta \Rightarrow |s_n(x + \alpha) - s_n(x)| < \epsilon.$$

(2) Since the series converges at x,

$$\exists N \, \forall n > N \quad |r_n(x)| < \epsilon.$$

(3) Since the series converges at $x + a$,

$$\exists N \, \forall n > N \quad |r_n(x + a)| < \epsilon.$$

Thus:

$$|s(x + a) - s(x)| = |s_n(x + a) + r_n(x + a) - s_n(x) - r_n(x)|$$
$$\leq |s_n(x + a) - s_n(x)| + |r_n(x)| + |r_n(x + a)|$$
$$\leq 3\epsilon.$$

Hence the function s is continuous. [It is pedantic to insist on ending with precisely ϵ. Why make the mathematics even more mysterious for the student.] Q.E.D.

By careful planning, the class ended just as the proof did, and I was relieved that there were no questions after class. The next day the students were hopping mad, for they had done their homework (this was a good class) and observed that some of the examples I had given (and left the graphs as exercises) contradicted Cauchy's Theorem. But they were ready to do mathematics. I asked them about one of the counterexamples they had discovered in their homework:

$$\sin(x) - \frac{1}{2}\sin(2x) + \frac{1}{3}\sin(3x) - \frac{1}{4}\sin(4x) + \cdots$$

Were all the terms of the series continuous functions? Did the series converge? Was the limit function really discontinuous? "Yes, yes, yes." they said. Well then, what about the theorem? Cauchy published it in his *Cours d'Analyse*, so it must be correct, right? "Yes," they readily agreed. They had also accepted the proof when it was presented in class, for it seemed correct to them. They were puzzled. Something was wrong, but what?

I asked if they had examined the proof to see if anything was wrong with it. No, that had not occurred to them. So I suggested that we should look at the proof carefully. Imre Lakatos makes the argument [12] that it was in the mid-nineteenth century that mathematicians made the same advance as my students were now making: When a proof is wrong, do not just abandon it, but analyze it carefully to see if there are any "hidden hypotheses" that would make it correct. Lakatos took this phrase from Philipp Ludwig von Seidel (1821–1896), who used it in 1847 when he took the steps that my students were now ready to do.

We shall now analyze Cauchy's proof. In step (1) we need to realize that δ depends on ϵ, x, and n. To make this explicit, we shall write $\delta(\epsilon, x, n)$. Now, in step (2), N depends on ϵ and x, so we write $N(\epsilon, x)$. However, in step (3), N depends on ϵ, x, and ALSO on a. Using the same notation, we express this by $N(\epsilon, x + a)$. Now comes the critical observation. To make Cauchy's proof work, we need an integer M bigger than $N(\epsilon, x)$ and simultaneously bigger than each $N(\epsilon, x + a)$, where $|a|$ is less than $\delta(\epsilon, x, n)$. Thus we must know that

$$M = \max_t N(\epsilon, t)$$

exists for all ϵ; i.e., that M does not depend on x. Consequently, the additional hypothesis that we need is the following:

$$\forall \epsilon > 0 \quad \exists M \ \forall n > M \ |r_n(x)| < \epsilon \ \forall x.$$

What I have done here is to motivate the definition of uniform convergence. It is precisely the additional hypothesis that is needed to make Cauchy's theorem correct and proof work. Had I just written it down in the usual definition-theorem-proof style of modern mathematics, it would appear to be very much ad hoc. The historical presentation allows the student to see the true origin of the concept. As Lakatos has observed, the correct concept is generated by the incorrect proof. This is one case where I feel that a historical presentation is absolutely necessary to the understanding of the material.

You may object that this type of presentation takes too much time, for it did take two whole class periods. But that is not so. The time was well spent. Presenting the wrong proof and then analyzing it to see what additional hypothesis is needed takes far less time than presenting an ad hoc definition, trying (probably unsuccessfully) to explain it, and then finally giving the proof. With my presentation there is no need to give a correct proof after the definition has been discovered; that is an easy exercise for the student. In fact, my students said there was no reason for me to write out a new proof, for they had a deep understanding of how it works. Moreover, with this presentation the students have also learned more. The opportunity to analyze an incorrect proof builds both confidence and skepticism (students must learn that books contain numerous errors). More importantly, it shows them where theorems come from: We make conjectures, attempt proofs, analyze them, and refine them. It also shows the importance of definitions, showing that they are carefully chosen, not things arbitrarily written down just before a proof. I trust you will agree with my assessment that without giving this historical presentation, the student's understanding of the concept of uniform convergence would be severely hampered.

In this example, the history stayed in the background, but by the time I had finished, the students were anxious to have some details. Since this whole issue has been extensively and hotly debated in the literature over many years, I shall refrain from giving the historical details here. Many of them are in Lakatos's book. For a current entry into the literature, see [13]. Nonetheless, I must end with a historical point. In 1826, a mathematician wrote "it seems to me that the theorem admits of exceptions" and then provided the first counterexample, the same counterexample that my students had done for homework [1]. The mathematician was the Norwegian, Niels Henrik Abel.

Conclusion

The history of mathematics plays many valuable roles in the classroom. The examples above show that an incomplete historical note can captivate students, that anecdote and biography make problems much more interesting than they would be without it, that history motivates students to want to do more mathematics, and that, on occasion, a historical approach is absolutely vital to understanding.

Exercises

1. Find the equation of the tractrix by evaluating the integral

$$-\int \frac{\sqrt{a^2 - x^2}}{x}\, dx.$$

There are at least three substitutions which will work: (1) $t^2 = a^2 - x^2$, (2) $x = a\sin(\theta)$, and (3) the nonstandard trigonometric substitution $x = a\cos(\theta)$. Do the problem all three ways. Which of these substitutions do you think provides the easiest solution?

2. Why was the cute diagram of Steinhaus flipped about the line $x = y$ before setting up the differential equation for the tractrix? The hard way to answer this is to start from Steinhaus's picture and solve the problem anew. The easy way is to think.

3. Find the equation of the curve of constant subnormal. You will need to know that the subnormal to a point on a curve is the line segment from the abscissa of the point to the place where the normal at that point intersects the x-axis. Also find the equation of the curve of constant subtangent. It is instructive to draw a careful sketch of these curves using only their definitions and then try to guess the solutions before you do any computation.

4. Eliminate a from the equations $x = k\ln|\sec a + \tan a|$ and $y = k\sec a$ to obtain $y = k\cosh(x/k)$, an equation of a catenary.

5. Use computer graphics to draw a parabola and a catenary. Can you tell the difference? To be faithful to the bridge problem, suppose a cable of length L is hung between two towers. Now add a roadway. Superimpose these two graphs on one coordinate axis. How much difference is there between the two curves? How do engineers decide how long the suspenders should be?

6. Work out the details of the counterexample to Cauchy's "Theorem."

7. Examine some old calculus books to see how they differ from our current texts both in the way they explain the fundamental ideas and in their choice of problems.

8. Investigate the history of suspension bridges and try to learn who first proved their cables are parabolic. Please write me if you find out.

Bibliography

1. Abel, Niels Henrik: "Recherches sur la série . . .," *Journal für die reine und angewandte Mathematik*, 1 (1826), 311–339. Reprinted in *Oeuvres Complètes* (1881), 1, 219–250. Also in Ostwald's *Klassiker*, #71.

2. Bernoulli, Johann: *Opera omnia*, 4 volumes, Lausanne and Geneva, 1742. Reprinted Hildesheim: Georg Olms, 1968.

3. Bernoulli, Johann: *Der Briefwechsel von Johann Bernoulli, (1667–1748)*. Basel: Birkhäuser, 1955.

4. Bronowski, J.: *The Ascent of Man*. Boston: Little, Brown, 1973.

5. Bos, H. J. M.: "Huygens, Christiaan," *Dictionary of Scientific Biography*, 6, 597–613.

6. Bos, H. J. M.: "Tractional Motion and the Legitimation of Transcendental Curves," *Centaurus* 34 (1988), 9–62.

7. Cauchy, Augustin: *Cours d'Analyse*, 1821. Reprinted *Oeuvres* II, 3.

8. Edwards, C. H., Jr. and Penney, David E.: *Calculus and Analytic Geometry*. Engelwood Cliffs, NJ.: Prentice-Hall, 1982.

9. Huygens, Christiaan: "Lettre l'auteur de l'Historire des Ouvrages des Sçavans," February 1693, 244–257. Reprinted *Oeuvres Complètes*, 10, 407–422.

10. Keller, A. G.: "Perrault, Claude," *Dictionary of Scientific Biography*, 10, 519–521.

11. Kline, Morris: *Mathematical Thought from Ancient to Modern Times*. New York: Oxford University Press, 1972.

12. Lakatos, Imre: *Proofs and Refutations*. Cambridge: Cambridge University Press, 1976.

13. Laugwitz, Detlef: "Infinitely Small Quantities in Cauchy's Textbooks," *Historia Mathematica* 14 (1987), 258–274.

14. Leibniz, Gottfried Wilhelm: "Supplementum geometriae dimensoriae sue generalissima omnium tetragonismorum effectio per motum; similiterque multiplex constructio lineae ex data tangentium conditione," *Acta eruditorum*, September 1693, 385–392. Reprinted *Matematische Schriften*, 5, 294–301.

15. Lowry, Bates: "Perrault, Claude," *Encyclopedia of World Art*, 11, 182.

16. Newton, Isaac: *The Mathematical Papers of Isaac Newton* (D. T. Whiteside, ed.), 8 volumes. Cambridge: Cambridge University Press, 1967–1981.

17. Pycior, Helena M.: "Biography in the Mathematics Classroom," in I. Grattan-Guinness, ed., *History in Mathematics Education*. Paris: Belin, 1987, 170–186.

18. Rickey, V. Frederick, and Tuchinsky, Philip M.: "An Application of Geography to Mathematics: History of the Integral of the Secant," *Mathematics Magazine* 53 (1980), 162–166.

19. Spivak, Michael: *Calculus*. New York: W. A. Benjamin, 1967.

20. Steinhaus, Hugo: *Mathematical Snapshots*. New York: Oxford University Press, 1960.

21. Thomas, George B., and Finney, Ross L.: *Calculus and Analytic Geometry*, 7th edition. Reading, MA: Addison-Wesley, 1988.

Improved Teaching of the Calculus
Through the Use of Historical Materials

Michel Helfgott

The School of Mathematical Sciences at the Universidad Nacional Mayor de San Marcos, Lima, Peru, offers an annual course on differential and integral calculus intended for first year students. The course, intimately linked to the natural sciences, is taught from a historical-cultural perspective and obtains its inspiration from the genetic method as expounded by Otto Toeplitz [18].

In this paper, I will relate how we at San Marcos face the challenge of introducing calculus using historical materials as a pedagogical device. San Marcos is the oldest university in the New World. Founded in 1551, when Peru was a Spanish colony, it has thirty thousand students enrolled in seventeen schools. I teach at the School of Mathematical Sciences, where two thousand undergraduate and seventy-five graduate students major in one of the following specialties: computer science, statistics, operations research, or mathematics.

The introductory course on calculus of one variable is required for all entering students. This annual course comprises thirty-two weeks of work (four hours of lectures and four hours devoted to problem discussion per week), not including the time allotted to exams. It covers the essential aspects of one variable differential and integral calculus.

Some Problems in Calculus Teaching

Probably every teacher of first year calculus must answer questions such as the following:

1. How can one ease the transition from high school mathematics to that taught at the University? This problem is particularly acute in countries where high school mathematics is not taught at an appropriate level.
2. What should be the place of mathematical rigor? What should be proved, and which proofs should be left for courses on analysis?
3. How should theory and applications be balanced within the course?
4. How can one make students feel that the subject is part of human history?

It is difficult to give definite answers to these questions. At our University, we have approached this task from several different angles. Our solution to the problem of transition is that students admitted to the School of Mathematical Sciences must take a three week intensive course whose main focus is Cartesian geometry and the rudiments of calculus. Since they haven't studied any calculus at high school, we believe that they should initially learn some basic techniques of differentiation and integration.

Derivatives linked to tangents, integrals as areas, and the unexpected connection between both concepts constitute the main topics of the introductory short course. In this way, students know from the very beginning where they are heading. We present nontrivial problems of maxima and minima that can be solved using only algebra and geometry. At the same time, we introduce problems whose solutions require techniques beyond high school mathematics. We also review the elements of Cartesian geometry of circles, lines, and parabolas, topics that are included in the high school programs. Several sections of the book by W. M. Priestley [14], with its pleasant style, come nearest to the material and approach we follow in the thirty hours of lectures (two per day). In addition, students get together every day under the guidance of a teaching assistant to discuss problems and their solutions. Once the short course comes to an end, an anonymous exam is given. Teachers can thus detect some of the students' deficiencies. Afterwards, the students enroll in the regular first year calculus course.

In the first year calculus course, with regards to "rigor," we try to strike a balance between intuitive arguments, proofs of some results, and the development and discussion of several models from the natural sciences. A historical point of view is used throughout the course. Our presentation is very different from the lemma-theorem-corollary style; we often use theorems whose geometrical meaning is taken to be evident, such as the mean value theorem. Proofs are presented later in the course. The fact that most students take courses in analysis in their later studies makes our job easier. In advanced courses, they see proofs of results such as the Bolzano–Weierstrass theorem.

In teaching the calculus, it is useful to supply many examples of applications from the natural sciences. We go further than usual, developing classical examples from mechanics and chemical kinetics. The latter are particularly helpful because they are simple, require few prerequisites, and use a significant amount of readily available data. We stress the meaning of the scientific method in its different stages of building models, obtaining consequences, and contrasting them with data. Applications appear everywhere, not necessarily at the end of a section.

Finally, historical aspects crop up frequently in the course; we always try to show that mathematics is intimately linked to the development of science and culture. Besides presenting biographies of great mathematicians, we include historical commentaries, such as those relating to the birth of logarithms and the Newton–Leibniz controversy. We do not convert the calculus course into a history of calculus course; our School offers, at the third year level, a one semester course on Foundations and History of Mathematics. Rather, we use historical developments to impress upon the students the view of mathematics as a discipline that has gradually evolved over many centuries. We try to follow the philosophy of Otto Toeplitz when he asserts:

> It is not history for its own sake in which I am interested, but the genesis at its cardinal points, of problems, facts, and proofs. [18, p. v]

Otto Toeplitz and the Genetic Method

The main source for the course is the book by Toeplitz mentioned above. Our classroom experience has taught us that from the pedagogical point of view it is very useful to go to the genesis, the origins, of mathematical theories. In this way, students achieve knowledge closer to the facts, and almost always this genetic approach is the most natural for beginning calculus students.

We use Toeplitz's book as a guide, a reference frame over which we build the course. As Alfred L. Putnam asserts in the preface to the English edition:

This book is not a substitute for a text, but it should be a most valuable supplement for those students who seek to know how the calculus arose and how it has come to its present form.

Everything that appears in *The Calculus: a Genetic Approach* is discussed in class, though not necessarily in the same order. Sometimes, we include some ε-δ techniques in order to compare how a statement was conceived and how it can be analyzed using our present standards of rigor.

For instance, in the discussion of logarithms [18, p. 90], we add the following considerations: Suppose there exists a differentiable function $f: R^+ \to R$ such that $f(xy) = f(x) + f(y)$. Let $x > 0$. For $u > 0$,

$$\frac{f(u) - f(x)}{u - x} = \frac{f(x \cdot u/x) - f(x)}{x \cdot (u/x) - x} = \frac{1}{x} \frac{f(u/x)}{(u/x) - 1}.$$

Hence

$$x \cdot \frac{f(u) - f(x)}{u - x} = \frac{f(u/x)}{(u/x) - 1}.$$

Therefore

$$\lim_{u \to x} \frac{f(u/x)}{(u/x) - 1} = c_x, \quad \text{where} \quad c_x = xf'(x).$$

Next we prove that $c_x = c_y \ \forall x, y > 0$. For this it is necessary to show that

$$\lim_{v \to y} \frac{f(v/y)}{(v/y) - 1} = c_x.$$

Given $\varepsilon > 0$, there exists $\delta > 0$ such that

$$\left| \frac{f(u/x)}{(u/x) - 1} - c_x \right| < \varepsilon \quad \text{for} \quad 0 < |u - x| < \delta.$$

Assume $0 < v - y < \rho\delta$, $\rho = y/x$. Then

$$0 < \left| \frac{v}{\rho} - \frac{y}{\rho} \right| < \delta, \quad \text{or} \quad 0 < \left| \frac{v}{\rho} - x \right| < \delta.$$

Thus

$$\left| \frac{f\left[\frac{(v/\rho)}{x} \right]}{\frac{(v/\rho)}{x} - 1} - c_x \right| < \varepsilon, \quad \text{or} \quad \left| \frac{f(v/y)}{(v/y) - 1} - c_x \right| < \varepsilon.$$

It follows that $xf'(x) = c$ or

$$f'(x) = c \cdot \frac{1}{x} \quad \text{for} \quad x > 0,$$

where the constant c depends solely on f.

Afterwards, we discuss how Gregory of Saint Vincent (1584–1667) discovered, in effect, that $A(xy) = A(x) + A(y)$, where $A(x)$ is the area under the curve $u = 1/t$ between 1 and x. On the other hand, a simple calculation shows that

$$\frac{1}{x} \leq \frac{A(x) - A(1)}{x - 1} \leq 1.$$

Hence $A(x)$ is differentiable with $A'(1) = 1$. But $A'(x) = c(1/x)$, $x > 0$. So $c = 1$, and $A'(x) = 1/x$. In other words, the area function $A(x)$ is a primitive of $1/x$.

The process that led us to the equality $A'(1) = 1$ suggests the fundamental theorem of the calculus (FTC) for monotonic continuous functions. Next, we develop the mean value theorem for integrals and the FTC for continuous functions through the use of sequences. At this point we give two historical notes:

(i) the role played by Isaac Barrow in the development of the FTC.

(ii) The Newton–Leibniz controversy.

In many calculus textbooks, the log is simply defined as

$$\int_1^x \frac{dt}{t}.$$

Then it is shown, using the method of substitution, that $\log xy = \log x + \log y$. This unmotivated development disturbs students because of its artificiality and makes them feel that mathematicians are, in a sense, magicians. The genetic method clarifies the historical development of logarithms and their relation with the FTC and takes us, without startling surprises, past the obstacles that the predecessors of Newton and Leibniz had to overcome. We note that the proof of the FTC for continuous monotonic functions is easy and does not present any complications. Mathematicians studied this case first; in it we can find the genesis of this famous and useful theorem. Later this theorem was extended to all continuous functions.

Overview of the Course

We build the whole edifice of the calculus on the concept of sequence and its limit, and thus proceed slowly in the beginning. Cavalieri's determination of the area under a cubic parabola is analyzed in detail, stressing the role played by Arab mathematicians. We use geometric series to shed light on Zeno's paradox.

A close look at the real number system and its rich history provides a good opportunity to define the number e and thus solve the problem of compound interest.

Fermat's method of calculating areas is used as a motivation for the limit of a function of a real variable. The computations are easy and give the student a good idea of one of Fermat's achievements [2]. Thereafter, we discuss how Descartes tried to solve the problem of tangents, the way Fermat found the tangent to a parabola [1, p. 168], and finally the modern approach to these issues. We begin the study of limits with limits of sequences, since in this way we can use all we know about the theory of convergence already covered. Only later do the ε-δ techniques make their appearance.

As for continuity, we proceed quite rapidly until we reach the intermediate value theorem (IVT). We analyze how Cauchy and Bolzano met the challenge of proving the IVT [4, pp. 564, 580], and present two different proofs that meet modern standards of rigor: an existence proof and an easily computer programmed proof [16, p. 67]. The IVT supplies a good opportunity to talk about what "rigor" has meant to different generations of mathematicians. The theorem on existence of a maximum and a minimum value of a continuous function on a closed interval and the theorem on the continuity of the inverse are carefully stated, accompanied by many illustrative examples, but are not proved in class. Nevertheless, we hand out to the students, as optional material, sheets with carefully developed proofs that make use of the Bolzano–

Weierstrass theorem. [10, p. 114] Most students will really understand these proofs only when they proceed to analysis in their second year, but it is important that they should be aware that the calculus can be founded on a proper definition of real number just as geometry is founded on its axioms.

After nine weeks, the ground is prepared for a systematic development of derivatives. We then emphasize calculation and physical applications, the latter in an historical context: the works of Galileo, Newton, and the rise of mechanics. Since we need to know that functions with equal derivatives differ by a constant, we freely use the mean value theorem (MVT) on the basis of its geometrical interpretation. Two weeks later a proof is provided.

Fermat, Rolle, and MVT provide us with the opportunity to show the students how a mathematician plays with the hypotheses of a theorem. Once the MVT is developed and its importance as a tool is recognized, we stop in order to say something about the scientific method [9, p. 280], the role played by Galileo, and how this method is used by scientists in daily practice.

In the short course, we dealt with some problems of maxima and minima of functions, using only algebra. Now, it is time to employ the calculus. We start this section by analyzing one of Fermat's original problems [4, p. 358]. Thereafter, the historical development of the phenomena of reflection in flat and curved mirrors and refraction play a dominant role. Many other well-known problems of maxima and minima are considered.

The sketching of curves is an appealing activity. Newton's ingenious method of approximation of roots of an equation makes things even more interesting.

We keep treating areas on an intuitive basis. The FTC for continuous monotone functions is easily obtained. After defining the logarithm as an integral, we develop most of its properties. Since this definition lacks motivation, we trace in detail the historical development of the logarithm. We do not use the FTC, which was not known at the time of Napier. Moreover, we show how the work on logarithms could have paved the way for the FTC.

Once we have proved the FTC for continuous functions, we can look for primitives of a large set of functions; in this connection we take a first look at the methods of integration by parts and substitution as they were developed at the end of the seventeenth century.

Our experience with students has shown us that one has to be very careful about introducing the integral as a limit. Here, probably more than in almost any other place in elementary calculus, historical perspective is pedagogically sound. We retrace several of the key steps developed by Cauchy and Riemann [8, p. 317] and then develop the integral, keeping in mind that this is not a course in mathematical analysis. Students learn about uniform continuity and about the integrability of continuous functions. Other topics are stated carefully but not proved. Applications to geometry and physics give the student an intuitive feeling for the significance of the theory they have just studied.

The next five weeks are devoted mainly to the exponential function and to an introduction to linear differential equations of first and second order. This is a point where students find multiple applications of the calculus and once more, realize the power and usefulness of the methods they have learned. Whenever possible, we develop models from first principles, try to place each problem in a historical context, and emphasize the necessity to work with data and the ability to formulate predictions [6, p. 139]. By treating differential equations (DE) in a first course in calculus we accomplish two things:

(i) Students realize that differential equations have historically always gone hand in hand with the calculus; to completely separate calculus from differential equations is like doing elementary algebra without solving linear equations.

(ii) In the second year, there is time to develop a qualitative theory of differential equations more completely.

L'Hospital's rules appear after the exponential function, since we need to know logarithms and the exponential function in order to give enough relevant examples. On the other hand, inverse hyperbolic functions are justified on the basis of their usefulness as primitives of well-known functions.

Taylor's theorem is derived as Taylor himself did [3, p. 287], as well as by using simple integration by parts. Once again, students profit from the comparison and get a sense of the discovery process in mathematics. The idea of approximation of functions and integrals is stressed throughout.

From time to time series appear naturally, be it when approximating $\log(x + 1)$ by successive integrations or when discussing Taylor polynomials under the assumption of uniform boundedness of all derivatives. But we do not develop the elementary theory of series in this course. Improper integrals and the theory of series constitute the first topic to be discussed in second-year calculus. Time is short and an important topic such as series has to be omitted. In their first year, students get a feeling for some concrete problems involving series and thus are ready for future work with series.

Time is ripe for a systematic development of the main methods of integration, emphasizing the contributions of late eighteenth-century mathematicians. Series representations of π show how a mathematical discovery makes its first appearance and how its proof is simplified afterwards [15, p. 715]. The catenary problem gives us the opportunity to deduce a differential equation from primary physical considerations and to solve it through a non-elementary substitution [15, p. 716]. We discuss the historical significance of the solution of this problem. On the other hand, we don't go into the intricate problems of integration that were commonly taught thirty years ago. We concentrate on the essential methodology, bearing in mind the availability of computers [8, p. 205].

When we fail to find elementary primitives by means of the useful methods of integration, then it is time to look at the trapezoidal and Simpson's method. Not much time is devoted to these topics, since most students will enroll afterwards in a numerical analysis course. However, we present the theoretical background underlying both methods so that they can be programmed and used on a computer [12].

The year's work ends with a study of curves, polar coordinates and related subjects. Our last lecture is on Kepler and Newton, with emphasis on Newton's discovery of the inverse square law on the basis of Kepler's laws and his deduction of these laws starting from the inverse square law [18, p. 150]. We could hardly find a more dramatic application of mathematics with which to conclude our first year calculus course. For a more complete description of the course contents, see the Appendix.

Some Student Needs

Usually, mathematics students are surprised that we are able to construct rigorously logarithmic and exponential functions but not trigonometric functions. We rely on their knowledge of high school trigonometry. We refer them to Spivak's book on calculus [17].

The more impatient and talented students, who can't wait until they enroll in the analysis sequence in order to find a satisfactory construction of the reals, are advised at the end of the year to read what A. H. Lightstone has to say about this matter [11]. They get a copy of the paper and take a look at its contents; the more tenacious ones start studying it. At the very least, they become aware of the complexity of the problem of the real numbers.

Besides presenting historical notes of universal interest about the origins and development of the calculus, we try to show our students the evolution of calculus teaching and of mathematics in Peru. For this purpose, the *Revista de Ciencias,* an almost ninety-year-old scientific journal published by our university, is an excellent source of information.

The arrival at San Marcos, in 1936, of the Polish mathematician Alfred Rosenblatt radically changed Peruvian mathematics. Before him, Peruvian mathematics was isolated from twentieth century European developments. In ten years of tenacious work, until his unexpected death in 1946, he trained a whole generation of young mathematicians who, in turn, prepared others for graduate work abroad. The productive effect of this one outstanding scholar is remarkable.

Professor Rosenblatt brought with him from Europe the latest information on functional analysis, topology and other mathematical areas that was not known in Peru at that time. Besides producing over 225 papers [13], he taught many important mathematical topics.

It is pertinent to mention the book *Analisis Algebraico* written in collaboration with Godofredo Garcia [5], a Peruvian mathematical physicist. It is a work filled with historical references and acute observations.

Our students are happy to learn about the evolution of their own school, and they are proud to belong to an institution of learning that throughout nearly 450 years has produced a considerable part of the Peruvian mathematical and cultural heritage.

Textbooks and Papers

As usually happens in course development projects, we are not completely satisfied with any textbook, and so have written class notes [7]. Nevertheless, the book by G. F. Simmons comes closest to our vision of how the calculus should be taught. This book is a serious attempt to blend techniques with applications and to provide a historical background.

Our best students are happy to have the opportunity to read and discuss some topics outside of the formal syllabus, topics that broaden their cultural view of mathematics. We plan to get more students interested in these matters; this project may not work for all since time is short and the amount of material they have to assimilate in the five regular courses (Calculus, Algebra, Computer Programming, Physics, and Humanities) is quite large. It should be mentioned that students "Read the Masters" once they enroll in the course on Foundations and History of Mathematics corresponding to the third year of studies. For us, it is still an open question whether they could do so in their first year.

Conclusion

We feel that, from a pedagogical standpoint, history is a valuable tool in mathematics teaching. By history, we mean history of mathematics linked with the history of science. It is also obvious that techniques, conceptual developments, computers, applications, and modelling are important. But we must not forget that providing a cultural background for a calculus course is a fundamental necessity.

In our experience, we have found that students welcome an approach that considers mathematics as an intellectual achievement of a certain age and place, whose main concepts have been evolving through many centuries.

Bibliography

1. Baron, M.: *The Origins of the Infinitesimal Calculus.* Oxford: Pergamon Press, 1969.
2. Campbell, W.: "An Application from the History of Mathematics," *Mathematics Teacher* 70 (1977), 538–540.
3. Edwards, C. H.: *The Historical Development of the Calculus.* New York: Springer Verlag, 1979.
4. Fauvel, J. and Gray, J.: *The History of Mathematics: A Reader.* Milton Keynes: The Open University, 1987.
5. Garcia, G. and Rosenblatt, A.: *Analisis Algebraico.* Lima: UNMSM, 1955.
6. Helfgott, M.: "Mathematics and the Natural Sciences: a Fructiferous Relation." *Proceedings of the Fourth Colloquium of the Peruvian Mathematical Society,* 1986.
7. Helfgott, M. and Nuñez, T.: *Notas de Clase de Calculo.* Facultad de Ciencias Matematicas, UNMSM, 1988.
8. Kemeny, J. G.: "Finite Mathematics—then and now," in *The Future of College Mathematics,* (A. Ralston and G.S. Young, eds.). New York: Springer-Verlag, 1983.
9. Kline, M.: *Mathematics for the Liberal Arts.* Reading, MA.: Addison-Wesley, 1967.
10. Kuratowski, K.: *Introduction to Calculus.* Oxford: Pergamon Press, 1961.
11. Lightstone, A. H.: "A Simple Alternative to Dedekind Cuts," *Scripta Mathematica* 26 (1963), 347–351.
12. Motter, W. I.: *Elementary Techniques of Numerical Integration and their Computer Implementation.* UMAP, Unit 379. Arlington, MA: COMAP.
13. Nuñez Bazalar, T.: "Vida y Obra de A. Rosenblatt," *Revista de la Facultad de Ciencias Matematicas de la UNMSM,* 2 (1988).
14. Priestley, W. M.: *Calculus: An Historical Approach.* New York: Springer-Verlag, 1979.
15. Simmons, G.: *Calculus with Analytic Geometry.* New York: McGraw Hill, 1985.
16. Smith, D.: *Interface: Calculus and the Computer.* Boston: Houghton Mifflin, 1961.
17. Spivak, M.: *Calculus.* New York: W. A. Benjamin, 1967.
18. Toeplitz, O.: *The Calculus: A Genetic Approach.* Chicago: The University of Chicago Press, 1963.

Appendix: Outline of Calculus Course

1. Sequences. How to determine the area under a parabola. The idea of the Greeks. J. Bernoulli's problem of compound interest. Limit of a sequence. Basic theory of convergence. The area under a cubic curve; the key ideas of the Arabs and Cavalieri. Geometric Series. Achilles and the tortoise and how we can look at this problem nowadays. Further properties of sequences. The problem of completeness. The number e as the limit of a sequence and its connection with compound interest. (4 weeks)

2. Limits of functions of a real variable.   How Pierre de Fermat calculated the area under $y = x^n$. Use of Fermat's method to determine the area under the square root function. The problem of tangents from the Greeks to Descartes; an analysis of Fermat's original method. Definition of limits through sequences. Development of the main results of the theory of limits. Limits of polynomials, square root functions, trigonometric functions, rational functions. The ε-δ definitions. One-sided limits. (2 weeks)

3. Continuity.   Algebra of continuous functions. Interchange of limits. Cauchy, Bolzano and the intermediate value theorem; roots of polynomials. Modern electronic devices. Justification of "obvious" results: The existence of a maximum and a minimum, and the continuity of the inverse. (1 week)

4. Derivatives.   Once again the problem of drawing tangents. Reflection in parabolic mirrors: telescopes and automobile headlights. Main properties of derivatives. Derivatives of polynomial, square root, trigonometric and rational functions. The derivative as a velocity. Statement of the mean value theorem and proof of the fact that two functions with equal derivatives differ by a constant. Applications to mechanics (how brakes work, ascent and descent.) The derivative of the composite of functions. Chain rule and Newton's escape velocity. The derivative of the inverse of a function. Derivatives of the inverse trigonometric functions. (3 weeks)

5. Theorems of Fermat, Rolle and MVT.   Weakening and strengthening of hypotheses. Approximation of the values of a function through the concept of derivative. Analysis of errors. Primitives. Simple differential equations make their appearance for the first time. The scientific method. Some models that appear as differential equations that can be solved readily. (2 weeks)

6. Maximum and minimum problems.   Criteria of the first and second derivative. How Fermat solved extremum problems. The problem of reflection in flat and curved mirrors since Hero of Alexandria. Refraction of light. Snell, Fermat and the way a scientific theory is built. Various applications. (2 weeks)

7. Concavity, convexity and points of inflection.   A special case of Taylor's theorem ($n = 2$). Sketching curves, especially polynomials. How Newton's method helps to localize roots. (1 week)

8. A new look at the problem of areas.   The problem of calculating areas and the fundamental theorem of calculus for continuous monotone functions. An important case: the logarithmic function. Deduction of the most important properties of logarithms. Napier, Gregorius, Sarasa and the way the concept of logarithm developed in the seventeenth century before Newton and Leibniz. How logarithms helped to prepare the ground for the fundamental theorem of calculus. Primitives and several ways to find them. Areas between curves. (3 weeks)

9. The integral as a limit.   Integrable functions. The contributions of Cauchy and Riemann. Why are we interested in limits of sample sums? The calculations of volumes, areas of surfaces of revolution and several problems in physics. (2 weeks)

10. Exponential function and related topics.   The exponential as the inverse of the logarithm. Main properties and an equivalent definition of e. Approximation of exponential functions. Exponential to any base and power functions. Linear differential equations of first order. Radioactivity, interchange of temperatures, electric circuits, chemical kinetics, dilution problems, the parachute problem. Complex-valued exponential functions and their use in electricity. The problem of drawing and interpreting results. A short analysis of complex-valued functions and how we find their derivatives and integrate them. Linear differential equations of second order and constant coefficients. General solution, initial value problems. Applications to mechanics, electricity, and enzyme kinetics from a historical perspective. The role played by Leibniz and the Bernoulli brothers in the early stages of the theory of differential equations. Hyperbolic cosine, sine and tangent and inverse hyperbolic function: what are they good for? J. Bernoulli and the Marquis de L'Hospital. Cauchy's mean value theorem. Rules of L'Hospital and their uses. (5 weeks)

11. Taylor polynomials.   Can we approximate functions through polynomials? Proof of Taylor's theorem. How Taylor got his results. Analysis of the remainder in some concrete problems. Lagrange's remainder formula. Taylor polynomials for the exponential, logarithmic, and trigonometric functions. Approximation of some integrals. (2 weeks)

12. A systematic development of several methods of integration.   Integration by parts. Approximation of π through the inverse tangent function and how Leibniz approximated π using, among other tools, integration by parts. Substitution methods. The catenary problem. Integration of rational functions. Differential equations with separable variables. Population models and problems of chemical kinetics. Trigonometric integrals. (2 weeks)

13. Numerical methods of integration.   The Trapezoidal and Simpson's method. Analysis of errors. (1 week)

14. Curves and polar coordinates.   Length of a curve. Length and area in polar coordinates. Parametric equations. The cycloid and Huygens' clock. Ellipse and hyperbola. Polar equations of conics. Kepler's laws and how Newton obtained his inverse square law. (2 weeks)

Euler and Heuristic Reasoning

Man-Keung Siu

Who, What and Why?

The title of this article begs answers to the following questions: Who was Euler? What is heuristic reasoning? Why are the two related? I shall answer them in the next three sections. But my aim in this discussion goes far beyond that, indeed even far beyond what my capability will allow me to achieve; for I wish very much to convey the message that what we shall discuss here constitutes a correct way to do mathematics, to study mathematics and to teach mathematics. I shall return to the last point towards the end of the article.

Who was Euler?

The first question is easy to answer since biographies of Euler can be found in many books, such as [3, 5, 14, 17]. Leonhard Euler was in the opinion of many the greatest mathematician (and physicist) of the eighteenth century. He was born in Basel, Switzerland on April 15, 1707. At the age of 13 he entered the University of Basel where he had the good fortune to study mathematics under the eminent mathematician Johann Bernoulli (1667–1748). Later in his life, he was fond of recollecting this pleasurable experience and acknowledging a debt to his teacher. He said [14, p. 342], "I soon found an opportunity to gain introduction to the famous professor Johann Bernoulli, whose good pleasure it was to advance me further in the mathematical sciences ... and wherever I should find some check or difficulties, he gave me free access to him every Saturday afternoon and was so kind as to elucidate all difficulties, which happened with such greatly desired advantage that whenever he had obviated one check for me, because of that ten others disappeared right away, which is certainly the way to make a happy advance in the mathematical sciences." At the age of 15, Euler received his first university degree and two years later his master's degree in philosophy. In 1727, he competed for the chair of physics at University of Basel and lost. Having had the good fortune not to win the chair of physics at Basel, Euler went to the Academy of St. Petersburg in Russia and spent thirteen very productive years there until 1741. In 1738, three years before leaving Russia, a violent fever destroyed the sight of his right eye. At the age of 34, Euler left Russia and moved to the Academy of Berlin in Prussia. He stayed there until 1766, in which year he returned to the Academy of St. Petersburg. At about the same time he lost sight of the other eye. An unsuccessful operation performed in 1771 resulted in near total blindness in the remaining years of his life. On September 18, 1783, Euler was working as usual. He spent that afternoon calculating the law of ascent of balloons. After dinner he outlined the calculation of the orbit of the newly-discovered planet Uranus. Then he played with his grandson. While playing with the child and drinking tea, he suffered a stroke. According to the eulogy written by his younger contemporary Marquis de Condorcet (1743–1794), "Euler ceased to live and to calculate" [3, p. 152].

Euler was the most prolific writer in the history of mathematics. Approximately one third of the research on mathematics, mathematical physics and engineering mechanics published in the last three-quarters of the eighteenth century was authored by him. From 1729 onward his work filled about half of the pages of the publications of the Academy of St. Petersburg, not only until his death in 1783, but continuing on over the next fifty years! From 1746 to 1771 he filled about half the pages of the publications of the Academy of Berlin. Shortly after his death, Nicolas Fuss (1755–1825, husband of a granddaughter of Euler) compiled Euler's publications collecting 756 articles, of which 355 were written in the last ten years of Euler's life when he was nearly totally blind! The modern revision of Euler's collected works began in 1911, and is not yet complete. By that time, 866 of his published articles had been collected, and it is estimated that over seventy large quarto volumes, each containing 300 to 600 pages, will be required to print them. And this collection does not yet include some 3000 pages of manuscripts and notes he left behind in Russia, about 3000 letters of personal correspondence, and some 25 volumes of expository books or treatises he wrote, several of which became important textbooks which nurtured generations of mathematicians who came after him.

Euler's contribution to mathematics can perhaps be glimpsed in the numerous terms, formulae, equations and theorems that bear his name. In 1983, the November issue of *Mathematics Magazine*, published as a tribute to Euler, contained a glossary of 44 such items [18, pp. 316–325]. Marquis de Condorcet observed in his eulogy of Euler, "All the noted mathematicians of the present day [late eighteenth century] are his pupils: there is no one of them who has not formed himself by the study of his works, who has not received from him the formulas, the method which he employs; who is not directed and supported by the genius of Euler in his discoveries" [18, p. 258].

What is Heuristic Reasoning?

This question is harder to answer. Fortunately someone else has already written much on it. Of course I am referring to the famous mathematician-mathematics educator and great mathematics teacher George Pólya (1887–1985), whose three fascinating books [11, 12, 13] should be on the reading list of every teacher of mathematics. Those who love a formal definition of "heuristic reasoning" will be disappointed. The very term itself connotes an air of many-sidedness and informality. Perhaps the best way to explain is to illustrate its many aspects via examples. Nevertheless I shall still quote two instructive passages from Pólya:

> Mathematical thinking is not purely "formal"; it is not concerned only with axioms, definitions, and strict proofs, but many other things belong to it: generalizing from observed cases, inductive arguments, arguments from analogy, recognizing a mathematical concept in, or extracting it from a concrete situation. [13, vol.2, pp. 100–101]

> Heuristic reasoning is reasoning not regarded as final and strict but as provisional and plausible only, whose purpose is to discover the solution of the present problem. . . . Heuristic reasoning is good in itself. What is bad is to mix up heuristic reasoning with rigorous proof. What is worse is to sell heuristic reasoning for rigorous proof. [12, p. 113]

Why Euler and Heuristic Reasoning?

I plan to illustrate the many aspects of heuristic reasoning via examples taken out of Euler's works. But why Euler? This brings us to the third question. Every mathematician practices heuristic reasoning to some extent. But unlike most authors who only present the final product in a neat and polished form which may invite awe and admiration but not necessarily add to understanding, Euler explained how he proceeded in his reasoning and described, sometimes in illuminating details, his process of discovery. Marquis de Condorcet noted:

> He [Euler] preferred instructing his pupils to the little satisfaction of amazing them. He would have thought not to have done enough for science if he should have failed to add to the discoveries, with which he enriched science, the candid exposition of his ideas that led him to those discoveries. [11, vol. 1, p. 90]

Pólya said:

> Naturally enough, as any other author, he [Euler] tries to impress his readers, but, as a really good author, he tries to impress his readers only by such things as have genuinely impressed himself.... We can learn from it a great deal about mathematics, or the psychology of invention, or inductive reasoning. [11, vol. 1, p. 90]

In this respect Euler's works are particularly instructive.

Example I. We choose as our first example sections 133–140 in Euler's book *Introductio in analysin infinitorum* (1748) [7, pp. 106–113], which was hailed by C. B. Boyer [14, p. 346] as "the foremost textbook of modern times." Euler started with the formula

$$(\cos z \pm i \sin z)^n = \cos nz \pm i \sin nz$$

(which is a result due to Abraham de Moivre (1667–1754) in 1730) to obtain

$$\cos nz = \frac{(\cos z + i \sin z)^n + (\cos z - i \sin z)^n}{2}$$

$$\sin nz = \frac{(\cos z + i \sin z)^n - (\cos z - i \sin z)^n}{2i}.$$

He developed them as binomial series to obtain

$$\cos nz = (\cos z)^n - \frac{n(n-1)}{1 \cdot 2}(\cos z)^{n-2}(\sin z)^2$$
$$+ \frac{n(n-1)(n-2)(n-3)}{1 \cdot 2 \cdot 3 \cdot 4}(\cos z)^{n-4}(\sin z)^4 - \cdots$$
$$\sin nz = \frac{n}{1}(\cos z)^{n-1}(\sin z) - \frac{n(n-1)(n-2)}{1 \cdot 2 \cdot 3}(\cos z)^{n-3}(\sin z)^3 + \cdots.$$

He then let z be infinitely small and n be infinitely large, but keeping nz of finite magnitude, say equal to v. He used the facts that $\sin z = z = v/n$ and $\cos z = 1$ to rewrite the two formulas as

$$\cos v = 1 - \frac{v^2}{1 \cdot 2} + \frac{v^4}{1 \cdot 2 \cdot 3 \cdot 4} - \cdots$$

$$\sin v = v - \frac{v^3}{1 \cdot 2 \cdot 3} + \frac{v^5}{1 \cdot 2 \cdot 3 \cdot 4 \cdot 5} - \cdots.$$

These are of course the correct power series for the sine and cosine functions. But the argument used is of the heuristic sort, viz. by *formal manipulation*. This technique, based on a trust in the power of symbols, was a prominent feature of eighteenth century mathematics.

Then in section 138, Euler "derived" his famous formula for e^{iv}. By the same reasoning outlined above, he obtained

$$\cos v = \frac{(1 + iv/n)^n + (1 - iv/n)^n}{2}$$

$$\sin v = \frac{(1 + iv/n)^n - (1 - iv/n)^n}{2i}.$$

In a preceding chapter he had already proven that $(1 + z/n)^n = e^z$. Hence the two formulas could be rewritten as

$$\cos v = \frac{e^{iv} + e^{-iv}}{2}$$

$$\sin v = \frac{e^{iv} - e^{-iv}}{2i},$$

which gave

$$e^{iv} = \cos v + i \sin v, \quad e^{-iv} = \cos v - i \sin v.$$

He then went on to obtain

$$v = \frac{1}{2i} \log_e \left[\frac{\cos v + i \sin v}{\cos v - i \sin v} \right] = \frac{1}{2i} \log_e \left[\frac{1 + i \tan v}{1 - i \tan v} \right].$$

In an earlier section he had proved the infinite series

$$\log_e \left(\frac{1 + x}{1 - x} \right) = \frac{2x}{1} + \frac{2x^3}{3} + \frac{2x^5}{5} + \frac{2x^7}{7} + \cdots$$

(due to James Gregory (1638–1675) in 1668), so he had

$$v = \frac{\tan v}{1} - \frac{(\tan v)^3}{3} + \frac{(\tan v)^5}{5} - \frac{(\tan v)^7}{7} + \cdots.$$

Letting $t = \tan v$, he got

$$v = \frac{t}{1} - \frac{t^3}{3} + \frac{t^5}{5} - \frac{t^7}{7} + \cdots.$$

Putting $t = 1$, so that $v = \pi/4$, he obtained the infinite series

$$\frac{\pi}{4} = 1 - \frac{1}{3} + \frac{1}{5} - \frac{1}{7} + \cdots,$$

which was a well-known series discovered by Gottfried Wilhelm Leibniz (1646–1716) and published in his "De vera proportione circuli" (1682). This technique of *partial confirmation* is another feature of heuristic reasoning. If the method yields a result which has been proved to be correct through other means, then the former result sounds more convincing, even though the method is still questionable.

Example II (See [2, 11, 16]).   While we are discussing infinite series, it is unlikely that we can omit that brilliant achievement of Euler concerning the computation of $1 + \frac{1}{2^2} + \frac{1}{3^2} + \frac{1}{4^2} + \cdots$.

For convenience of exposition we shall adopt a modern notation and write

$$\zeta(s) = \sum_{n=1}^{\infty} \frac{1}{n^s} \qquad \text{(zeta function)}.$$

Pietro Mengoli (1625–1686) asked for the value of $\zeta(2)$ in 1650. John Wallis (1616–1703) computed $\zeta(2)$ to three decimal places in 1655, but did not recognize the significance of 1.645. This problem withstood the efforts of the Bernoulli brothers. In 1731, Euler computed $\zeta(2)$ to six decimal places. In view of the slow convergence rate of the series, even numerical evaluation is no small task. For instance, Euler's computation motivated the discovery of what is known today as the Euler–MacLaurin summation formula. In 1735 Euler sharpened his calculation to obtain an answer

$$\zeta(2) = 1.64493406684822643647.\ldots$$

But he was not satisfied, for he wanted the *exact* value. Laboring on this task, he succeeded in 1735 by "generalizing" the factorization of polynomials to transcendental functions represented as power series. Although this story is probably familiar to many as it occurs in several books, it is worth repeating.

Let $\alpha_1, \ldots, \alpha_n$ be roots of the equation

$$a_n X^n + a_{n-1} X^{n-1} + \cdots + a_1 X + a_0 = 0 \qquad \text{where} \quad a_0 \neq 0, a_n \neq 0$$

(so that $\alpha_1, \ldots, \alpha_n$ are all nonzero). Then we have

$$a_n X^n + a_{n-1} X^{n-1} + \cdots + a_1 X + a_0 = a_0 (1 - x/\alpha_1) \cdots (1 - x/\alpha_n),$$

and hence $a_1 = -a_0 \left(\frac{1}{\alpha_1} + \cdots + \frac{1}{\alpha_n} \right)$. Euler treated a power series as a polynomial, only with more terms! He noted that

$$\sin v = v - \frac{v^3}{1 \cdot 2 \cdot 3} + \frac{v^5}{1 \cdot 2 \cdot 3 \cdot 4 \cdot 5} - \cdots = 0$$

has roots $0, \pm\pi, \pm 2\pi, \pm 3\pi, \ldots$, so that

$$\frac{\sin v}{v} = 1 - \frac{v^2}{1 \cdot 2 \cdot 3} + \frac{v^4}{1 \cdot 2 \cdot 3 \cdot 4 \cdot 5} - \cdots = 0$$

has roots $\pm\pi, \pm 2\pi, \pm 3\pi, \ldots$, i.e.,

$$1 - \frac{x}{1 \cdot 2 \cdot 3} + \frac{x^2}{1 \cdot 2 \cdot 3 \cdot 4 \cdot 5} - \cdots = 0$$

has roots $\pi^2, (2\pi)^2, (3\pi)^2, \ldots$. From the relation discussed above, he obtained

$$-\frac{1}{1 \cdot 2 \cdot 3} = -\left(\frac{1}{\pi^2} + \frac{1}{2^2 \pi^2} + \frac{1}{3^2 \pi^2} + \cdots \right),$$

i.e.,

$$\frac{\pi^2}{6} = 1 + \frac{1}{2^2} + \frac{1}{3^2} + \frac{1}{4^2} + \cdots.$$

Euler applied the same technique to the equation $1 - \sin x = 0$ which has (double) roots $\pi/2, \pi/2, -3\pi/2, -3\pi/2, 5\pi/2, 5\pi/2, -7\pi/2, -7\pi/2, \ldots$, i.e.,

$$1 - \frac{x}{1} + \frac{x^3}{1 \cdot 2 \cdot 3} - \frac{x^5}{1 \cdot 2 \cdot 3 \cdot 4 \cdot 5} + \cdots = 0 \qquad \text{with these roots}.$$

He found

$$-1 = -\left(\frac{4}{\pi} - \frac{4}{3\pi} + \frac{4}{5\pi} - \frac{4}{7\pi} + \cdots\right),$$

i.e.,

$$\frac{\pi}{4} = 1 - \frac{1}{3} + \frac{1}{5} - \frac{1}{7} + \cdots, \quad \text{the famous series of Leibniz.}$$

He said, "For our method which may appear to some as not reliable enough, a great confirmation comes here to light. Therefore, we should not doubt at all of the other things which are derived by the same method" [11, vol. 1, p. 21]. Again, *partial confirmation* is at work!

The prominent feature of heuristic reasoning we discern in Euler's argument is that of *analogy*. Besides helping to discover an answer, analogy can sometimes lead to new theory. In this case, the analogy between factorization of polynomials and that of power series opened up the theory of infinite product and partial fraction decomposition of transcendental functions. A rigorous version of the argument outlined above lies in the expression

$$\frac{\sin x}{x} = \prod_{n=1}^{\infty} \left(1 - \frac{x^2}{n^2\pi^2}\right),$$

which was proved by Euler in 1742. Another famous instance of an infinite product is the Euler Identity (presented to the Academy of St. Petersburg in 1737)

$$\zeta(s) = \prod_p (1 - 1/p^s)^{-1}, \quad \text{where } p \text{ runs through all primes.}$$

This analytic version of the fundamental theorem of arithmetic is the starting point of Riemann's theory of zeta functions.

Euler returned to this problem of evaluating $\zeta(s)$ many times. In particular, he was aware of the heuristic nature in the 1735 argument. He later proved that $\pi^2/6$ was the correct answer, and computed $\zeta(2n)$ more generally in 1739. Investigations on $\zeta(n)$ for odd n led him in 1749 to a discovery which was equivalent to the functional equation of the zeta function, subsequently forgotten for over a century until resurrected by Bernhard Riemann (1826–1866) in 1859! It is of some interest to note that the irrationality of $\zeta(3)$ was established only recently, by Roger Apéry in 1978 [15].

Example III (See [8, 9, 11, 13]). For a change, let us leave analysis and go to geometry. In a letter of November, 1750 to Christian Goldbach (1690–1764), Euler mentioned some results he noticed in his investigation on polyhedra:

> Recently it occurred to me to determine the general properties of solids bounded by plane faces, because there is no doubt that general theorems should be found for them, just as for plane rectilinear figures, whose properties are: (1) that in every plane figure the number of sides is equal to the number of angles, and (2) that the sum of all the angles is equal to twice as many right angles as there are sides, less four. Whereas for plane figures only sides and angles need to be considered, for the case of solids more parts must be taken into account, namely
>
> I. the faces, whose number $= H$;
> II. the solid angles, whose number $= S$;

III. the joints where two faces come together side to side, which, for lack of an accepted word I call 'edges', whose number $= A$;

IV. the sides of all the faces, the number of which all added together $= L$;

V. the plane angles of all faces, the total number of which $= P$. [4, p. 76]

Again, *analogy* is at work. It is natural to ask what are analogies of facts we know about a polygon in the case of a polyhedron. "Sides" become "faces," for they both serve to bound the object under investigation. What about "sides of a face" then? Euler distinguished between "side" (*latus*) and "edge" (*acies*) and even emphasized this in his letter. For a polygon, we need only know the number of sides (E), which is equal to the number of vertices (V). It follows as a theorem that the sum of all interior angles of a convex polygon is equal to $(V - 2)\pi$. For a polyhedron, we need to know more parameters. What Euler denoted by H, S, A are today usually written respectively as F (number of faces), V (number of vertices), E (number of edges). It is no longer true that the number of faces must be equal to the number of vertices. However, there is an analogous result for the sum of interior angles. Euler stated this as Theorem 11 in his letter: The sum of all plane angles is equal to four times as many right angles as there are solid angles, less eight, that is $= 4S - 8$ right angles. Using contemporary notation, $\sum \alpha = (2V - 4)\pi$ where α runs through all interior angles of all faces. In this connection it is extremely interesting to look at Theorem 6 mentioned in that same letter: In every solid enclosed by plane faces the aggregate of the number of faces and the number of solid angles exceeds by two the number of edges, or $H + S = A + 2$. Again in contemporary notation, it says that $V - E + F = 2$, the famous Euler formula. Today we know that this formula is valid for a certain class of polyhedra only. At that time, Euler did not yet see the subtlety, but apparently he was talking about a convex polyhedron without explicitly stating the fact.

After illustrating his theorems with an example, Euler concluded [4, p. 77], "I find it surprising that these general results in solid geometry have not previously been noticed by anyone, so far as I am aware; and furthermore, that the important ones, Theorem 6 and 11, are so difficult that I have not yet been able to prove them in a satisfactory way." His statement is correct as far as ancient Greek mathematics is concerned, but it is incorrect in that René Descartes (1596–1650) had found similar results in 1639. However, Descartes' manuscript was discovered and published in 1860, so Euler could not have known about Descartes' work! Today we honor both mathematicians by referring to that strikingly beautiful formula as the Euler–Descartes formula. It is interesting to note that Theorem 6 and Theorem 11 are equivalent since $\sum \alpha = 2(E - F)\pi$. In the form of Theorem 11, which is comprehensible to any ancient Greek mathematician, the result looks like one that should not have escaped the attention of Greek mathematics. But throughout the centuries in which Greek geometry flourished, the result did not appear anywhere. In view of the fact that Theorem 11 is equivalent to Theorem 6, the reason is quite simple. Theorem 6 concerns the combinatorial properties of a polyhedron rather than its metrical properties and so lies completely outside the Greek mathematicians' field of interest. No wonder it never found its way into Greek mathematics. Indeed, this formula opened up a new page in the history of mathematics and motivated the new branch of mathematics called topology.

In the years after Euler wrote the letter, he devoted two memoirs to those two important theorems. He gave a proof, later found to be insufficient. Augustin Louis Cauchy (1789–1857) gave a proof in 1811 which met the standard of rigor of his day. (It is still nowadays presented in most popular accounts as a proof of the formula.) Quite a number of counterexamples were discovered after Cauchy's proof was given. They indicated inadequacy not only in the proof, but even in the formulation, viz. What is a polyhedron? The proof now usually offered in a topology text is that due to Karl George von Staudt (1798–1867) and produced in 1847, already a whole century after Euler discovered it! An instructive and enlightening dialogue with generous historical footnotes about this formula, written by Imre Lakatos (1922–1974) [9], is strongly recommended for additional reading. Following Pólya's idea [13, vol. 2, section 15.6], let us try to reconstruct Euler's trend of thought with the aid of historical documents. Suppose the goal is to find an analogue of the formula $\sum \alpha = (V - 2)\pi$ for a polygon with V vertices. As possible choices we can investigate $\sum \alpha$ where α runs through: (i) dihedral angles of the polyhedron, (ii) solid angles of the polyhedron, (iii) plane angles of all faces of the polyhedron. As an exercise, readers can convince themselves that (i) is not a good candidate since even for a tetrahedron, $\sum \alpha$ will depend on the shape of the tetrahedron as evidenced by the two tetrahedra illustrated in Figure 1a (while $\sum \alpha$ for a triangle does not depend on the shape of the triangle). For the same reason, (ii) is not a good candidate either, as evidenced by the two tetrahedra illustrated in Figure 1b.

We are left with (iii) as our choice. Let us collect some data from the polyhedra illustrated in Figure 2,

Polyhedron	(a)	(b)	(c)	(d)	(e)	(f)
F	6	4	8	7	9	5
$\sum \alpha$	12π	4π	8π	16π	14π	8π

The pattern appears erratic! We need some guiding principle in examining experimental data so as to elicit valuable information which will enable us to make an informed guess. (However, we should guard against preconceived ideas that can bias our thinking. We should keep an open,

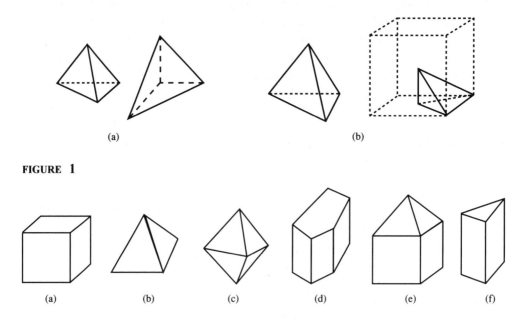

(a)   (b)

FIGURE 1

(a)   (b)   (c)   (d)   (e)   (f)

FIGURE 2

objective attitude.) Note that

$$\sum \alpha = \sum_f \sum \alpha_f = \sum_v \sum \alpha_v$$

where $\sum_f$ means summation over all faces and α_f runs through all plane angles of a face; $\sum_v$ means summation over all vertices, α_v runs through all plane angles at a vertex. But $\sum \alpha_v < 2\pi$ for each vertex, which is a theorem for a convex polyhedron, proved in Euclid's *Elements* as Proposition 21 of Book 11. A heuristic geometric argument is obtained by "flattening out" that polyhedral angle onto the plane. Hence we see that $\sum \alpha < 2V\pi$. Why not look at the discrepancy $2V\pi - \sum \alpha$ for those data in the table above? If you do, you will see immediately a conjectured formula for $\sum \alpha$, which is nothing other than Theorem 6!

Let us further apply two usual techniques in heuristic reasoning. First *specialize*: Is the conjectured formula an analogue of that for a polygon? Consider a (convex) polygon with V vertices. Make two identical copies and join corresponding vertices by vertical edges to form a prism. The conjectured formula tells us that

$$\sum \alpha = (4V - 4)\pi = 4V\pi - 4\pi.$$

But we also know that $\sum \alpha = 2S + 2V\pi$ where S is the sum of all interior angles of the polygon. Hence, we obtain $S = (V - 2)\pi$. Next we *generalize*: Can we use the formula for a polygon to derive the conjectured formula for the polyhedron? We shall flatten the given polyhedron "in a special way" (so that the base polygon is convex and has N vertices). Since $\sum \alpha = 2(E - F)\pi$ (explained earlier on), the angle sum is invariant under the flattening provided E, F remain unaltered. Since

$$\sum \alpha = (N - 2)\pi + (N - 2)\pi + (V - N)2\pi,$$

we see that it simplifies to $(2V - 4)\pi$. Although there are quite a number of objections one can raise against this "proof," it makes the result even more convincing.

One result leads to another. In the same month that Euler wrote his letter to Goldbach, he also presented a paper titled "Elementa Doctrinae Solidorum" to the Academy of St. Petersburg in which he tried to classify polyhedra. He noted, "While in plane geometry polygons can be classified very easily according to the number of their sides, which of course is always equal to the number of their angles, in stereogeometry the classification of polyhedra represents a much more difficult problem, since the number of faces alone is insufficient for this purpose" [9, p. 6]. Everybody can easily see why F alone is not enough. The three polyhedra shown below in Figure 3 all have $F = 6$. But nobody likes to say they belong to the same type. For one

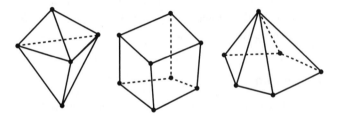

FIGURE 3

thing, each polyhedron has a different V, viz. 5, 8, 6 respectively. How about including both F and V? Still that is not enough, as the two polyhedra shown in Figure 4 demonstrate, since the faces are of different shapes.

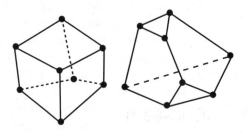

FIGURE 4

Euler invented the term "edge" for polyhedra, which he distinguished from "side," a concept pertaining to polygons. It is noteworthy that he emphasized the novelty of this new term, possibly because he had hoped at first that it might help in the classification of polyhedra. Again, let us collect data from the polyhedra shown in Figure 5,

Polyhedron	(a)	(b)	(c)	(d)	(e)	(f)
F	6	6	6	6	7	7
V	5	8	6	8	10	10
E	9	12	10	12	15	15

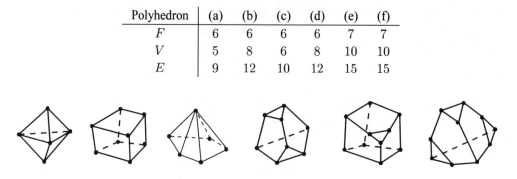

FIGURE 5

What do you observe? Polyhedra with the same F, V seem to have the same E as well. Thus, it seems that E, the number of edges, contributes nothing to the classification problem; it does not give extra information over what can be gathered from F and V. Does that mean disappointment? No, it means triumph! It suggests that E is a function of F and V. Indeed, it looks like E increases with F, V jointly. Why not try $(V + F) - E$? If you do, you will immediately obtain that famous formula of Euler!

Induction and Deduction

Induction is a kind of heuristic reasoning. It is the process of discovering general laws by the observation and combination of particular instances. (In this respect, "mathematical induction" is not induction; it is *deduction*.) It helps us to discover an answer, but it cannot yield the final say, which has to be gained by deductive reasoning. Euler was very clear about this point. He once said:

> It will seem not a little paradoxical to ascribe a great importance to observations even in that part of the mathematical sciences which is usually called Pure Mathematics, since the current opinion is that observations are restricted to physical objects that make impression on the senses. ... The kind of knowledge which is supported only by observations and is not yet proved must be carefully distinguished from the truth; it is gained by induction, as we usually say. Yet we have seen cases in which mere induction led to error. [11, vol. 1, p. 3]

I shall illustrate his warning with examples again taken from his works. In a letter dated December, 1729 to Euler, Goldbach asked, "Is Fermat's observation known to you, that all numbers $2^{2^n} + 1$ are primes? He said he could not prove it; nor has anyone else done so to my knowledge" [16, p. 172]. Euler's reception was at first cool, but in June, 1730 he suddenly caught fire and started to read Fermat's work seriously, and this began his life-long interest in number theory. The numbers $F_n = 2^{2^n} + 1$ referred to in the letter are now known as Fermat numbers. Around 1640 Pierre de Fermat (1601–1665) mentioned the conjecture that all Fermat numbers were prime. Indeed, we see that

$$F_1 = 5, \quad F_2 = 17, \quad F_3 = 257, \quad F_4 = 65537$$

are all prime. In 1732, Euler by studying the factors of $a^{2^n} + b^{2^n}$, showed that

$$F_5 = 4294967297 = 641 \times 6700417$$

and showed the conjecture to be false. As another example, take the curious property of the polynomial $X^2 + X + 41$ that Euler discovered in 1772, viz. it yields a prime number for $X = 0, 1, 2, \ldots, 39$. Can we conclude from these forty consecutive affirmative answers that it will always produce prime numbers for all values of X? No; it is false for $X = 41$. However, coincidence is rare in mathematics. The existence of coincidence demands, and implies, explanation. In this very case, the coincidence is related to the discriminant of the quadratic polynomial, viz. -163. For a more startling example, let us look at this question: Is $1 + 1141y^2$ ever a square for $y \neq 0$? It can be rephrased as the diophantine equation $x^2 - 1141y^2 = 1$, one particular instance of the so-called "Pell equation" (which was misnamed by Euler in 1730 although it has nothing to do with John Pell (1611–1685); in fact it was considered in India as early as the seventh century!). It so happens that the smallest $y \neq 0$ which gives an affirmative answer is 30,693,385,322,765,657,197,397,208. Even with a supercomputer, experimental evidence will always indicate a negative answer! But actually there are infinitely many y's which supply an affirmative answer!

However, Euler, being fallible, did commit errors at this game of guessing. He once made the following conjecture which generalized "Fermat's Last Theorem": $x_1^n + \cdots + x_m^n \neq y^n$ if $1 < m < n \ (n \geq 3)$ for integral values $x_1, \ldots, x_m, y$. This was refuted by L. J. Lander and T. R. Parkin in 1967, almost two centuries later. Their counterexample is

$$27^5 + 84^5 + 110^5 + 133^5 = 144^5.$$

Recently, N. Elkies found a counterexample for the case $n = 4$,

$$2682440^4 + 15365639^4 + 18796760^4 = 20615673^4.$$

(The conjecture is true for $n = 3$.) Another famous misjudgement of Euler is his 1779 conjecture on the nonexistence of orthogonal Latin squares of order $2n$, n odd. It was refuted by Roy Chandra Bose, Ernest Tilden Parker and S. S. Shrikhande in 1958.

Example IV (See [1, 11, 16]).   The final example I shall discuss is a profound discovery of Euler in number theory, which appeared in a 1747 memoir. It offered a "most extraordinary law of the number concerning the sum of their divisors" [11, vol. 1, p. 91]. For ease of exposition we shall adopt today's notation $\sigma(n) = $ sum of all divisors of n. For instance, $\sigma(6) = 1 + 2 + 3 + 6 = 12$ and $\sigma(n) = 1 + n$ if and only if n is a prime. At the beginning of the memoir Euler said, "Till now the mathematicians tried in vain to discover some order in the sequence of the prime numbers and we have every reason to believe that there is some mystery which the human mind shall never penetrate. . . . I am myself certainly far from this goal, but I just happened to discover an extremely strange law governing the sums of the divisors of the integers which, at the first glance, appear just as irregular as the sequence of the primes, and which, in a certain sense, comprise even the latter. This law, which I shall explain in a moment, is, in my opinion, so much more remarkable as it is of such a nature that we can be assured of its truth without giving it a perfect demonstration" [11, vol. 1, p. 91]. The last sentence sounds paradoxical to someone trained in mathematics. How can one be assured of a theorem without proving it? Let us see how Euler explained this phenomenon.

Euler devised a table of $\sigma(n)$ for n in the range $1 \leq n \leq 99$. It does look pretty erratic:

n	0	1	2	3	4	5	6	7	8	9
0		1	**3**	**4**	7	**6**	12	**8**	15	13
10	18	**12**	28	**14**	24	24	31	**18**	39	**20**
20	42	32	36	**24**	60	31	42	40	56	**30**
30	72	**32**	63	48	54	48	91	**38**	60	56
40	90	**42**	96	**44**	84	78	72	**48**	124	57
50	93	72	98	**54**	120	72	120	80	90	**60**
60	168	**62**	96	104	127	84	144	**68**	126	96
70	144	**72**	195	**74**	114	124	140	96	168	**80**
80	186	121	126	**84**	224	108	132	120	180	**90**
90	234	112	168	128	144	120	252	**98**	171	156

(The table is self-explanatory. For instance, the entry in the row labelled 40 and column labelled 7 is $\sigma(47) = 48$. Entries in boldface print correspond to primes.) He then gave the rule, viz. the recurrence relation

$$\sigma(n) = \sigma(n-1) + \sigma(n-2) - \sigma(n-5) - \sigma(n-7)$$

$$+ \sigma(n-12) + \sigma(n-15) - \sigma(n-22) - \sigma(n-26)$$

$$+ \sigma(n-35) + \sigma(n-40) - \sigma(n-51) - \sigma(n-57) + \cdots$$

where (i) the signs $+$ and $-$ each arise twice in succession, (ii) the sequence continues as long as the number under the sign σ is nonnegative (so the sequence stops somewhere), (iii) if $\sigma(0)$ turns up, it is to be interpreted as n, (iv) the sequence 1, 2, 5, 7, 12, 15, 22, 26, 35, 40, 51, 57, . . . follows the pattern in which differences between consecutive terms are 1, **3**, 2, **5**, 3, **7**, 4, **9**, 5, **11**, 6, ... As illustration, Euler computed a few examples to convince the reader of the validity of his rule. He then said, "The examples that I have just developed will undoubtedly dispel any qualms which we might have had about the truth of my formula." He continued, "I confess that I did not hit on this discovery by mere chance, but another proposition opened the path to this beautiful property—another proposition of the same nature which must be accepted as true although I am unable to prove it" [11, vol. 1, p. 95].

What Euler referred to is his investigation on the infinite product $\prod_{n=1}^{\infty}(1 - x^n) = (1 - x)(1 - x^2)(1 - x^3) \cdots$ in 1741. This investigation was motivated by a combinatorial problem concerning the partitions of an integer raised in 1740 by Philipp Naudé (1684–1745). By actually computing the product, Euler observed that the pattern came out as

$$1 - x - x^2 + x^5 + x^7 - x^{12} - x^{15} + x^{22} + x^{26} - x^{35} - x^{40} + x^{51} + \cdots.$$

To an untrained eye this pattern may look irregular. But Euler noticed that alternate exponents formed two sequences, viz.,

$$1, \ 5, \ 12, \ 22, \ 35, \ 51, \ldots, \qquad \text{and} \qquad 2, \ 7, \ 15, \ 26, \ 40, \ 57, \ldots.$$

The first sequence is that of pentagonal numbers of the general form $n(3n - 1)/2$ (so called by the Pythagoreans (c. fifth century B.C.) since they are the numbers of vertices of pentagons of proportionately increasing sizes as illustrated in Figure 6).

The second sequence is obtained from the first by adding respectively 1, 2, 3, 4, $\ldots$, i.e., with the nth term being $n(3n + 1)/2$. Thus, Euler observed that the remarkable formula might hold:

$$\prod_{n=1}^{\infty}(1 - x^n) = 1 + \sum_{n=1}^{\infty}(-1)^n x^{n(3n+1)/2} + \sum_{n=1}^{\infty}(-1)^n x^{n(3n-1)/2}$$
$$= \sum_{n=-\infty}^{\infty}(-1)^n x^{n(3n+1)/2}$$

According to Euler, "this is quite certain, although I cannot prove it" [1, p. 279]. However, he did prove it ten years later. He could not possibly guess that both series and product would be part of the theory of elliptic modular functions developed by Carl Gustav Jacob Jacobi (1804–1851) eighty years later! Let us return to his 1747 memoir. He said, "As we have thus discovered that those two infinite expressions are equal even though it has not been possible to demonstrate their equality, all the conclusions which may be deduced from it will be of the same nature, that is, true but not demonstrated. Or, if one of these conclusions could be demonstrated, one could reciprocally obtain a clue to the demonstration of that equation; and it was with this purpose in mind that I maneuvered those two expressions in many ways" [11, vol. 1, p. 96].

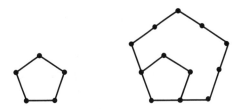

FIGURE 6

Perhaps his most salient feature is the extraordinary promptness with which he always reacted even to casual suggestions or stimuli ... . Every occasion was promptly grasped; each one supplied grist to his mill, often giving rise to a long series of impressive investigations. Hardly less striking is the fact that Euler never abandoned a problem after it had once aroused his insatiable curiosity. ... All his life, even after the loss of his eyesight, he seems to have carried in his head the whole of the mathematics of his day, both pure and applied. Once he had taken up a question, not only did he come back to it again and again, little caring if at times he was merely repeating himself, but also he loved to cast his net wider and wider with never failing enthusiasm, always expecting to uncover more and more mysteries, more and more "herrliche proprietates" lurking just around the next corner. Nor did it greatly matter to him whether he or another made the discovery. [16, pp. 283–284]

Euler was a genius, whose height very few can hope to reach. But we can all learn from his enthusiasm to work, his insatiable curiosity to probe, and his determination to procure deeper and deeper understanding. As students of mathematics we should strive for these characteristics. As teachers of mathematics we should influence our students to also strive for them.

Bibliography

1. Andrews, G. E.: "Euler's Pentagonal Number Theorem," *Mathematics Magazine* 56 (1983), 279–284.
2. Ayoub, R.: "Euler and the Zeta Function," *American Mathematical Monthly* 81 (1974), 1067–1086; Addendum 82 (1975), 737.
3. Bell, E. T.: *Men of Mathematics*. New York: Simon and Schuster, 1965 (first published in 1937).
4. Biggs, N. L., Lloyd, E. K., and Wilson, R. J.: *Graph Theory*. Oxford: Clarendon Press, 1976.
5. Burckhardt, J. J., Fellmann E. A., and Habicht, W. (eds.): *Leonhard Euler 1707–1783, Beiträge zu Leben und Werk*. Boston: Birkhäuser, 1983.
6. Chow, S. C., Siu, F. K., and Siu, M. K.: "Some Remarks on the Proposed Secondary Mathematics Curriculum in People's Republic of China—About the Role of Mathematics in School Curriculum" (in Chinese), *Dousou Bimonthly* 38 (1980), 76–83 (also in *Essays on Mathematics Teaching From the Dousou Bimonthly*. Hong Kong: Commercial Press, 1981, 57–64).
7. Euler, L.: *Introduction to Analysis of the Infinite*, Book I (J. D. Blanton, trans.). New York: Springer-Verlag, 1988.
8. Federico, P. J.: *Descartes On Polyhedra*. New York: Springer-Verlag, 1982.
9. Lakatos, I.: *Proofs and Refutations*. Cambridge: Cambridge University Press, 1976.
10. Pólya, G.: *Collected Papers*, Vol. IV (G.-C. Rota, ed.). Cambridge: MIT Press, 1984.
11. Pólya, G.: *Mathematics and Plausible Reasoning*, Vol. 1 and 2. Princeton: Princeton University Press, 1954.
12. Pólya, G.: *How To Solve It*, 2nd Edition. Princeton: Princeton University Press, 1957.
13. Pólya, G.: *Mathematical Discovery*, Vol. 1 and 2. New York: John Wiley, 1962 (combined edition, 1981).
14. Truesdell, C.: "Leonhard Euler, Supreme Geometer," in Truesdell, C: *An Idiot's Fugitive Essays on Science*. New York: Springer-Verlag, 1984, 337–379.
15. Van der Poorten, A.: "A proof that Euler missed ... ," *Mathematical Intelligencer* 1 (1979) 195–203.
16. Weil, A.: *Number Theory: An Approach Through History From Hammurapi to Legendre*. Boston: Birkhäuser, 1984.
17. Youschkevitch, A. P.: "Euler," in *Dictionary of Scientific Biography*, Vol. 4. New York: Scribner's, 1970–1980.
18. *Mathematics Magazine* 56 (1983), 258–325.

Converging Concepts of Series: Learning from History

Joel P. Lehmann

The job of a teacher is to facilitate student learning, where "learn" is an active verb whose subject is "student." If we are to do more than just present material, we must understand the learning process, identify areas of particular difficulty, and develop teaching strategies that will help the student overcome the difficulty. Learning is not a uniformly continuous process. It happens sometimes incrementally, sometimes in more of a quantum leap. Some topics seem to build slowly on a foundation already in place, others require a major shift in thinking. This appears to be the case both for the individual student of mathematics and for the collective culture of mathematicians. The leaps required for the students are very likely to coincide with what historically was a leap for the collective culture. Using the history of such a topic to facilitate the students' leap to understanding of a topic is more than just an enjoyable diversion; it can be a sound pedagogical technique.

What I propose is a model for the development of mathematical concepts. No claim is made for its being a deep psychological model for learning, nor a major step in cognitive theory. It is a model that provides a point of view, both to understand how mathematical concepts developed historically and to plan pedagogy. The model's value will be to help identify where ideas require a major shift in understanding (which I suspect we generally categorize as "mathematical maturity"). When we are aware of where in our history the major shifts occurred, we will be more sensitive to the students' problems of trying to grasp the material. If we can understand how mathematicians accommodated a radical idea when it was first developed, we can better plan the strategy to help the student learn and accommodate it.

Model

The model I propose posits several levels of successive abstraction. The first level is people just doing things—finding the length of a side of a triangle, or circumference of a circle, or area or volume, etc.—solving a single, concrete problem, with no indication of generalizable method or theory. For example, the Old Testament passage cited for the use of 3 as a working equivalent of π [2] is:

> Also, he made a molten sea of ten cubits from brim to brim, round in compass, and
> five cubits the height thereof; and a line of thirty cubits did compass it round about.
> [I Kings vii. 23]

This is specific, the numbers (arrived at by guess or measurement or approximation or whatever it takes) are given and that's the end of it. An answer is all that is sought, and if the presented value works, there is no need to explain how the answer was determined. This is the level at which the student accepts as authority the answers given in the back of the book, or tables of values, or the instructor's word.

The second level is then a method or algorithm for obtaining a solution of a specific type of problem. The method of presentation may still be in the form of a single worked example, but there is an understanding that the numbers used are just representative, that similar steps with other numbers will also work. Most of the problems in the Rhind Papyrus (c. 1650 B.C.) [6] are of this level, especially its area problems, the division of loaves or grain, and the mixture ("pesu") problems. These show that some abstracting has occurred, some common element identified that makes the problems "all the same, with only the numbers changed," rather than a collection of different problems. A solution is, in a sense, recycled.

The third level represents a higher level of abstraction, a removal away from actual physical objects; objects under discussion become more of what we would think of as defined terms rather than objects having a physical existence. The "aha" problems in the Rhind Papyrus are of this sort, for example:

Problem 24: A quantity and its 1/7 added together become 19. What is the quantity? Assume 7. $1\frac{1}{7}$ of 7 is 8. As many times as 8 must be multiplied to give 19, so many times 7 must be multiplied to give the required number. [6, p. 66]

The quantity sought has an abstract quality different from that of applied problems, with no hint of its being a quantity *of* anything in particular. And the method of false position used to solve it represents a fairly sophisticated mathematical technique. Any convenient value is assumed for the desired quantity and is substituted into the expression, then appropriate adjustments are made to obtain the correct value.

While most of the problems may convey a sense of physical reality, that there is an actual square object, round field or pool under consideration at this third level, you begin to get concepts of idealized circles or squares. A classic example of this situation would be Greek geometry, which viewed physical observation as only the first process towards understanding reality, which in itself was a mental construct. Areas of polygons and volumes of pyramids might be used by engineers and architects, but that was not the intent of Euclid's work.

By this third level, the process is sufficiently non-obvious so that some kind of instructions are needed beyond a single numerical example. Archimedes (287–212 B.C.), approximating the area of a circle with inscribed and circumscribed polygons, must give instructions on how to proceed [9, vol. ii, pp. 50–52]. Antiphon the Sophist (c. 430 B.C.), attempting to square the circle using inscribed polygons, supplies instructions for doubling the number of sides of the polygon and how to proceed [9, vol. i, pp. 221–22].

The mention of Antiphon brings up a problem, a sign of difficulties to come. Antiphon made a leap from inscribed polygons with a finite (though large) number of sides to an infinite number of sides, from "approximate as closely as you like" to "be exactly the same," from polygonal sides *approximating* the curve to polygonal sides *coinciding* with the curve. The rejection of Antiphon's faulty reasoning [9, vol. 1, p. 221] still left the question, "How do you find areas of figures with curves for boundaries?" Special cases led to classes of figures for which areas could be determined, which led to abstract ideas of area based on the use of inscribed triangles or rectangles. Nothing worked for circles. We have the third level maturing into something different, a fourth level.

The fourth level is perhaps a maturation of the third level, a look at more sophisticated questions or problems that occur. There is a separate level of realization here, of distinctions that have to be made in the abstract objects being studied, of differences that need to be considered. The definitions and the objects under study force a reconsideration of the concepts. Archimedes used inscribed and circumscribed polygons to approximate areas of circles successfully, but recognized that they were approximations. Antiphon had mistakenly extended a similar procedure from finite to infinite number of sides, failing to realize that areas of polygons, no matter how large the number of sides, will never exactly coincide with that of a circle. A failure of the method or process used has brought up a problem. In most cases, resolution of such problems will require an increase in the level of abstraction and sophistication. In this case, a general method for finding areas of figures with curved boundaries required the development of calculus.

An analogy and pictorial representation of the model might be useful. (See Figure 1.) The analogy is a mountain path, with "altitude" signs along the way. The gentle first slope at Level 1 leads to a somewhat steeper (though straight) path up to Level 2. Level 3 looks steeper and winds a bit. When you hit Level 4, you encounter a crevasse. Frequently you are unaware of this abyss until you have almost stepped into it. To continue your ascent requires alternative action. The action may be lateral movement trying to get around it, or backtracking, or even returning to the starting point and beginning over; it may also be building a bridge across the crevasse. And what do you discover on the other side? Usually you are at another Level 1, which gives you a brief respite but which has a path leading upwards again. The mathematical version would be something like this:

Level 1: A single, concrete problem is solved, using methods improvised for this single problem. Usually there is no indication of generalizable method or theory.

Level 2: In some cases the method or algorithm used for solving a single problem can be used as or adapted as a solution for other problems of the same type. A solution is, in a sense, recycled. This recycling is the value of abstraction, which we quite frequently forget to mention to students.

Level 3: Particularly useful techniques may be applied beyond the original object of study. By isolating and abstracting key features, the method may be adapted to a larger class of

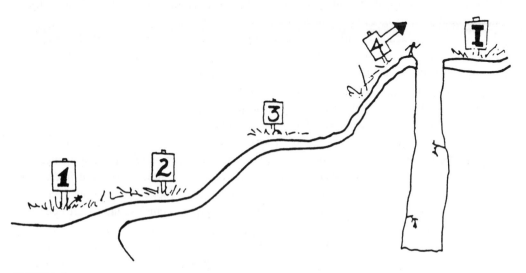

FIGURE 1

problems. There is a widening of the field of applicability. The objects under discussion become more defined terms rather than objects having physical existence; they are removed from the "intuitive" understanding of the object that gave rise to the algorithm.

Level 4: Level 3 then may lead to an exuberant and confident application to all manner of problems, until difficulties or contradictions arise. Thus Level 4 is perhaps a maturation of Level 3, a look at more sophisticated questions or problems that occur. There is a separate level of realization here, of distinctions that have to be made in the abstract objects being studied, of differences that need to be considered. A crisis may be reached where accepted methods fail or produce contradictory results—the method has been extended beyond its applicability, and some accommodation must be made. The definitions and the objects under study force a reconsideration of the concepts. This failure is illustrated by the crevasse which must be bridged.

Level 1 is a straightforward, "intuitive" stage of development, generally a relatively easy concept for students to comprehend. Getting to Level 2 requires some small mental adjustment, while Level 3 is a much larger (and to some, perhaps, impossible) cognitive shift. Level 4 "crisis" states may have required years or generations or even centuries of coping by the mathematical community, and it is not surprising that our students will have difficulty with them. Moreover, the levels are not linear, and there may be not a single path but many paths all crossing and recrossing, so you have no clear direction in which to move. There will be loops and recursion, with strategic retreats to former positions and reinstatement of formerly excluded ideas. Levels will rarely be clearly delineated; transition boundaries are very fuzzy things indeed. The same topics can be viewed as levels in the development of different concepts; we can use another metaphor and see the development as an interweaving of threads into a tapestry that makes following a single thread as difficult as sorting out this sentence.

An Example of the Model

As an example of this intricate development in both historic and pedagogical terms, I would like to use the concept of the *sum of the series*. The standard presentation of infinite series in calculus courses as taught in the United States is the following:
(1) A short introduction to infinite sequences, to prepare for sequences of partial sums;
(2) Abstract definitions of infinite series, with convergence defined in terms of limits of sequences of partial sums;
(3) Theorems and convergence tests for positive term series;
(4) Theorems and convergence tests for alternating series;
(5) Theorems and convergence tests for general series;
(6) Definition of and theorems about power series.

The instructional emphasis is on convergence and especially on tests of convergence. We spend our time finding out whether series converge or not, but little or no time finding out what the series converge *to*. And since most series that arise in applications are relatively well behaved, examples and exercises for testing convergence often comprise highly artificial and pathological cases.

That an unending collection of numbers may have a finite sum is not an easy concept to grasp. Since series are generally presented without history and separate from applications, the student must wonder not only "What are these things?" but also "Why are we doing this?"

The preoccupation with determining convergence but not the sum makes the whole process seem artificial and pointless to many students—and instructors as well. The fact is that series are greatly different from anything encountered before; and they are made harder to understand because we present solutions to Level 4 problems without first convincing the students that there is a problem. It's as if in our model we air-lifted them to the far side of the crevasse instead of leading them up the path and helping them construct a bridge. Recall that as mathematicians were first developing the concepts of series, they did not know in advance the final forms that these concepts would take. This is exactly the position our students are in: they too are developing concepts and do not know what the final form of those concepts should be. If we select topics that not only illustrate the concepts, but also trace the historical progression of ideas, we can help students make the transition to understand the Level 4 concerns we are addressing.

Let us look at a teaching strategy and a plan for sequencing topics and building around the model, looking at the several cognitive shifts which must be made. We will see that there is a shift (Level 2) in going from adding up specific numbers to finding the sum of a finite series. A shift to Level 3 can be illustrated with the move to more algebraic techniques. The first big shift in definition of "sum" will come with the move to infinite series, where the rules of arithmetic may no longer apply. By thus preparing the stage for identified cognitive shifts, the questions of "Why do we need to do things this way?" or "Why do we need to concern ourselves about this?" do not arise because the students will see the necessity for themselves.

Figurate Numbers.   We may take as a starting point the Greek figurate numbers [9]. Determining the triangular numbers and their sums is a Level 1 development that students can understand immediately. Triangular numbers and their sums are tangible and concrete: they can be found by counting dots on a diagram.

Triangular Numbers	N	Sum
.	1	1
. .	2	3
. . .	3	6
. . . .	4	10
. . . . .	5	15
. . . . . .	6	21
. . . .	n	$\dfrac{n(n+1)}{2}$

We make a minor shift (Level 2) by looking at the nth triangular number as a sum,

$$1 + 2 + 3 + \cdots + n = \frac{n(n+1)}{2}.$$

It is still familiar, though the "$\cdots$" and the "n" take a little getting used to. There is no great change from simple addition, and we can put in numbers to check the formulas and reassure ourselves that all is well.

A small step takes us to square numbers:

Square Numbers	N	Sum
	1	1
	2	4
	3	9
	4	16
	5	25
	6	36

Obtaining the nth square number in a similar fashion leads to the sum of odd numbers,

$$1 + 3 + 5 + \cdots + (2n + 1) = (n + 1)^2.$$

As other forms of figurate numbers were considered, the second level determination of properties common to all polygonal numbers appears, an organizing principle for triangular numbers, square numbers, pentagonal numbers, and so on. Certain sums go together, add up to the same kind of numbers which have some defining property. Nicomachus of Gerasa (c. A.D. 100) noted the pattern of numbers rather than shapes, and made a step forward in abstraction to Level 3 [13]. For Nicomachus, triangular numbers represented the sum of an arithmetic sequence with common difference 1, the square numbers a sequence with common difference of 2, pentagonal and hexagonal numbers 3 and 4 respectively. He generalized to a class of objects only abstractly similar. In this we see a move toward abstraction and away from the concrete, which can help the student make the same step. This is illustrated further by the table Nicomachus presents, which retains its touch with its geometrical roots only in the names [13, pp. 248–9]:

Triangles	1	3	6	10	15	21	28	36	45
Squares	1	4	9	16	25	36	49	64	81
Pentagonals	1	5	12	22	35	51	70	92	117
Hexagonals	1	6	15	28	45	66	91	120	153
Heptagonals	1	7	18	34	55	81	112	148	189

Nicomachus notes that "each polygonal number is the sum of the polygonal in the same place in the series with one fewer angle, plus the triangle, in the highest row, one place back in the series." This clearly represents a cognitive shift from the figures with which we began, for here we have interrelationships between pure numbers, with the defining characteristic being the generating algorithm rather than a geometrical arrangement of dots.

This shift to numbers and formulas and away from geometric figures made possible different discoveries relative to sums. Having seen that the sum of successive odd numbers was always a square, the Greeks turned to investigating cubes. When it was discovered that 2^3 is $3 + 5$, 3^3 is $7 + 9 + 11$, 4^3 is $13 + 15 + 17 + 19$, and so on, they were in a position to sum the series $1^3 + 2^3 + 3^3 + \cdots + r^3$: it was only necessary to find out how many terms of the series $1 + 3 + 5 + \cdots$ this sum of r cubes includes. The number of terms being $1 + 2 + 3 + \cdots + r$, the desired sum of the first r cubes is

$$\left[\frac{r(r+1)}{2}\right]^2$$

[9, p. 109]. This is moving into Level 3 activity. As we parallel the historical development in our presentation, we can add and hold interest in the development of formulae which are otherwise confusing and unnatural in a mere bare-bones direct presentation.

A Level 4 crisis occurred with Zeno (c. 450 B.C.), whose paradox of Achilles and the tortoise brought into question the meaning of adding an infinite number of terms. The response to Zeno was not a resolution of the problem but a backing away from any questions of the infinite or infinite processes, an avoidance of the question. It would be almost two thousand years before the Level 4 crisis of Zeno would be fully appreciated and confronted.

Algebraization of Finite Series. We are solidly on Level 3 with the continued algebraization of finite series, in the pursuit of formulas for sums of powers of successive integers. Two examples that might help ease the transition to this more abstract form come from Alhazen and one, following a slightly different developmental strand, from Jakob Bernoulli (1654–1705).

In the eleventh century, the Arab mathematician al-Haitham (c. 965–1039), known in the West as Alhazen, computed the volume of a segment of a parabola revolved about its base. This required formulas for sums of cubes and fourth powers of integers. Alhazen found sums of cubes in terms of squares and of fourth powers in terms of cubes. [7] A geometric model which depends only on equating areas of rectangles can help the modern student (and instructor) understand Alhazen's arithmetic argument and the otherwise confusing expression that results.

The rectangle has height $n+1$, and width $1+2+\cdots+n$. The area is $(n+1)(1+2+\cdots+n)$. The area can also be seen as the sum of the areas of the individual pieces, the squares and rectangles inside. The total area of squares is

$$1^2 + 2^2 + \cdots + n^2,$$

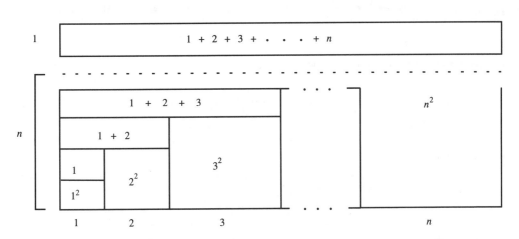

FIGURE 2
Total area equals the sum of the parts

and of the rectangles

$$1 + (1+2) + (1+2+3) + \cdots + (1+2+\cdots+n).$$

We then have, in modern notation,

$$(n+1)\sum_{i=1}^{n} i = \sum_{i=1}^{n} i^2 + \sum_{p=1}^{n}\sum_{i=1}^{p} i$$

from which is found

$$\sum_{i=1}^{n} i^2 = \frac{n(n+1)(2n+1)}{6}.$$

Now, we can use this result to get the sum of cubes:

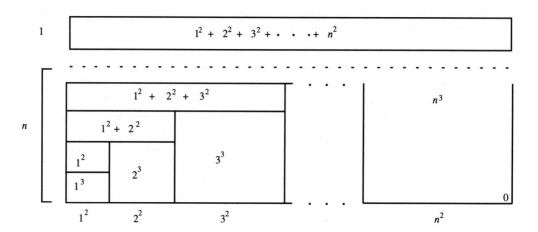

FIGURE 3

Proceeding in the same fashion as above, we get

$$(n+1)\sum_{i=1}^{n} i^2 = \sum_{i=1}^{n} i^3 + \sum_{p=1}^{n}\sum_{i=1}^{p} i^2$$

from which it follows that

$$\sum_{i=1}^{n} i^3 = \frac{[n(n+1)]^2}{4}.$$

We can continue this process to give a more concrete framework on which to secure abstract concepts.

Moving from the specific cases above to a general form for the sum of kth powers:

$$(n+1)\sum_{i=1}^{n} i^k = \sum_{i=1}^{n} i^{k+1} + \sum_{p=1}^{n}\sum_{i=1}^{p} i^k.$$

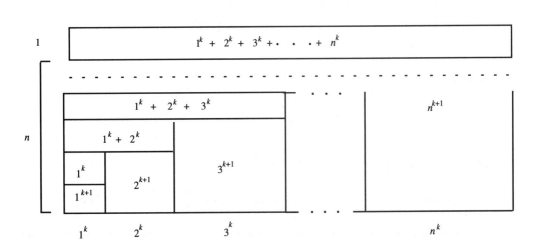

FIGURE 4

An otherwise overpowering formula can be related to familiar ideas and made not only more understandable but also more meaningful and "intuitive." Intuition requires careful development and nurturing. At least, it does if we hope to develop it in weeks rather than years or centuries!

A further historical note is appropriate, from another strand of historical development that is interwoven inextricably at this point. By the time of Pierre de Fermat (1636) the geometrical origins of figurate numbers had developed into a more algebraic form along a somewhat different path than the polygonal numbers used by Nicomachus. The first type of figurate numbers were for Fermat as the triangular numbers were for Nicomachus, as noted above. The nth triangular number was the sum of the first n integers. But in this development, the second type of figurate numbers were not square numbers but pyramidal numbers, where the nth pyramidal number was formed by summing the first n triangular numbers. In general, figurate numbers of type k were formed by summing figurate numbers of type $k-1$. These figurate numbers form the columns of the arithmetic triangle, usually called Pascal's triangle, because of his extensive investigation of its properties.

$$
\begin{array}{ccccccc}
1 & & & & & & \\
1 & 1 & & & & & \\
1 & 2 & 1 & & & & \\
1 & 3 & 3 & 1 & & & \\
1 & 4 & 6 & 4 & 1 & & \\
1 & 5 & 10 & 10 & 5 & 1 & \\
. & . & . & . & . & . & .
\end{array}
$$

A general formula for the nth figurate number of type k, given by Fermat in 1636 without proof [12, pp. 229–232] was

$$\sum_{i=1}^{n} \frac{i(i+1)(i+2)\cdots(i+k-1)}{k!} = \frac{n(n+1)(n+2)\cdots(n+k)}{(k+1)!}.$$

Jakob Bernoulli (1654–1705) used a variation of the Arithmetic Triangle, making it a square in which the general form for the nth element in column k is $\dfrac{(n-1)(n-2)\cdots(n-k)}{k!}$, and the first k entries in column k are zeroes.

n	C_1	C_2	C_3	C_4	C_5	C_6	
1	1	0	0	0	0	0	0
2	1	1	0	0	0	0	0
3	1	2	1	0	0	0	0
4	1	3	3	1	0	0	0
5	1	4	6	4	1	0	0
6	1	5	10	10	5	1	0
7	1	6	15	20	15	6	1
8	1	7	21	35	35	21	7

He observed [15, p. 86] that "the series of figurate numbers, supplied with the corresponding zeros, have a submultiple ratio to the series of equals." This is a chance to expose the student to the value of symbolic expressions when compared to the difficulty of working solely with words. Bernoulli is saying that the ratio of the sum of the first n terms in column k (of which the first k terms will be zero) to n times the last term, will be $1/(k+1)$. For example, from C_3:

$$\frac{0+0+0+1+4}{4+4+4+4+4} = \frac{0+0+0+1+4+10}{10+10+10+10+10} = \frac{0+0+0+1+4+10+20}{20+20+20+20+20+20+20} = \frac{1}{4}.$$

This gave him an equation relating sums of powers of integers to a known quantity. He now proceeds in a quite modern fashion to solve for the sum with the highest exponent in terms of sums of lower powers, which are already known. This is the same thing Alhazen did, but Bernoulli's development proceeds from a more algebraic perspective.

From column C_2 we have

$$\frac{0+0+1+\cdots+\frac{(n-1)(n-2)}{2}}{\frac{(n-1)(n-2)}{2}+\frac{(n-1)(n-2)}{2}+\cdots+\frac{(n-1)(n-2)}{2}} = \frac{\sum_{i=1}^{n}(i^2-3i+2)}{n(n^2-3n+2)} = \frac{1}{3}$$

$$\sum_{i=1}^{n}(i^2-3i+2) = \frac{n^3-3n^2+2n}{3}$$

$$\sum_{i=1}^{n} i^2 - \frac{3n(n+1)}{2} + 2n = \frac{n^3-3n^2+2n}{3}$$

and solving,

$$\sum_{i=1}^{n} i^2 = \frac{n^3}{3} + \frac{n^2}{2} + \frac{n}{6}.$$

Bernoulli then used this result to derive from column C_3 that

$$\sum_{i=1}^{n} i^3 = \frac{n^4}{4} + \frac{n^3}{2} + \frac{n^2}{4}.$$

Except for notation, as he uses our integral sign instead of our capital sigma to denote sum, his treatment is very modern, and he gives instructions for continuing the process as far as desired.

But there is a higher level of abstraction to come. Moving from the table to the derived expressions, he establishes a pattern from the sequence of sums. Looking first at the sums of powers, he computes:

$$\sum i = \frac{1}{2}n^2 + \frac{1}{2}n$$

$$\sum i^2 = \frac{1}{3}n^3 + \frac{1}{2}n^2 + \frac{1}{6}n$$

$$\sum i^3 = \frac{1}{4}n^4 + \frac{1}{2}n^3 + \frac{1}{4}n^2$$

$$\sum i^4 = \frac{1}{5}n^5 + \frac{1}{2}n^4 + \frac{1}{3}n^3 - \frac{1}{30}n$$

$$\sum i^5 = \frac{1}{6}n^6 + \frac{1}{2}n^5 + \frac{5}{12}n^4 - \frac{1}{12}n^2$$

$$\sum i^6 = \frac{1}{7}n^7 + \frac{1}{2}n^6 + \frac{1}{2}n^5 - \frac{1}{6}n^3 + \frac{1}{42}n$$

$$\sum i^7 = \frac{1}{8}n^8 + \frac{1}{2}n^7 + \frac{7}{12}n^6 - \frac{7}{24}n^4 + \frac{1}{12}n^2$$

$$\sum i^8 = \frac{1}{9}n^9 + \frac{1}{2}n^8 + \frac{2}{3}n^7 - \frac{7}{15}n^5 + \frac{2}{9}n^3 - \frac{1}{30}n$$

$$\sum i^9 = \frac{1}{10}n^{10} + \frac{1}{2}n^9 + \frac{3}{4}n^8 - \frac{7}{10}n^6 + \frac{1}{2}n^4 - \frac{3}{20}n^2$$

$$\sum i^{10} = \frac{1}{11}n^{11} + \frac{1}{2}n^{10} + \frac{5}{6}n^9 - n^7 + n^5 - \frac{1}{2}n^3 + \frac{5}{66}n$$

He then goes on to observe:

> Whoever will examine the series as to their regularity may be able to continue the table. Taking c to be the power of any exponent, the sum of all n^c or
>
> $$\sum i^c = \frac{1}{c+1}n^{c+1} + \frac{1}{2}n^c + \frac{c}{2}An^{c-1} + \frac{c(c-1)(c-2)}{4!}Bn^{c-3}$$
> $$+ \frac{c(c-1)(c-2)(c-3)(c-4)}{6!}Cn^{c-5}$$
> $$+ \frac{c(c-1)(c-2)(c-3)(c-4)(c-5)(c-6)}{8!}Dn^{c-7} + \cdots$$
>
> and so on, the exponents of n continually decreasing by 2 until n or n^2 is reached. The capital letters $A, B, C, D \ldots$ denote in order the coefficients of the last terms in the expressions for $\sum i^2, \sum i^4, \sum i^6, \sum i^8, \ldots$. With the help of this table it took me less than half of a quarter of an hour to find that the tenth powers of the first 1000 numbers being added together will yield the sum
>
> $$91,409,924,241,424,243,424,241,924,242,500$$

From this it will become clear how useless was the work of Ismael Bullialdus spent on the compilation of his voluminous *Arithmetica Infinitorum* in which he did nothing more than compute with immense labor the sums of the first six powers, which is only a part of what we have accomplished in the space of a single page. [15, p. 90]

In this last paragraph, Bernoulli clearly describes the cognitive shift that has occurred, the shift from special cases to a general abstract solution, and stresses the value of the latter. This is a point teachers cannot overemphasize to their students.

Infinite Series. A Level 4 shift occurs with the move from finite sums to infinite series. So far we have reached a level where we have general forms for finite sums of powers of integers, and while there have been some major shifts—especially in treating finite sums as objects in equations—things still have close ties to arithmetic that is familiar and "normal." Adding an infinite number of terms can bring us around to Zeno again: after a thousand years there is still coping and adjusting to be done. The move to infinite series was not a single step, and there were many paths intermingled to produce the change. The one I wish to follow requires us to back up a bit first, back to the fourteenth century.

A Level 1 in infinite sums existed in medieval times. In the second quarter of the fourteenth century the natural philosophers at Merton College in Oxford, including Thomas Bradwardine (1290–1349) and Richard Swineshead (the Calculator) (fl. c. 1350), were involved with the problem of quantifying change. Their investigations into the "latitude of forms," though pursued in rhetorical fashion, presented a break from the tradition of unchanging geometric figures moving in uniform motion. One problem considered by Swineshead is relevant to our current endeavor:

> If a point moves throughout the first half of a certain time interval with a constant velocity, throughout the next quarter of the interval at double the initial velocity, throughout the following eighth at triple the initial velocity, and so on ad infinitum; then the average velocity during the whole time interval will be double the initial velocity. [7, p. 91]

In modern notation, this is equivalent to

$$\frac{1}{2} + \frac{2}{4} + \frac{3}{8} + \cdots + \frac{n}{2^n} + \cdots = 2.$$

Swineshead used a geometrical argument in his proof, to show that a sum of an infinite number of terms can have a finite value.

The fourteenth century also produces the seeds of Level 4 difficulties when, in about 1350, the natural scientist Nicole Oresme (c. 1323–1382) showed that the harmonic series $1 + \frac{1}{2} + \frac{1}{3} + \cdots$ diverged, that if the successive terms were added the whole would become infinite. He proved this by noting that $\frac{1}{3} + \frac{1}{4}$ is greater than $\frac{1}{2}$, as is the sum of the next four terms ($\frac{1}{5}$ through $\frac{1}{8}$), and the next 8 terms, and so on. We thus have an early example of a convergent infinite series with its sum, and one of a divergent series [3]. Knowledge of the latter seems not to have inhibited subsequent users of infinite series.

We can progress further onto Level 2 for infinite sums by looking at Leibniz's harmonic triangle, used in a somewhat similar fashion to the way Bernoulli used the arithmetic triangle. In 1672, Gottfried Leibniz (1646–1716) had recognized that if you take a given sequence and form a new finite sequence by taking consecutive differences of the terms of the given sequence, and then sum the terms of the new sequence, the result is the difference of the first and last terms of the original sequence. Thus, if $a_0, a_1, \ldots, a_n$ is the original sequence, and $d_i = a_i - a_{i+1}$, then

$$d_0 + d_1 + \cdots + d_{n-1} = (a_0 - a_1) + (a_1 - a_2) + \cdots + (a_{n-1} - a_n) = a_0 - a_n.$$

Leibniz realized that if the nth term of the original sequence went to zero as n increased, then the derived infinite series of differences would sum to the original first term, i.e.,

$$\sum_{i=0}^{\infty} d_i = \sum_{i=0}^{\infty} (a_i - a_{i+1}) = a_0.$$

Leibniz used this idea to solve a problem posed to him by Christiaan Huygens (1629–1695), the problem of finding the sum of the series

$$\frac{1}{1} + \frac{1}{3} + \frac{1}{6} + \frac{1}{10} + \cdots + \frac{1}{n(n+1)/2} + \cdots$$

Combining his insight on sums of differences with a familiarity with the Arithmetic Triangle for figurate numbers, Leibniz formed what he called his Harmonic Triangle. Starting in the first row with the reciprocals of the integers, subsequent rows were formed from differences of consecutive terms from the preceding row.

$$\frac{1}{1} \quad \frac{1}{2} \quad \frac{1}{3} \quad \frac{1}{4} \quad \frac{1}{5} \quad \frac{1}{6} \quad \frac{1}{7} \quad \cdots$$

$$\frac{1}{2} \quad \frac{1}{6} \quad \frac{1}{12} \quad \frac{1}{20} \quad \frac{1}{30} \quad \frac{1}{42} \quad \cdots$$

$$\frac{1}{3} \quad \frac{1}{12} \quad \frac{1}{30} \quad \frac{1}{60} \quad \frac{1}{105} \quad \cdots$$

$$\frac{1}{4} \quad \frac{1}{20} \quad \frac{1}{60} \quad \frac{1}{140} \quad \cdots$$

$$\frac{1}{5} \quad \frac{1}{30} \quad \frac{1}{105} \quad \cdots$$

$$\cdot \qquad \cdot \qquad \cdot$$

Just as the Arithmetic Triangle provided formulae for sums of a finite number of terms, the Harmonic Triangle provided formulae for sums of an infinite number of terms. In particular, the sum of each row is equal to the first term of the preceding row. Thus,

$$\frac{1}{2} + \frac{1}{6} + \frac{1}{12} + \frac{1}{20} + \cdots = 1$$

$$\frac{1}{3} + \frac{1}{12} + \frac{1}{30} + \frac{1}{60} + \cdots = \frac{1}{2}$$

$$\frac{1}{4} + \frac{1}{20} + \frac{1}{60} + \frac{1}{140} + \cdots = \frac{1}{3}$$

Double the first of these gives Huygens' requested sum, which is of course 2.

What we have with Leibniz and his Harmonic Triangle is an example of Level 2, with a method applicable to a restricted class of infinite series. Though it still retains the idea of sum being related to addition as usual, yet the case of sums of differences has the look of partial sums about it. It has the added benefit of actually finding the sum of many infinite series, in a logical way that removes the mystery from the whole process.

A Level 3 move comes when series with numbers are extended to series containing variable expressions. Isaac Newton (1642–1727), working with series in *A Treatise on the Methods of Series and Fluxions* (1671, first published 1736), gives a good description of the transition that is beginning:

> Since the operations of computing in numbers and with variables are closely similar—indeed there appears to be no difference between them except in the characters by which quantities are denoted, definitely in one case, indefinitely so in the latter—I am amazed that it has occurred to no one (if you except N. Mercator with his quadrature of the hyperbola) to fit the doctrine recently established for decimal numbers in similar fashion to variables, especially since the way is then open to more striking consequences. For since this doctrine has the same relationship to Algebra that the doctrine in decimal numbers has to common Arithmetic, its operations of Addition, Subtraction, Multiplication, Division and Root-extraction may easily be learnt from the latter's provided the reader be skilled in each, both Arithmetic and Algebra, and appreciate the correspondence between decimal numbers and algebraic terms continued to infinity ... It is the advantage of infinite variable-sequences that classes of more complicated terms (such as fractions whose denominators are complex quantities, the roots of complex quantities and the roots of affected equations) may be reduced to the class of simple ones: that is, to infinite series of fractions having simple numerators and denominators and without the all but insuperable encumbrances which beset the others. [16, pp. 33–34]

Newton is asserting that any legal operation that can be performed in arithmetic on numbers can likewise be performed in algebra on variable expressions. Just as arithmetic operations produce highly useful infinite decimal expressions, so the same operations may produce highly useful infinite series in algebra. This is a point most if not all students would agree with, and they require proof that there is anything wrong with this reasoning.

This represents a move into Level 4, however, though it is not yet recognized. The question of convergence has been blithely ignored. We see Newton working with the two series

$$\frac{1}{1+x} = 1 - x + x^2 - x^3 + \cdots$$

and

$$\frac{1}{x+1} = \frac{1}{x} - \frac{1}{x^2} + \frac{1}{x^3} - \cdots,$$

and no distinction is drawn between them. If we substitute 1 for x in either expression, we have

$$\frac{1}{2} = 1 - 1 + 1 - 1 + 1 - \cdots$$

in both cases. The partial sums of the expression on the right alternate between 0 (if an even number of terms are added) and 1 (if an odd number of terms are added). Trying to add together more and more terms does not lead us to any appropriate sum, certainly not $\frac{1}{2}$.

There are several ways to handle the dilemma. One is to ignore the difficulty and merely accept the result:

In former times—before the strict foundation of infinite series—mathematicians found themselves fairly at a loss when confronted with paradoxes such as this.

And even though the better mathematicians instinctively avoided arguments such as the above, the lesser brains had all the more opportunity of indulging in the boldest speculations. Thus, e.g., Guido Grandi believed that in the above erroneous train of argument which turns 0 into 1, he had obtained a mathematical proof of the possibility of creation of the world from nothing! [11, p. 133]

A somewhat less metaphysical alternative which also avoids a cognitive shift is to define the sum of a series in terms of limits of sequences of partial sums (as we do now), and dismiss as "divergent" any series that does not satisfy this convergence requirement. This is a backtrack that avoids the crevasse that has come before us.

Yet another alternative is somehow to redefine the concept of "sum" in such a way that these aberrant series can be reclaimed, building a bridge and thus making a Level 4 accommodation. Among those making the attempt to save the divergent series for analysis was Leonhard Euler (1707–1783). He attempted to redefine the meaning of "sum" in a significantly more abstract fashion, further from the then common understanding of "sum" as "to add up." In his 1760 paper on divergent series, Euler gives reasons for trying to cope with divergent series rather than dismissing them out of hand:

> Whenever in analysis we arrive at a rational or transcendental expression, we customarily convert it into a suitable series on which the subsequent calculations can more easily be performed. Therefore infinite series find a place in analysis inasmuch as they arise from the expansion of some closed expression, and accordingly in a calculation it is valid to substitute in place of the infinite series that formula from which the series came. Just as with great profit rules are usually given for converting expressions closed but awkward in form into infinite series, so likewise the rules, by whose help the closed expression, from which a proposed infinite series arises, can be investigated, are to be thought highly useful. Since this expression can always be substituted without error for the infinite series, both must have the same value. [1, p 148]

Euler summarizes Leibniz's arguments for $1 - 1 + 1 - 1 + \cdots$ being assigned the value $\frac{1}{2}$. The first is that the series is the expansion by division of $\frac{1}{1+a} = 1 - a + a^2 - a^3 + \cdots$ with a replaced by 1. The second rests on the fact that the sum of a finite number of terms is 0 for an even number of terms, 1 for an odd:

> Now if, therefore, the series is taken to infinity and (consequently) the number of terms cannot be regarded as either even or odd, it cannot be concluded that the sum is either 0 or 1, but we ought to take a certain median value which differs equally from both, namely $\frac{1}{2}$. [1, p 145]

Euler wanted to make a Level 4 accommodation, refine the definition of sum to a more abstract form, thus making the concept applicable to a wider range of series than the partial sum definition. He proposed the following definition for "sum":

> Understanding of the question is to be sought in the word "sum"; this idea, if thus conceived—namely, the sum of a series is said to be that quantity to which it is brought closer as more terms of the series are taken—has relevance only for convergent series, and we should in general give up this idea of sum for divergent series. Wherefore,

those who thus define a sum cannot be blamed if they claim they are unable to assign a sum to a series. On the other hand, as series in analysis arise from the expansion of fractions or irrational quantities or even of transcendentals, it will in turn be permissible in calculations to substitute in place of such a series that quantity out of whose development it is produced. For this reason, if we employ this definition of sum, that is, to say the sum of a series is that quantity which generates the series, all doubts with respect to divergent series vanish and no further controversy remains on this score, inasmuch as this definition is applicable equally to convergent or divergent series. [1, p 144]

The intuitive formation of this definition of "sum" reflects an attitude still current among applied mathematicians and physicists: problems that arise naturally (i.e., from nature) do have solutions, so the assumption that things will work out eventually is justified experimentally without the need for existence sorts of proof. Assume everything is okay, and if the arrived-at solution works, you were probably right, or at least right enough. Emil Borel (1871–1956) noted this, observing that

the older mathematicians had sufficiently good experimental evidence that the use of such series as if they were convergent led to correct results in the majority of cases when they presented themselves naturally. [4, p. 320]

This is exactly the attitude of many students—all the real problems ("real" meaning "arising in physical reality") have things working out, so why bother with the details that only show up in homework problems?

Examples for which this intuitive definition leads to problems can help meet student objections. They are instructive for discussion, even if they have already been resolved. An objection made to Euler's definition giving $\sum(-1)^n = \frac{1}{2}$, made by Jean-Charles Callet (1744–1799) in an unpublished memorandum submitted to J. L. Lagrange (1736–1813) [10], pointed out that the same series can arise from the expansion of different functions, for example,

$$\frac{1+x}{1+x+x^2} = \frac{1-x^2}{1-x^3} = 1 - x^2 + x^3 - x^5 + x^6 - x^8 + \cdots$$

which at $x = 1$ gives $\frac{2}{3} = \sum(-1)^n$, instead of Euler's $\frac{1}{2}$. Lagrange considered this objection and argued that Callet's example was incomplete. When the missing terms were included, the series should have been written

$$1 + 0x^1 - x^2 + x^3 + 0x^4 - x^5 + x^6 + 0x^7 - x^8 + \cdots$$

so that what was summed was

$$1 + 0 - 1 + 1 + 0 - 1 + 1 + 0 - 1 + \cdots,$$

a series whose partial sums are 1, 1, 0, 1, 1, 0, ... with an average sum of $\frac{2}{3}$. [4, pp. 319–20]

A less *ad hoc* solution to this problem was offered by G. Frobenius (1848–1917) [4, p. 319] by defining

$$\lim_{n \to \infty} \left(\sum a_n x^n \right) = \lim_{n \to \infty} \frac{S_0 + S_1 + \cdots + S_n}{n+1},$$

that is, you average the partial sums. This redefinition of "sum" is a Level 4 activity, a bridge over the crevasse.

Conditional and Absolute Convergence. We have reached the beginnings of Level 4 difficulties and the first attempts at coping. The special case of the alternating series could be explained away, but other and greater problems were arising.

Consider the following series,

$$S = 1 - \frac{1}{2} + \frac{1}{3} - \frac{1}{4} + \frac{1}{5} - \frac{1}{6} + \cdots$$

which converges to $S = \ln 2$. We have

$$\frac{1}{2}S = \frac{1}{2} - \frac{1}{4} + \frac{1}{6} - \frac{1}{8} + \frac{1}{10} - \frac{1}{12} + \cdots = 0 + \frac{1}{2} + 0 - \frac{1}{4} + 0 + \frac{1}{6} + 0 - \frac{1}{8} + 0 + \cdots$$

since the series and its convergence are unaffected by the addition of zero terms. "Adding" the original series S to this latter series term by term, we have

$$\frac{3}{2}S = 1 + 0 + \frac{1}{3} - \frac{1}{2} + \frac{1}{5} + 0 + \frac{1}{7} - \frac{1}{4} + \frac{1}{9} + 0 + \cdots = 1 + \frac{1}{3} - \frac{1}{2} + \frac{1}{5} + \frac{1}{7} - \frac{1}{4} + \frac{1}{9} + \frac{1}{11} - \frac{1}{6} + \cdots$$

when the zeroes are dropped. Comparison shows this series contains the same terms, although in different order, as the original series, but its sum is $\frac{3}{2} \ln 2$. By rearranging the terms in the series, we have changed its sum. Thus, the series does not behave as a traditional sum. [5, p. 168]

Recognition that rearranging the terms of an infinite series could change its sum, first noted by A. L. Cauchy (1789–1857) in *Resumés Analytiques* (Turin, 1833) [11, p. 138], was a clear Level 4 challenge. Here is the demonstrated need for a distinction between those convergent series for which a rearrangement made no difference and those for which it did—in today's terms, between absolutely and conditionally convergent series. Riemann's Rearrangement Theorem [11, pp. 318–9], which shows that it is possible to rearrange the terms of a conditionally convergent series so that the derived series converges to any desired value, may be the final word but it ought not to be the first (or only) indication of the problem.

The above example, and others like it, could be used already in a first semester calculus class to motivate and explain the need for the distinction between conditionally and absolutely convergent series. Such examples clearly demonstrate the need for a precise definition for "sum" and the need to distinguish series for which rearrangement makes no difference and those for which it does.

Final Thoughts on Series

What I have tried to present is a way to lead a student gradually to understanding what infinite series are, based on the developmental model. The historical development of the concepts and the difficulties that forced revision and redefinition show why we have to be precise in how we define sums and how we work with them. The aim is to promote understanding while at the same time defusing the "Why are we doing this?" question. The presentation of answers without the questions is avoided, the solutions to Level 4 crevasses developed rather than presented full blown.

It would perhaps be wise to keep in mind that in the process of adapting, things that in one age are excluded from respectable mathematics may return and be reconciled. Though the accepted usage now relegates every series that does not converge (in the sense of limits of partial sums) to the category of "divergent series" and then ignores it, there is a rigorous theoretical base for summability of divergent series. Following Euler's example, the definition of sum can be made so that all previously convergent series still converge and to the same sum, but the extended definition can now apply to some divergent series. Rigorous definitions can exclude the ambiguity of the "natural" forms, and $\frac{1}{2}$ is the acceptable "sum" for the alternating series. As J. E. Littlewood wrote in the preface to Hardy's book *Divergent Series*:

> The title holds curious echoes of the past, and of Hardy's past. Abel wrote in 1828: 'Divergent series are the invention of the devil, and it is shameful to base on them any demonstration whatsoever.' In the ensuing period of critical revision they were simply rejected. Then came a time when it was found that something after all could be done about them. This is now a matter of course, but in the early years of the century the subject, while in no way mystical or unrigorous, was regarded as sensational, and about the present title, now colourless, there hung an aroma of paradox and audacity. [8, p. i]

Perhaps it would also be well to interject a cautionary note. It is frequently pointed out that in reading primary sources we should try to avoid judging them by "today's standard" or with the hindsight of knowledge of intervening development. But the student is coming to the material fresh, without knowing what "today's standard" is nor what the intervening developments were. Thus, if topics are developed historically the students won't—they can't—have this biased hindsight.

A major value in stressing this search for cognitive shifts through our history is that we become aware of the context of the shift and the ways of thought. We see that what was appropriate before the shift may not be appropriate after it. Minor shifts may involve slightly different methods, or slight variations in technique or ways of viewing. Major shifts involve not just totally new concepts but radically shifted bases and foundations of thought. Newton didn't just cause a change in mathematical notation; it was a change in the basis of what is a valid argument, of what is and is not "proof." The full Level 4 shift represents in its fruition more than just redefining a single term, it represents a change in our understanding of the whole concept of sum. As G. H. Hardy (1877–1947) said:

> It does not occur to a modern mathematician that a collection of mathematical symbols should have a "meaning" until one has been assigned to it by definition. It was not a triviality even to the greatest mathematicians of the eighteenth century. They had not the habit of definition: it was not natural to them to say, in so many words, "by X we mean Y." There are reservations to be made, ... but it is broadly true to say that mathematicians before Cauchy asked not "How shall we define $1 - 1 + 1 - \cdots$" but "What is $1 - 1 + 1 - \cdots$," and this habit of mind led them into unnecessary perplexities and controversies which were often really verbal.
>
> It is easy now to pick out one cause which aggravated this tendency, and made it harder for the older analysts to take the modern, more "conventional," view. It generally seems that there is only one sum which it is "reasonable" to assign to a divergent series: thus all "natural" calculations with the series $[1 - 1 + 1 - \cdots]$ seem to point to the conclusion that its sum should be taken to be $\frac{1}{2}$. We can devise arguments leading to a different value, but it always seems as if, when we use them, we are somehow "not playing the game." [8, p. 5]

This is the kind of adjusting our students are also trying to make, and that we are trying to help them make.

For Further Consideration

What I have presented is a teaching strategy for the concept of "sum of a series," based on my model of historical development. The time spent on each transition should increase with the level, so that a transition from level 3 to level 4 should be allowed more time than from level 1 to level 2. Other concepts that might be successfully developed according to the model in the same fashion are given below. These are intended as suggestions for possible development, not as a hard and fast formulation of how the development should be done. Any concept has many paths leading to its current form, of which one is indicated here.

1. Area:
   Early concepts in Egypt and Babylon
   Greek quadrature
   Problem of arcs as boundary—Antiphon and Archimedes
   Infinitesimals for quadrature—special techniques
   Calculus and integration—Riemann integration, Lebesgue integration

2. The concept of "proof":
   Single example
   Verbose argument
   Assumed "obviousness"
   Appeal to authority
   Epsilon-delta
   Formal proofs—Russell *et al.* [Paradoxes, Theory of classes (blind alley, back up),
       Constructivists]
   Gödel—we are still trying to cope

3. Convergence:
   Sequences
   Series of terms
   Sequences of functions
   Sums of functions
   Uniform convergence

4. Number:
   One-to-one correspondence
   Counting numbers
   Rationals
   Irrationals (definitely Level 4)
   Imaginary and complex
   Infinitesimals—Wallis, Newton *et al.*
   Negative numbers (another and surprisingly recent Level 4)
   Infinitesimals again, reborn and perhaps legitimate

Bibliography

1. Barbeau, E. J. and Leah, P. J.: "Euler's 1760 Paper on Divergent Series," *Historia Mathematica* 3 (1976), 141–160. (Translation of Euler's paper)
2. Beckmann, Petr: *A History of Pi*. New York: St. Martin's Press, 1971.
3. Boyer, Carl B.: *The History of the Calculus and its Conceptual Development*. New York: Dover, 1959.
4. Bromwich, T. J. I.: *An Introduction to the Theory of Infinite Series*, 2nd edition. London: MacMillan, 1942.
5. Buck, R. Creighton and Ellen F.: *Advanced Calculus*, 2nd edition. New York: McGraw-Hill, 1965.
6. Chace, Arnold Buffum: *The Rhind Mathematical Papyrus*. Reston, VA: The National Council of Teachers of Mathematics, 1979.
7. Edwards, C. H., Jr.: *The Historical Development of Calculus*. New York: Springer-Verlag, 1979.
8. Hardy, G. H.: *Divergent Series*. Oxford: Clarendon Press, 1949.
9. Heath, Sir Thomas: *A History of Greek Mathematics*. New York: Dover, 1981.
10. Kline, Morris: *Mathematical Thought from Ancient to Modern Times*. New York: Oxford University Press, 1972.
11. Knopp, Konrad: *Theory and Application of Infinite Series*. London: Blackie and Son, 1954 (Reprint of 1928 edition).
12. Mahoney, M. S.: *The Mathematical Career of Pierre de Fermat*. Princeton: Princeton University Press, 1973.
13. Nicomachus of Gerasa, *Nicomachus of Gerasa: Introduction to Arithmetic* (Martin Luther D'Ooge, trans.). New York: Macmillan, 1926. (Limited edition reprint of Part II from *Nicomachus of Gerasa*, vol. XVI of the *Humanistic Series* by the University of Michigan Press, for the members of St. John's College.)
14. Reiff, R.: *Geschichte der unendlichen Reihen*. Wiesbaden, 1969 (reissue of 1889 edition).
15. Smith, David Eugene: *A Source Book in Mathematics*. New York: Dover, 1959.
16. Whiteside, D. T., ed.: *The Mathematical Papers of Isaac Newton*, vol. III. Cambridge: Cambridge University Press, 1969.

Historical Thoughts on Infinite Numbers

Lars Mejlbo

Introduction

For some years now I have given a one-semester course on the concept of infinity in mathematics, aimed at pre-service teachers in the middle of their education. This course has several aims: to introduce the mathematical concept of infinity itself, to show how the development of a mathematical topic depends in part on social factors as well as on its more "internal" history, to present an introduction to the philosophy of mathematics, and, not least, to give the students an idea of how to talk about mathematics.

Here I discuss only a portion of the course material, namely some different approaches to operations with infinitely large numbers. Our students generally come with the common sense view that there is only one infinity, and so the question of manipulating it cannot arise. To be presented with the idea that there are different sizes of infinite number, and with no less than three different approaches to operating on them, is thus a fascinating, if somewhat mind-boggling, experience for them.

The Infinite as Operationally Similar to the Finite: Leonhard Euler

The first view of infinity I discuss is about the contrast, and the relation, between infinitely large and infinitely small quantities. This has a long history, but it was probably G. W. Leibniz (1646–1712) who had the first reasonably coherent view, in the context of the development of the calculus [2]. But even Leibniz was fairly relaxed about how such quantities should be interpreted, as can be seen from his letter to the French mathematician Pierre Varignon:

> And to this effect I have given once some lemmas on incomparables in the Leipzig *Acta*, which one may understand as one wishes, either as rigorous infinities, or as quantities only, of which the one does not count with respect to the other. [2, p. 56]

Leibniz remarked in the same letter, that you can calculate with infinite numbers in just the same way as with finite numbers, because the rules that are satisfied for the finite are satisfied for the infinite and vice versa. But it was really the later Swiss mathematician Leonhard Euler (1707–1783) who used to greatest effect the comparability of rules for manipulating finite and infinite quantities.

Euler's views on the infinitely small and infinitely large are explained in chapter 3 of his 1755 monograph on differential calculus, *Institutiones calculi differentialis* ([5] or [8]): he states that the infinitely small has no physical existence, and whether there is an infinitely large magnitude in nature depends on whether God created the Universe infinitely large or not. But in theoretical mathematics, both infinitely large and infinitely small quantities are useful and necessary. Admittedly, there are difficulties in Euler's position, or so it seemed to later mathematicians. He identifies infinitely small quantities with zero when he chooses, but in practice, often considers infinitely small quantities as reciprocals of infinitely large quantities.

It is interesting that Euler's views were formed in a context of religious and philosophical disputation. As a sincere Christian with a simple faith in divine revelation, Euler was passionately opposed to the more rationalistic religion promoted by such German thinkers as Leibniz and Christian Wolff (1679–1754). His remarks about the infinite were probably polemically aimed at the influential philosophical school of Wolffians.

Euler's views on the infinite developed and ripened during his lifetime, but there is no fundamental disagreement between his early and later opinions. The main characteristic for us is the great operational fluidity which the concept acquired in his hands, with a profound intuition that almost defies logic. A good example is his handling of infinite series: throughout his life he treated the infinite series

$$u_1 + u_2 + u_3 + \cdots \tag{1}$$

as though it were the infinite sum

$$u_1 + u_2 + u_3 + \cdots + u_{i-1} + u_i \tag{2}$$

where i is an infinitely large natural number. This is a concept not now used in mathematics, but in Euler's hands, it proved remarkably versatile.

In a paper written in 1734 [6], where the subject is the harmonic series

$$\frac{c}{a} + \frac{c}{a+b} + \frac{c}{a+2b} + \cdots \quad (a,\ b,\ c > 0), \tag{3}$$

Euler stated without proof a result which looks like the so-called Cauchy condition. In modern language, he said that the infinite series (1) has a finite sum if and only if for all finite natural numbers n and all infinite natural numbers i, the sum

$$u_{i+1} + u_{i+2} + u_{i+3} + \cdots + u_{ni} \tag{4}$$

is infinitely small.

We may see from two examples how the condition works. First we show that the harmonic series (3) is infinite. Because each of the $ni - i = (n-1)i$ terms in the "tail" of (4) is greater than the last term, we have

$$\frac{c}{a+ib} + \cdots + \frac{c}{a+(ni-1)b} > \frac{(n-1)ic}{a+(ni-1)b}.$$

Since i is infinite, we may discard the a in the denominator. Therefore,

$$\frac{c}{a+ib} + \cdots + \frac{c}{a+(ni-1)b} > \frac{(n-1)ic}{(ni-1)b} > \frac{(n-1)ic}{nib} = \frac{(n-1)c}{nb},$$

which is not infinitely small.

On the other hand, the series

$$\frac{c}{a} + \frac{c}{a+b} + \frac{c}{a+2^\alpha b} + \cdots + \frac{c}{a+k^\alpha b} + \cdots \tag{5}$$

is finite for $\alpha > 1$, since

$$\frac{c}{a + i^\alpha b} + \cdots + \frac{c}{a + (ni-1)^\alpha b} < \frac{(n-1)ic}{a + i^\alpha b} = \frac{(n-1)ic}{i^\alpha b} = \frac{(n-1)c}{i^{\alpha-1}b},$$

and the final quantity is infinitely small.

As Pringsheim pointed out in [11], Euler's condition is only necessary, not sufficient. A counterexample is the series

$$\frac{1}{2\ln 2} + \frac{1}{3\ln 3} + \cdots$$

for which the condition holds, but the series does not, in fact, have a finite sum. It is important in Euler's condition that n should be finite—this is used, for example, in the proof of the inequality (6). If n can take infinite values, then Euler's condition is equivalent to the Cauchy condition.

These examples are just some of those which show how adeptly Euler used ordinary finite conceptions in place of infinite numbers, and calculated as usual.

Toward the end of his life, in 1778, Euler published a paper [7] where he concerned himself with the order structure of the infinitely large and infinitely small. I will give just some of his results on the domain of the infinitely large, without going into detail of his proofs (for which, see the discussion in the Appendix to [2]). His results for the infinitely small can be derived immediately from the infinitely large case by division, using the normal rules for inequalities.

Using modern symbolism, we can put Euler's problem this way. He was investigating the order relation $\ll$ defined by

$$x \ll y \quad \text{iff} \quad \frac{y}{x} \quad \text{is infinite.}$$

Various results follow quite speedily. If x is infinitely large, then

$$x \ll x^2 \ll x^3 \ll \cdots \tag{7}$$

and

$$\cdots \ll \sqrt[3]{x} \ll \sqrt{x} \ll x. \tag{8}$$

But (7) is bounded above, since $x^n \ll a^x$ for n finite and $a > 1$; and (8) is bounded from below by an infinite number since $\ln x \ll \sqrt[n]{x}$ for n finite. Also, for all finite n and $0 < \epsilon < 1$, we have

$$x^n \ll x^{n+\epsilon} \ll x^{n+1}.$$

So we can be fairly confident that Euler knew the order structure of the infinite very well; he knew that the order is dense with no first or last element, and that there is no unbounded sequence of infinite numbers in this order structure. But of course he did not say this in modern language—after all, the theory of abstract ordered sets was only created about a century later.

Accepting the Actually Infinite: Bernard Bolzano

Let us jump ahead some years now, to the Czech mathematician and philosopher Bernard Bolzano (1788–1848), one of the first in modern times who wholeheartedly accepted actually infinite sets and magnitudes, set apart from God. His work *Paradoxien des Unendlichen (Paradoxes of the infinite)*, which he wrote in his last years, was published in 1851 after his death. To the age-old conundrum of whether a finite human being could imagine a collection of infinitely many things, he replied that he could talk about the collection of the inhabitants of Prague, or Peking, without imagining each of them, and was able to make true statements about it—for example, that there are between 100,000 and 120,000 inhabitants of Prague. [1, §14]

Bolanzo accepted that infinite sets harbor some paradoxes. He was probably the first to emphasize that an infinite set can always be mapped one-to-one onto a proper subset, but this does not mean—in his opinion—that an infinite set and a proper subset can have the same number of elements. He explicitly rejected that possibility, in fact [1, §21 and §41–42], and accepted the applicability to this case of the old axiom the whole is greater than the part. Thus he took the view that a line of length 12 has more points than a line of length 5, despite the map $5y = 12x$ establishing a one-to-one correspondence between the two sets [1, §20].

To explain this paradox, Bolzano points out that while 3 and 4 (say), and their images $7\frac{1}{5}$ and $9\frac{3}{5}$ have the same geometric relation, their arithmetic relation is quite different, and so this pair of points does not play the same role. [1, §23]

The following points are fundamental in Bolzano's calculations with infinite numbers.

4. If the distance between the points a and b equals that between α and β, then the set of points between a and b must be put equal to the set of points between α and β.

5. Extended figures having equal sets of points are also equal in magnitude. The converse, that extensions equal in magnitude have equal sets of points, is not true.

6. When a pair of spatial objects are completely similar [in the geometrical sense], their two sets of points must have the same ratio as their two magnitudes. [1, §41]

From these and certain other, not too explicit, principles he derives the following argument.

1. The first point to which we would direct the reader's attention is this: the set of points in an arbitrarily short straight line az must be deemed infinitely greater than the infinite set of points obtained by taking out of it first the terminal point a, then b at a suitable distance towards z, then c at a shorter distance from b, and so forth in such wise that the sum of the infinitely many distances ab, bc, cd, ... comes to be less than or equal to az. For each of the infinitely many pieces ab, bc, cd, ... into which az is partitioned is itself a finite straight line, and can be handled just as az has been handled, that is to say: in each of these pieces another such infinite set of points can be designated, all of which belong simultaneously to az itself. Such an infinite set of points must consequently be contained in the whole line az infinitely often.

3. Let E denote the set of points lying between a and b, both included, and let the straight line ac have the integer n for its length measured by ab as unit. The set of points in the straight line ac, both terminals included, will then be equal to $nE - (n-1)$.

4. The set of points in the surface of a square of side 1 (the usual unit of area) including the periphery will be equal to E^2.

5. The set of points in the surface of a rectangle with one side m and the other side n and including the periphery will be $mnE^2 - [n(m-1) + m(n-1)]E + (m-1)(n-1)$.

6. The set of points in the volume of a cube of side 1 (the usual unit of volume) including the periphery will be E^3.

In a sense these calculations are quite normal and not really surprising; but there are difficulties, and Bolzano did not seek to avoid them.

9. Every particular point on such a line is situated in an exactly similar manner with regard to the parts of the line on either side of itself, and its situation offers no other conceptually expressible features than does the situation of any other point on the line. Yet for all that, we are not entitled to say that the point partitions the line into two portions equally long; for if we could say this of any point a, we could say it of any other point b for the same reason, and so we come to the contradiction that aR and aS were equal, but bR and bS unequal—the latter following from bR equal to aR plus ab, together with bS equal to aS minus ab. We ought rather to make the assertion that a bilaterally unterminated straight line simply has no midpoint at all—no point, that is, determinable solely by its conceptually expressible relation to this line. [1, §49]

Bolzano did not develop his calculations with infinite numbers further than this; I do not know whether he considered these paradoxes an insurmountable obstacle, or whether he just felt himself temporarily stumped.

I think that Bolzano's struggles form an important case study to be explored with students in the teaching of mathematics today, because they show a brilliant man caught up in difficulties he cannot easily resolve. Students can see that mathematics is not always a story of continual success and smooth development. There are of course a number of such stumblings, in the prehistory and history of set theory in particular; the book by Hawkins [9] tells an interesting story about this.

Of course, Bolzano's views cannot be considered "mistakes" in a straightforward way. It is perhaps more that they have not, in this area, been directly fruitful in subsequent mathematics. His *Paradoxes of the infinite* was, however, known to Cantor and Dedekind, at least from 1880, and some parts inspired Dedekind. Cantor, too, was pleased to find support from Bolzano's audacious belief that actual completed infinities could be contemplated in mathematics, though he was critical of Bolzano's treatment when analyzed in detail [4, p. 124]. It is to Cantor's work that we now turn.

The Arithmetic of Infinite Numbers: Georg Cantor

Georg Cantor (1845–1918) is often regarded as the founder of set theory, but as with other widespread popular beliefs this view has to be modified somewhat in the light of modern historical research. (See [9].) But here we are concerned with Cantor's work in founding a modern theory of infinite numbers, which can be safely attributed to him without such reservations.

The natural numbers $1, 2, 3, 4, 5, \ldots$ have two prominent aspects; as cardinal numbers they record the number of elements in a finite set, as ordinal numbers they give a number label to each element in an ordered finite set. The difference in everyday usage is quite subtle, since one often counts—i.e., orders and labels—a small finite set in order to find the number of elements in it. Experience shows that the number of elements is independent of the ordering, and we continue to believe that even for large sets, where we do not count but calculate the number of elements. So for a finite set, that is, where the natural numbers are involved, one can choose to emphasize the cardinal or the ordinal aspect, and the result is much the same. This is not so for infinite numbers.

At the start, Cantor used the cardinal aspect to consider infinite numbers. In contrast to Bolzano, he defined two sets A and B to have the same number of elements if and only if there is a bijection of A on B. (So in the case of Bolzano's example Cantor would judge that lines of length 12 and 5 do have the same number of points, because of the bijection $5y = 12x$.) In the case of finite sets this is the usual cardinality mapping: you know that a set of cups and a set of saucers have the same cardinal number if and only if you can put each cup in a saucer with no cups or saucers left over.

Cantor did not use the word number but power (*mächtigkeit*). He showed in the 1870s that most infinite sets we meet in real or complex analysis have either the power of the natural numbers or the power of the real numbers (the "continuum"), and that these two powers are different—still a very striking result.

In the 1880s he developed further ideas which we meet in higher analysis, where we use topological notions. These concern well-ordered sets; Cantor was interested in counting these, so he used the ordinal aspect of both finite and infinite numbers. To get at ordinal numbers he used two principles of generation (*Erzeugungsprincipien*):

1. Each ordinal number α has an immediate successor $\alpha + 1$. That is, $\alpha < \alpha + 1$, and (this is the well-ordering) for all ordinals β, either $\beta \leq \alpha$ or $\alpha + 1 \leq \beta$.
2. A well-ordered set of ordinals without last element has an immediate successor, that is, an ordinal α such that all ordinals in the set are less than α, but no ordinal is between α and all ordinals in the set.

The ordinal numbers may readily be constructed, using these principles. The first principle gives the natural numbers

$$1, \; 2, \; 3, \; 4, \; \ldots, \; \nu, \; \nu + 1, \ldots$$

Then the second principle gives the first infinite ordinal number ω, say, and we can use the first principle again to produce its successors

$$\omega + 1, \; \omega + 2, \; \omega + 3, \ldots, \omega + n.$$

We can continue this process, to produce ever larger ordinal numbers. It is a fascinating game to carry on as far as we can—but it is surprising, indeed shocking, to realize how difficult it is to produce more than a countable number of them!

I like to introduce the ordinal numbers in this way, because I think it is much more transparent and student-friendly than more formal approaches. But I like it also because it leads us to consider a paradox, called the Burali-Forti paradox after the Italians who announced it in 1897; is not the set of all ordinals itself a well-ordered set without a last element?—therefore by principle 2 it must have an immediate successor. It seems to follow that there is an ordinal larger than all ordinals. This is an interesting and accessible topic to raise with students. In fact, it now transpires that Cantor had anticipated this paradox ([4], [12], [13]), which helps to explain why he took this and the other paradoxes discovered about this time in his stride.

In summary, the path I have been sketching here is a useful way of sharing with pre-service teachers some exciting and thought-provoking ideas. We also discuss, of course, other ways of extending numbers such as from the natural numbers to the real numbers; but somehow the historical extensions from the finite to the infinite provide a richer source of surprising and counter-intuitive results, which the students enjoy and profit from thinking about.

Bibliography

1. Bolzano, Bernard: *Paradoxien des Unendlichen*, (F. Prihonsky, ed.). Berlin: Mayer and Müller, 1889. English version: *Paradoxes of the infinite*, (D. A. Steele, trans.). New Haven: Yale University Press, 1950.

2. Bos, Henk: "Differentials, higher-order differentials and the derivative in Leibnizian calculus," *Archive for history of exact sciences* 14 (1974), 1–90.

3. Cantor, Georg: *Gesammelte Abhandlungen mathematischen und philosophischen Inhalts*, (E. Zermelo, ed.). Berlin: Springer, 1932 (reprinted Hildesheim: Olms, 1966).

4. Dauben, Joseph Warren: *Georg Cantor: His Mathematics and Philosophy of the Infinite*. Cambridge: Harvard University Press, 1979.

5. Euler, Leonhard: *Institutiones calculi differentialis. Opera Omnia* (1) 10. Berlin: Teubner, 1913.

6. Euler, Leonhard: "De progressionibus harmonicis observationes," *Opera Omnia* (1) 14, Berlin: Teubner, 1925, 87–100

7. Euler, Leonhard: "De infinities infinities gradibus tam infinite magnorum quam infinite parvorum," *Opera Omnia* (1) 15, Berlin: Teubner, 1927, 298–313.

8. Euler, Leonhard: *Vollständige Anleitung zur Differenzial-Rechnung, Erster Theil*. Berlin, 1798 (repr. Wiesbaden, 1981) (German translation of 5).

9. Hawkins, Thomas: *Lebesgue's Theory of Integration: Its Origins and Development*. Madison: University of Wisconsin Press, 1970.

10. Leibniz, Gottfried Wilhelm: *Mathematische Schriften*, Vol. 4. Leipzig, 1859 (reprinted Hildeshiem: Olms, 1971).

11. Pringsheim, A.: "Euler und das 'Cauchysche' Konvergenskriterium." *Bibliotheca mathematica* (3) 6 (1905), 252–256.

12. Purkert, W.: "Georg Cantor und die Antinomien der Mengelehre," *Bulletin Societé Mathematique Belgique* (A) 38 (1986), 313–327.

13. Purkert, W. and Ilgauds, H. J.: *Georg Cantor: 1845–1918*. Basel: Birkhäuser, 1987.

Arthur Cayley

Historical Ideas in Teaching Linear Algebra

Victor J. Katz

As a subject in the undergraduate curriculum at American universities, linear algebra is relatively new. Though many of the individual topics now included under the name linear algebra have roots stretching back many centuries, their organization into a program of study is only a few decades old. Perhaps because of this, the typical structuring of a linear algebra course follows the historical order of development more closely than that of many other courses studied by undergraduates. For example, the first linear algebra topic usually encountered is that of the solution of systems of linear equations, a subject which to some extent was studied by the Babylonians nearly 4000 years ago. The next topics may well be determinants, which date from roughly 300 years ago, and the elements of vector geometry in 2-space and 3-space, a concern of the early nineteenth century. The more abstract notions of vector spaces and linear transformations, built upon concrete foundations, were not fully developed in the mathematical literature until the late nineteenth and early twentieth centuries and are typically studied toward the end of a linear algebra program.

It seems, then, that in these studies, unlike in some other undergraduate studies, there is no pedagogical reason to reorganize the course along more historical lines. Nevertheless, there are several topics included in a typical linear algebra course which can be enriched by associating them with their history. I will discuss three major parts of linear algebra in which historical ideas can be applied: the solution of systems of linear equations, basic ideas about matrices, and the concept of vectors along with the more abstract notion of a vector space. In each area I will outline certain ideas which one can present in class based on the appropriate historical sources.

Systems of Linear Equations

We begin by considering the ancient methods of solving systems of linear equations. Chinese scholars over two millennia ago developed a method for solving systems of two equations in two unknowns and a different one for solving larger systems. The method for solving systems of two equations, which starts with the "guessing" of possible solutions and then concludes by adjusting the guess to get the correct solution, shows clearly that the Chinese understood the basic idea of linearity. As such, its use today can also help our students with that understanding.

The Chinese method occurs in Chapter 7 of the *Jiuzhang suanshu* (*Nine Chapters in the Mathematical Art*) [20] which dates from the early Han period (c. 200 B.C.). The basic idea is as follows: Suppose one is given an affine function of one variable, say $f(x) = rx + s$, and wants to solve the equation $f(x) = b$. It may even be that b is not known. The Chinese tried two values for the unknown, say x_1 and x_2, and determined what they called the surplus b_1 and the

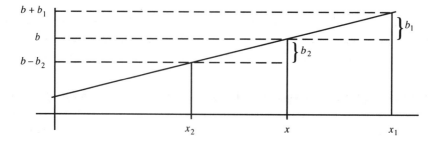

deficiency b_2. That is, b_1 and b_2 were determined so that $f(x_1) = b + b_1$ and $f(x_2) = b - b_2$. To find the desired value for x, then, one uses linearity:

$$\frac{b_1}{x_1 - x} = \frac{b_2}{x - x_2}.$$

An instructor can easily explain this proportion to a class by use of a straight line diagram (Figure 1). It then follows that $b_1 x - b_1 x_2 = b_2 x_1 - b_2 x$ or that

$$x = \frac{b_1 x_2 + b_2 x_1}{b_1 + b_2}.$$

The rule described by this formula is used by the Chinese author as he goes through some 20 examples of this procedure in various modifications. Sometimes he uses two surpluses or two deficiencies. It is doubtful that the Chinese discovered the rule by using the diagram of Figure 1, but they obviously understood the consequences of linearity. Here are two examples of the rule taken from the *Jiuzhang*:

> One pint of good wine costs 50 gold pieces, while one pint of poor wine costs 10. Two pints of wine are bought for 30 gold pieces. How much of each kind of wine was bought? [20, p. 76]

This problem can be set up as a system of two equations in two unknowns:

$$x + y = 2$$

$$50x + 10y = 30$$

It is the left side of the second equation, with $2 - x$ substituted for y, which is $f(x)$. That is, $f(x) = 40x + 20$. The two trial values are $x = \frac{1}{2}$ for which $f(\frac{1}{2}) = 30 + 10$, and $x = \frac{1}{5}$ for which $f(\frac{1}{5}) = 30 - 2$. Hence $b_1 = 10$, $b_2 = 2$, and by the formula above, $x = \frac{1}{4}$ is the number of pints of good wine purchased. The value for the number of pints of poor wine, $y = 1\frac{3}{4}$, is then easily calculated.

A second example shows that it is not necessary to know b:

> Several people buy some chickens in common. If each paid 9 coins, the surplus is 11; if each paid 6, the deficiency is 16. How many people were there and what is the price of the chickens? [20, p. 70]

In this case, let x represent the number of coins paid by each person. Then $f(x) = rx$ represents the total price paid, where r represents the number of people. The two trial values are $x_1 = 9$, for which $b_1 = 11$, and $x_2 = 6$, for which $b_2 = 16$. Again, the formula above gives $x = \frac{210}{27} = \frac{70}{9}$ as the price paid by each. The smallest integral solution to the question, that there were 9 people and the price was 70 coins, is easily checked to be correct. (A similar method for solving equations of this type, known as the Rule of False Position, was used in medieval Europe.)

The Chinese also developed the essence of what is today known as the method of Gaussian elimination to solve larger systems. Though as usual they did not provide any reasons for their method, the important idea to convey from this method is that given a system of linear equations one seeks an equivalent system in which one of the equations only contains one unknown, a second only two, etc. Though it is extremely unlikely that Gauss was influenced by the Chinese in his development of Gaussian elimination, it is important to tell a class that, in fact, the Chinese had discovered the method close to two millennia before the "Prince of Mathematicians."

One can also present the first Chinese example, from chapter 8 of the *Nine Chapters*:

> There are three classes of corn, of which three bundles of the first class, two of the second, and one of the third make 39 measures. Two of the first, three of the second, and one of the third make 34 measures. And one of the first, two of the second, and three of the third make 26 measures. How many measures of grain are contained in one bundle of each class? [20, p. 80]

This problem can be translated into a system of 3 equations in 3 unknowns:

$$3x + 2y + z = 39$$
$$2x + 3y + z = 34$$
$$x + 2y + 3z = 26$$

Using counting rods, the Chinese physically set up this problem on their counting board in the configuration of a matrix according to the directions:

> Arrange the 3, 2 and 1 bundles of the three classes and the 39 measures of their grains at the right. Arrange other conditions at the middle and at the left. [20, p. 80]

$$
\begin{array}{ccc}
1 & 2 & 3 \\
2 & 3 & 2 \\
3 & 1 & 1 \\
26 & 34 & 39
\end{array}
$$

The author then instructs the reader to multiply the middle column by 3 and subtract off twice the right column, then to multiply the left column by 3 and subtract off the right column. This reduces the matrix to:

$$
\begin{array}{ccc}
0 & 0 & 3 \\
4 & 5 & 2 \\
8 & 1 & 1 \\
39 & 24 & 39
\end{array}
$$

The next step is to multiply the left column by 5 and then subtract off 4 times the middle column, leaving a "column-reduced" matrix:

$$\begin{array}{ccc} 0 & 0 & 3 \\ 0 & 5 & 2 \\ 36 & 1 & 1 \\ 99 & 24 & 39 \end{array}$$

This is equivalent to the triangular system

$$3x + 2y + z = 39$$

$$5y + 2z = 24$$

$$36z = 99$$

from which one easily finds $z = \frac{99}{36} = 2\frac{3}{4}$ and then y and x by back-substitution.

Linear Systems in Europe

It was not until the necessity arose that European mathematicians began the search for new methods of solving linear systems. One such mathematical necessity came about from a question in the theory of curves. This led Gabriel Cramer (1704–1752) to publish what is now called Cramer's rule in his *Introduction à l'Analyse des Lignes Courbes Algébriques* (1750) [4]. Cramer had discovered that the equation of a plane curve of degree n is determined if one knows the coordinates of $\frac{1}{2}n^2 + \frac{3}{2}n$ points; for example, a second degree curve can be determined if one knows five points, a third degree curve if one knows nine. The question was then to determine the specific equation of the curve. Cramer demonstrated a method for degree two. To determine the curve represented by $A + By + Cx + Dy^2 + Exy + x^2 = 0$ which passes through the points (α, a), (β, b), (γ, c), (δ, d), and (ϵ, e), Cramer simply replaced the x and y by these five values in turn to get five linear equations for the five unknowns A, B, C, D, E. To find the solution, he referred to an appendix as he wrote,

> I believe I have found a rule very convenient and general to solve any number of equations and unknowns which are no more than first degree. [4, p. 60]

In the appendix, Cramer writes out explicitly the result for the solution of systems of one, two, and three equations respectively. For example, the system which he writes as

$$A = Z_1 z + Y_1 y + X_1 x$$

$$A = Z_2 z + Y_2 y + X_2 x$$

$$A = Z_3 z + Y_3 y + X_3 x$$

has the solution for z given by

$$z = \frac{A_1 Y_2 X_3 - A_1 Y_3 X_2 - A_2 Y_1 X_3 + A_2 Y_3 X_1 + A_3 Y_1 X_2 - A_3 Y_2 X_1}{Z_1 Y_2 X_3 - Z_1 Y_3 X_2 - Z_2 Y_1 X_3 + Z_2 Y_3 X_1 + Z_3 Y_1 X_2 - Z_3 Y_2 X_1} \qquad \text{[4, p. 657]}$$

with similar results given for y and x. Cramer does not explain how the result was achieved. What he does do is give a general rule. The value of each unknown is a fraction of which the common denominator contains $n!$ terms where n is the number of equations. Each term is composed of the coefficient letters, for example ZYX, always written in the same order, but the indexes are permuted in all possible ways. The sign is determined by the rule that if in any given permutation the number of times a larger number precedes a smaller number is even, then the sign is " + "; otherwise it is "− ". The numerator for each unknown is found in basically the same way except that every occurrence of the coefficient corresponding to the given variable is replaced by the constant term with the same index. Cramer further notes that a solution will not be found if the denominator turns out to be 0. [4, p. 659] In that case, if the numerators are also 0, then the problem is indeterminate, while if any of the numerators are not 0, then the problem is impossible.

Assuming that students are familiar with the basic elimination rules for solving systems of linear equations, they can be asked to derive Cramer's rule in the order two case as well as, perhaps, also in the order three case. This exercise will provide a good introduction to determinants; in fact, determinants first occurred historically in connection with the solution of systems of equations. In working out Cramer's rule using elimination the students will be following in the footsteps of Colin Maclaurin (1698–1746) who apparently discovered this same rule as early as 1729. Maclaurin's version was not published, however, until after his death; it occurs in his *A Treatise of Algebra* [15] of 1748. In chapter 12 of part I, Maclaurin derived Cramer's rule for the two-variable case by going through the elimination procedure, then derived the three-variable version by reducing it to the earlier case. That is, given three equations in three unknowns x, y, z, he solved the first and second equations for y (treating z as a constant), then the first and third equations, then equated the two values and found z. Maclaurin's method, students may find, is somewhat easier than a direct calculation using elimination.

Historically, it took some time for the ideas of Cramer and Maclaurin to spread. Even Leonhard Euler (1707–1783), in his *Elements of Algebra* of 1770, gave no such rule for solving systems of linear equations. His only suggestion was to solve each equation for the same unknown and then compare the solutions pairwise. If this procedure was too tedious, one might also look to introduce new unknowns, for example the sum of the given ones, which would simplify the calculations. On the other hand, the translator of Euler's work into French near the end of the eighteenth century did refer to Cramer's work in a footnote to Euler's own discussion [6, p. 211].

Neither Cramer nor Maclaurin gave any examples of particularly complicated systems of linear equations. Thus, there did not appear to be any reason in the eighteenth century to find a better way to solve such equations. But by early in the nineteenth century the needs of astronomy required an improved procedure. It was Carl Friedrich Gauss (1777–1855) who presented a systematic method of elimination in connection with the method of least squares in a paper of 1811 [8] dealing with the determination of the orbit of the asteroid Pallas. To determine the various parameters of this elliptical orbit, such as the eccentricity of the ellipse and the inclination of the plane of the orbit to the plane of the earth's orbit, Gauss made a series of observations over a number of years. Each observation, together with astronomical theory, resulted in a linear equation involving the various parameters. In fact, he determined altogether twelve linear equations involving six unknowns. Unlike the examples of Cramer and Maclaurin, however, the coefficients were not integers. For example, one of the equations is

$$0.79363x + 143.66y + 0.39493z + 0.95929u - 0.18856v + 0.17387w = 183.93$$

In a system such as Gauss's, with twelve equations relating six unknowns, one does not expect to find an exact solution, particularly since the coefficients occurring in each equation are subject to observational error. What one seeks, therefore, is the best approximation to the solution. In modern terminology, if the system is written as $Ax = b$, where A is a 12×6 matrix, x an unknown 6-dimensional vector, and b a known 12-dimensional vector, the goal is to find the value for x which minimizes the length $|Ax - b|$ of the error vector. This is the basic idea behind the method of least squares. It turns out that this solution x is given by the exact solution to the square system $(A^t A)x = A^t b$, where A^t is the transpose of the matrix A. Though Gauss does not use matrix notation he, in effect, uses the same procedure as the Chinese to reduce this system to a triangular one, from which the solution could be found using back substitution.

It would be difficult to go through Gauss's exact procedure in a class, since his notation is somewhat obscure, but one can certainly convince the class that there is a necessity for having such a procedure. One should also mention that Gauss's procedure was improved somewhat by the German geodesist Wilhelm Jordan (1842–1899) late in the nineteenth century. (See [14] for more details.) Jordan in his work on surveying had to use the method of least squares. In geodesy, as in astronomy, there is some redundancy in the various measurements taken. However, there are always various conditions connecting the given measurements, and these lead, as above, to an overdetermined system of linear equations to which the least squares method applies. To solve the linear systems resulting from such problems, Jordan developed what amounted to a systematic method of back substitution, which ultimately gave solutions for the unknowns in terms of formulas involving the original coefficients. Jordan then created the algorithm now known as the Gauss-Jordan method to reduce the system to a diagonal one from which the solutions can be immediately read.

The Rise of Matrix Theory

Although Jordan wrote the various editions of the text in which his method appears in the last quarter of the nineteenth century, he does not use matrices in the sense that they are used today [14]. Around the same time period, however, several mathematicians developed matrix notation and matrix algebra to solve problems related to obtaining solutions of systems of linear equations. It was not solely from this problem and the related notion of determinants that the basic ideas of matrix algebra grew. Another primary theme in the history of linear algebra was the idea of a linear substitution.

In Gauss's *Disquisitiones arithmeticae* of 1801 [7] there is a detailed treatment of the arithmetic theory of quadratic forms, in particular of binary quadratic forms. These are functions of two variables of the form $Ax^2 + 2Bxy + Cy^2$ where A, B, C are integers. Gauss, generalizing an older problem on the representation of integers as sums of squares, determined which integers M can be represented by a given form through an appropriate choice of integers x, y. His approach was to note that equivalent forms, those for which there is a linear substitution with determinant equal to 1 which transforms one to the other, represent the same integers. Thus he had to deal with the notion of a linear substitution, a substitution of the form

$$x = ax' + by', \qquad y = cx' + dy'$$

with a, b, c, d integers. Such a substitution transforms the given form F to a new form F'. It is from this idea of a linear substitution that the notion of matrix multiplication comes. For if a second substitution is taken

$$x' = ex'' + fy'', \qquad y' = gx'' + hy''$$

which transforms the form F' into the form F'', the composition of the two substitutions gives a new substitution transforming F into F'':

$$x = (ae + bg)x'' + (af + bh)y'', \qquad y = (ce + dg)x'' + (cg + dh)y''.$$

The coefficient matrix of the new substitution is the product of the coefficient matrices of the two original substitutions.

In his study of ternary quadratic forms, Gauss performed an analogous composition which in effect gave the rule for multiplying two 3×3 matrices together. But Gauss himself did not explicitly refer to this idea as a multiplication. He occasionally used single letters, such as S, to refer to a particular substitution but he did not write expressions such as SS' to refer to this composition. Although the idea of matrix multiplication is certainly implicit in Gauss's work, the actual notation is not yet there. Other mathematicians, including Gotthold Eisenstein (1823–1852) and Charles Hermite (1822–1901), studied linear substitutions as objects. The former, in fact, used the notation $S \times T$ to represent the substitution composed of S and T. (See [12] for details.) Eisenstein realized that substitutions could be added and multiplied in some sense as if they were single objects, but his untimely death prevented him from pursuing his investigations further.

The first mention of the term "matrix" to denote a rectangular array of numbers was in an 1850 paper of James Joseph Sylvester (1814–1897). [19] The standard non-technical meaning of that term is "a place in which something is bred, produced or developed." For Sylvester, then, a matrix, which was an "oblong arrangement of terms," was an entity out of which one could form various square arrays to produce determinants. Sylvester himself, however, made no use of the term at this time. It was left to his friend Arthur Cayley (1821–1895) to put the terminology to use in a paper of 1855 [2] and then write it with the appropriate notation in a paper of 1858 [3]. In 1855 he noted that one could use matrices to represent square systems of numbers. As he said, "it seemed to me very convenient for the theory of linear equations" [2, p. 186]. Thus, employing brackets, he wrote

$$(\xi, \eta, \zeta, \ldots) = \begin{pmatrix} \alpha & \beta & \gamma & \cdots \\ \alpha' & \beta' & \gamma' & \cdots \\ \alpha'' & \beta'' & \gamma'' & \cdots \\ \cdot & \cdot & \cdot & \cdots \end{pmatrix} (x, y, z, \ldots)$$

to represent the system of equations

$$\xi = \alpha x + \beta y + \gamma z + \cdots$$
$$\eta = \alpha' x + \beta' y + \gamma' z + \cdots$$
$$\zeta = \alpha'' x + \beta'' y + \gamma'' z + \cdots$$
$$\cdot = \cdots\cdots\cdots\cdots$$

Note here that Cayley, in fact, knows how to multiply matrices. Although initially he multiplies a row by a row rather than a row by a column, later in the paper he writes down the standard rules for multiplying two square matrices. He also employs the concept of the inverse of a matrix. For he next writes that the solution of the system of equations should be written as

$$(x, y, z, \ldots) = \begin{pmatrix} \alpha & \beta & \gamma & \cdots \\ \alpha' & \beta' & \gamma' & \cdots \\ \alpha'' & \beta'' & \gamma'' & \cdots \\ \cdot & \cdot & \cdot & \cdots \end{pmatrix}^{-1} (\xi, \eta, \zeta, \ldots). \qquad [2, \text{p.187}]$$

Of course, this representation comes from the basic analogy of the matrix equation to a simple linear equation in one variable. Cayley also knew from Cramer's rule how to solve the system. He therefore describes the terms of this new inverse matrix as fractions having for a common denominator the determinant of the original matrix while the numerators are the appropriate minor determinants. [2, p. 187]

In 1858, Cayley introduced single letter notation for matrices and notes not only how to multiply them but also how to add and subtract. Most of the explicit calculations of the paper are done for 2×2 or 3×3 matrices, but Cayley indicates that his conclusions in general apply to matrices of any degree. Once students have been introduced to matrices, then, it is probably useful to go through some of the ideas of Cayley's paper. The important idea here, as in the idea of the inverse mentioned above, is the analogy between ordinary algebraic manipulations and those with matrices. Cayley showed that various algebraic laws are satisfied. One should ask the students to try to demonstrate these also.

It was Cayley's use of the single letter notation for matrices which probably suggested to him the result which is today known as the Cayley-Hamilton theorem, namely that

$$\det \begin{pmatrix} a - M & b \\ c & d - M \end{pmatrix} = 0,$$

where M represents the matrix

$$\begin{pmatrix} a & b \\ c & d \end{pmatrix}.$$

Cayley first communicated this "very remarkable" theorem in a letter to Sylvester on November 19, 1857. In it and in the 1858 paper Cayley simply calculated with M as if it were an ordinary algebraic quantity and showed that

$$(a - M)(d - M) - bc = (ad - bc)M^0 - (a + d)M^1 + M^2 = 0.$$

Again, it is easy enough for students to show this. One simply needs to square M and look at the sum of the matrices involved. Note that by M^0 Cayley means the matrix which he generally wrote as **1** and which is today written as

$$I = \begin{pmatrix} 1 & 0 \\ 0 & 1 \end{pmatrix}.$$

Cayley's motivation in stating this result was to show that "any matrix whatever satisfies an algebraical equation of its own order," and therefore that "every rational and integral function [of the matrix] ... can be expressed as a rational and integral function of an order at most equal to that of the matrix, less unity" [3, p. 483]. Cayley went on to show that one can adapt this result even for irrational functions. In particular, he shows how to calculate $L = \sqrt{M}$, where M is a given 2×2 matrix. Suppose

$$M = \begin{pmatrix} a & b \\ c & d \end{pmatrix} \qquad L = \begin{pmatrix} \alpha & \beta \\ \gamma & \delta \end{pmatrix}.$$

Since L satisfies $L^2 - (\alpha+\delta)L + (\alpha\delta - \beta\gamma) = 0$ and $L^2 = M$, a substitution shows immediately that

$$L = \frac{1}{\alpha+\delta}[M + (\alpha\delta - \beta\gamma)\mathbf{1}].$$

The task is now to express $X = \alpha + \delta$ and $Y = \alpha\delta - \beta\gamma$ in terms of the entries in the given matrix M. Cayley solves this by pointing out that the above expression for L implies that

$$L = \begin{pmatrix} \frac{a+Y}{X} & \frac{b}{X} \\ \frac{c}{X} & \frac{d+Y}{X} \end{pmatrix}.$$

By noting that the sum of the entries on the main diagonal in this matrix is again X, while the determinant is Y, one gets two equations for the unknowns X and Y involving only the entries of the original matrix.

The other man for whom the Cayley-Hamilton theorem is named, William Rowan Hamilton (1805–1865), also did not give a general proof of the theorem. In addition, his statement of the theorem was in terms of linear transformations, not in terms of matrices [10]. In particular, he wrote out the result in dimension three in terms of a linear function of vectors and in dimension four in terms of a linear function of quaternions. His major interest, however, was in deriving an expression for the inverse of a linear function. For example, if the equation $M^2 + aM + bI = 0$ is valid, then $bI = -M(aI + M)$ and M^{-1} can be easily found. One can then exploit this idea to deal with other negative powers of a given matrix.

Cauchy's Work

Students may well ask, "Why should one consider the characteristic equation $\det(M - \lambda I) = 0$ of the matrix in the first place? What meaning do its roots have?" There are several sources for the idea of a characteristic value, or eigenvalue, including works of Jean d'Alembert (1717–1783), Joseph Louis Lagrange (1736–1813), and Pierre Simon Laplace (1749–1827) on the solution of certain systems of differential equations as well as that of Euler on the analytic geometry of quadric surfaces. The general idea is that quantities expressible by square arrays are often in some sense equivalent to quantities whose only nonzero terms are the diagonal ones. The entries along the diagonal form what is called the "spectrum of the matrix." In fact, it was probably only because of spectral theory that matrix algebra became an important part of the mathematics curriculum. (See [11] for more details.) The mathematician who contributed the most to the early development of this topic was Cauchy. It is therefore necessary to examine his contribution, in particular his fundamental paper of 1829 [1] in which he proved that symmetric matrices have real eigenvalues.

Cauchy, of course, was not dealing with matrices as such. What he was dealing with was quadratic forms, or, as he put it, real functions homogeneous of the second degree. His aim was to find a linear substitution of variables so that in terms of the new variables the quadratic form would be expressed as a sum of squares. In terms of geometry, for example, this means

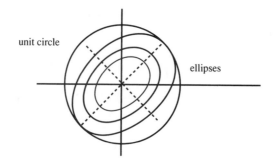

FIGURE 2

that one can find new orthogonal axes for a central conic section or quadric surface such that in terms of these axes the curve or surface is expressible solely as the sum (or difference) of square terms. The two-variable version of Cauchy's proof that this is always possible contains many of its important ideas and is simple enough to present in an elementary class.

Cauchy's idea was to find the maximum and minimum of the quadratic form $f(x, y) = ax^2 + 2bxy + cy^2$ subject to the condition that $x^2 + y^2 = 1$. If one considers the geometry of this question, one can understand that the point at which the extreme value occurs is that point on the unit circle which also lies on the end of one axis of one of the family of ellipses (or hyperbolas) described by the quadratic form. (See Figure 2.) If one then takes the line from the origin to that point as one of the axes and the perpendicular to that line as the other, the equation in relation to those axes will only contain the squares of the variables.

Cauchy was aware of the calculus procedures for finding extreme values subject to given conditions. He therefore simply stated the result, essentially the principle of Lagrange multipliers, that the extreme value occurs when the ratios $f_x/2x$ and $f_y/2y$ are equal. Setting each of these equal to λ gives the two equations

$$\frac{ax + by}{x} = \lambda \qquad \frac{bx + cy}{y} = \lambda,$$

which can be rewritten as the system

$$(a - \lambda)x + by = 0$$
$$bx + (c - \lambda)y = 0$$

[2, p. 175]

This system has nontrivial solutions only if its determinant equals 0; i.e., only if the equation $(a - \lambda)(c - \lambda) - b^2 = 0$. In matrix terminology, this equation is the characteristic equation $\det(A - \lambda I) = 0$, essentially the equation dealt with by Cayley.

It is easy to see in this quadratic case that the two roots of this equation are real and distinct except in trivial cases. Cauchy's proof of this fact modified somewhat from the general case will be sketched below. First, though, it is important to see how the roots of the characteristic equation enable the matrix to be diagonalized. Letting λ_1, λ_2 be the two roots and (x_1, y_1), (x_2, y_2) the corresponding solutions for x and y, we know from the original conditions that

$x_1^2 + y_1^2 = 1$ and $x_2^2 + y_2^2 = 1$. Furthermore, if in the equations

$$(a - \lambda_1)x_1 + by_1 = 0$$

$$(a - \lambda_2)x_2 + by_2 = 0$$

the first is multiplied by x_2, the second by x_1 and the second is then subtracted from the first, the result is the equation

$$(\lambda_2 - \lambda_1)x_1 x_2 + b(y_1 x_2 - x_1 y_2) = 0.$$

Similar operations involving the two equations with $c - \lambda_1$ give the equation

$$b(y_2 x_1 - y_1 x_2) + (\lambda_2 - \lambda_1)y_1 y_2 = 0.$$

Adding these two equations together gives $(\lambda_2 - \lambda_1)(x_1 x_2 + y_1 y_2) = 0$. Since $\lambda_1 \neq \lambda_2$, it follows that $x_1 x_2 + y_1 y_2 = 0$. In modern terminology, the two vectors (x_1, y_1), (x_2, y_2) are orthonormal. If one now makes the linear substitution

$$x = x_1 u + x_2 v \qquad y = y_1 u + y_2 v,$$

one easily calculates that the new quadratic form is $\lambda_1 u^2 + \lambda_2 v^2$ as desired. The original symmetric matrix has therefore been diagonalized by use of an orthogonal diagonalizing matrix.

Cauchy, in fact, gave a general proof that the eigenvalues are real by assuming the contrary. In the two variable case that would mean that the two eigenvalues would be complex conjugates of one another. By considering the equations satisfied by these two complex eigenvalues, however, one then finds that x_2 must be the complex conjugate of x_1 and y_2 that of y_1. But then $x_1 x_2 + y_1 y_2$ could not be 0.

The basic arguments of Cauchy's paper provided the beginnings to an extensive theory dealing with the properties of the eigenvalues of various types of matrices and with canonical forms. Much of this theory was developed in the second half of the nineteenth century by such mathematicians as Georg Frobenius (1849–1917), Camille Jordan (1838–1922), and Karl Weierstrass (1815–1897), but this subject is too complicated for a first course in linear algebra.

The Concept of a Vector

The concept of a vector in its earliest manifestation comes from physics. There is evidence of velocity being thought of as a vector, that is, as a quantity with magnitude and direction, in Greek times. For example, in the treatise entitled *Mechanica* attributed, though probably falsely, to Aristotle but dating probably from his time (fourth century B.C.), is written,

> When a body is moved in a certain ratio (i.e., has two linear movements in a constant ratio to one another), the body must move in a straight line, and this straight line is the diagonal of the figure (parallelogram) formed from the straight lines which have the given ratio. [13, v.1, p. 346]

Heron of Alexandria, in his *Mechanics* gives a proof of this idea [13, v.2, p. 348]. He assumes that the point at A moves uniformly over line AB while at the same time line AB moves uniformly along the lines AC and BD so that it always remains parallel to its original position. (See Figure 3.) Assuming further that the time it takes A to reach B is the same as the

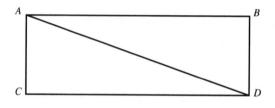

FIGURE **3**

time it takes AB to reach CD, Heron shows that in fact the point A moves along the diagonal AD.

This basic idea of two motions adding vectorially, that is, the so-called parallelogram of velocities, was generalized to physical forces in the sixteenth and seventeenth centuries. A readily accessible example of this practice is found in the beginning of Isaac Newton's *Principia* as Corollary I to the Laws of Motion:

> A body acted on by two forces simultaneously, will describe the diagonal of a parallelogram in the same time as it would describe the sides by those forces separately.
> [17, p. 14]

Newton proves this result by appealing to his Second Law: The change of motion is proportional to the motive force impressed; and is made in the direction of the right (straight) line in which that force is impressed. The proof then shows that since neither force alters the movement generated by the other, the body originally at A will end up both on the line BD and on the line CD. Therefore, its position will ultimately be at the intersection point D of these two lines, and it will have moved along the diagonal AD. (See Figure 4.) Newton continues by showing how any force can be resolved into components which, geometrically, form the sides of a parallelogram of which the original force is the diagonal.

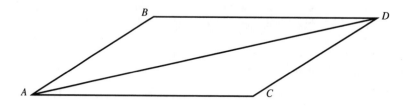

FIGURE **4**

Although he used the parallelogram of forces often in his work, Newton did not consider that he was "adding" vectors. He had no conception of any algebraic operations on these forces. Nevertheless, it would seem that the notion of combining forces by this method of diagonals is an excellent pedagogical tool for imparting the idea of a vector to students.

The merging of this physical idea with the mathematical idea of a system of coordinates is perhaps best explained by looking at the initial development of the geometrical representation of complex numbers as found in the 1797 work of the Norwegian surveyor, Caspar Wessel (1745–1818). Although his work was unfortunately not noticed by other European mathematicians, and therefore several of them independently discovered the same idea somewhat later, it is appropriate that Wessel's ideas form the basis of the discussion. Wessel's aim in his *On the Analytical Representation of Direction* [21] was to

> express right lines so that in a single equation involving one unknown line and others known, both the length and the direction of the unknown line may be expressed. [21, p. 55]

He made clear that these expressions had to be capable of being manipulated algebraically. In particular, he wanted to express algebraically a method of changing directions other than the simple method of using a negative sign to indicate a direction opposite to a given one. Wessel's analysis of his aims and his solution provide an excellent illustration for students of how one uses analogy to extend important concepts to new situations. His treatment is also excellent for use in class as an introduction to the study of vector spaces.

Wessel began by describing a technique of combining lines similar to that of Newton. The difference is that he thought of this technique as an algebraic operation analogous to addition and, in fact, called it by that name:

> Two right lines are added if we unite them in such a way that the second line begins where the first one ends and then pass a right line from the first to the last point of the united lines. This line is the sum of the united lines. [21, p. 58]

He next noted that one could add as many lines as one wanted in succession. In particular, for this "addition," "the order in which these lines are taken is immaterial;" that is, the commutative law holds.

Next, Wessel turned to multiplication. For this he set down a number of properties which he felt were essential. First, the product of two lines in the same plane with the positive unit was also in that plane. Secondly, the length of the product line had to be the product of the lengths of the two factor lines. Thirdly, if all directions are measured from the positive unit, which he called 1, the angle of direction of the product was to be the sum of the angles of direction of the two factors. Wessel then proceeded to designate the unit of length one perpendicular to the positive unit by ϵ and showed that his properties implied that $\epsilon^2 = (-\epsilon)^2 = -1$, or that $\epsilon = \sqrt{-1}$. The entire range of familiar properties of the complex numbers can then be derived. In terms of the discussion here, Wessel's most important contribution is the realization that any vector (in the plane of the unit), what Wessel called an "indirect line," can be represented as $a + b\epsilon$. In modern terminology, 1 and ϵ form a basis for the two (real) dimensional vector space of the complex numbers. For Wessel, these vectors can be added by adding the "coefficients," while they can be multiplied by the rule $(a + b\epsilon)(c + d\epsilon) = ac - bd + (ad + bc)\epsilon$. Wessel has thus described not only a vector space, but also an algebra. (For more information on Wessel, consult the paper by Otto Bekken elsewhere in this volume.)

In 1837, William Rowan Hamilton published an essay [9] in which he developed the theory of complex numbers anew by defining them as ordered pairs (a, b) of real numbers. He argued that, after all, the $+$ in the representation $a + b\sqrt{-1}$ does not have the same meaning as the $+$ in ordinary addition; $b\sqrt{-1}$ cannot be added to a. Thus it is well to describe these complex numbers without using that symbol. From another point of view, he simply replaced the representation of these numbers as "line segments" in the plane starting at the origin with that of the "coordinates" of the end point. Addition was then defined in the standard way, $(a, b) + (c, d) = (a + c, b + d)$, and multiplication by the rule $(a, b)(c, d) = (ac - bd, ad + bc)$.

Now there was no reason to worry about what $\sqrt{-1}$ meant. In this representation, that imaginary number was simply represented by the pair $(0,1)$, which was the square root of the pair $(-1,0)$. Given that all operations were defined in terms of those of real numbers, he noted that it was perfectly legitimate, in his theory of pairs, to write, if one desired, $a+b\sqrt{-1}$ in place of the pair (a,b). More importantly, however, it appears that Hamilton was the first to realize the importance of the basic laws of operation in his set of pairs, namely, the associative, commutative, and distributive laws. He clearly described them, though not by current names, in his discussion of the properties of the pairs.

Now that he had a reasonable set of operations for number pairs, Hamilton wanted to generalize it to three dimensions. After all, much of physics took place in three-dimensional space. As he said in a letter of 1841 to Augustus De Morgan,

> If my view of Algebra be just it must be possible, in some way or other, to introduce not only triplets but polyplets so as in some sense to satisfy the symbolical equation **a** $=(a_1, a_2, \ldots, a_n)$; **a** being here one symbol as indicative of one (complex) thought; and $a_1, a_2, \ldots, a_n$ denoting n real numbers positive or negative. [5, p. 27]

The struggle for Hamilton, of course, was not in the addition of his polyplets. That was easy. The struggle was in the multiplication. Knowing the basic laws for his pairs, he wanted his triplets also to satisfy the associative and commutative laws for multiplication as well as the distributive law. He also wanted division to be possible and lengths to multiply. Finally, he wanted his triplets to have a reasonable interpretation in three-dimensional space.

The story of Hamilton's discovery of the quaternions, that is, of a reasonable way of considering four-tuples, is well-known. (See [5] for details.) But the basic result as far as the teaching of linear algebra is concerned is that Hamilton gave an example of a four-dimensional vector space, which also turned out to be a four-dimensional algebra. Though Hamilton did not succeed in discovering a multiplication for triplets satisfying his criteria (because, of course, it was impossible), it turned out that the multiplication of quaternions led the mathematical physicists of the last half of the nineteenth century to develop the dot and cross products. In addition, it was Hamilton in one of his first papers on quaternions who gave the name vector to the trinomial expression $ix+jy+kz$ and noted that it represents a "right line having direction in space."

Though there are other early sources for the dot and cross products of vectors in three-space, particularly in the work of Hermann Grassmann (1809–1877), probably the source most helpful for their introduction in a first course in linear algebra is the work of Hamilton on quaternions as exemplified in James Clerk Maxwell's (1831–1879) *Treatise on Electricity and Magnetism*. [16] The basic idea is simple. Given two "vectors," $A = ai + bj + ck$, and $B = di + ej + fk$, multiply them together using the basic rules for quaternions. The result is $-(ad + be + cf) + (bf - ce)i + (cd - af)j + (ae - bd)k$. Hence the scalar part, written as $S.AB$, is the negative of the modern dot product, while the vector part, written as $V.AB$, is the modern cross product. (For more information on the use of vectors, see the paper by Karin Reich elsewhere in this volume.)

These vector operations, being useful in physics, were only studied in three-dimensional space. As noted, however, Hamilton in the 1840s had indicated that one should be able to consider n-tuples as individual objects for arbitrary n. Hamilton was chiefly interested in the multiplication of these objects. The notion of an n-dimensional vector space had to come primarily out of the desire to add such objects. The basic idea that there were objects that one could add together in a natural way and that there were a certain limited number of these objects by which any of them could be expressed by the use of addition and scalar multiplication came into the mainstream of nineteenth century mathematics from a wide variety of sources. These include the study of integrals in n-space by Mikhail Ostrogradskii (1801–1861) and Eugene Catalan (1814–1894), the work of Cauchy on line integrals, ultimately generalized into the study of homology by Henri Poincaré (1854–1912), and the study of the roots of polynomial equations and congruences by Evariste Galois (1811–1832) and others. Many of these sources can be adapted into examples accessible to beginning students.

Formalization: Vector Spaces

By the end of the nineteenth century there were sufficient examples available in the mathematical community from which someone could abstract the notion of a vector space. Apparently, the first such abstract definition was given by Giuseppe Peano (1858–1932) in his *Calcolo geometrico* [18] of 1888. Peano's aim in the book, as the title indicates, was to develop a geometric calculus. Such a calculus

> consists of a system of operations analogous to those of algebraic calculus but in which the objects with which the calculations are performed are, instead of numbers, geometrical objects. [18, p. 21]

Much of the book consists of such calculations dealing with points, lines, planes, and volumes. But in chapter IX Peano gives a definition of what he calls a linear system. Such a system consists of quantities provided with operations of addition and scalar multiplication. The addition satisfies the commutative and associative laws (though these are not named by Peano), while the scalar multiplication satisfies two distributive laws and an associative law as well as the law that $1a = a$ for every quantity a. In addition, Peano includes as part of his axiom system the existence of a zero quantity satisfying $a + 0 = 0$ for any a as well as $a + (-1)a = 0$. Peano also defined the dimension of a linear system as the maximum number of linearly independent objects in the system and noted that the set of polynomial functions in one variable forms a linear system of infinite dimension.

Curiously, Peano's work had no immediate effect on the mathematics scene. The definition was even forgotten. It did not enter the mathematical mainstream until Hermann Weyl (1885–1955) essentially repeated Peano's axioms for what Weyl called an "affine geometry" in his *Space-Time-Matter* [22] of 1918. Besides giving several examples satisfying the definitions, he also gave a philosophic reason for adopting such, a reason which should be discussed in class:

> Not only in geometry, but to a still more astonishing degree in physics, has it become more and more evident that as soon as we have succeeded in unravelling fully the natural laws which govern reality, we find them to be expressible by mathematical relations of surpassing simplicity and architectonic perfection. It seems to me to be one of the chief objects of mathematical instruction to develop the faculty of perceiving this simplicity and harmony, which we cannot fail to observe in the theoretical physics of the present day. It gives us deep satisfaction in our quest for knowledge. Analytical geometry [the axiom system he has presented] ... conveys an idea even if inadequate, of this perfection of form. [22, p. 23]

Weyl then brings the subject of linear algebra full circle, pointing out that by considering the coefficients of the unknowns in a system of linear equations in n unknowns as vectors, "our axioms characterize the basis of our operations in the theory of linear equations." [22, p. 24]

Conclusion

The teaching of linear algebra, like the teaching of most topics in the undergraduate curriculum, benefits from the use of historical materials. Not only do the historical materials help provide insight into the understanding of the topics themselves, but also their discussion helps enliven the class and show the students that linear algebra, like the rest of mathematics, grew up in a certain milieu and was developed in order to solve certain problems. Though there is no complete study of the history of linear algebra ([5] comes the closest), the interested teacher can find additional information about the topics covered here as well as about other topics in any of the standard history of mathematics texts.

For Further Consideration

1. In any discussion of the solutions of a system of linear equations, the question needs to be asked, "What happens if there are more unknowns than equations?" The Chinese authors of the *Jiuzhang* seem to have ignored this question. Problem 13 of chapter 8 provides an example:

Five families have a well in common. The two ropes of the first family do not reach to the bottom; they need one rope of the second family. Similarly, three ropes of the second family need one of the third; four ropes of the third need one of the fourth; five ropes of the fourth need one of the fifth; and six ropes of the fifth need one of the first. How deep is the well and what is the length of each family's rope? [20, p. 87]

Here there are five equations, but six unknowns (the length of each family's rope and the depth of the well). The Chinese author in this instance merely provides an answer. To find it, he says that the basic rule should be applied. If one follows his instructions, using the variables s to represent the depth of the well and v to represent the length of the fifth family's rope, one finally obtains an equation of the form $721v = 76s$. The length of the remaining ropes also depend on s. The Chinese, however, did not mention anything other than the one solution corresponding to $s = 721$, namely, $v = 76$.

Did the Chinese realize there were other solutions? Do our students? There is no discussion in the Chinese sources of obtaining infinitely many solutions to such problems. But that is not surprising. After all, they needed, and found, one solution to the given problem. Why would they want any other solutions? Such questions might provide a basis for some good class discussion on the solution of systems of linear equations.

2. Does Maclaurin's method of elimination generalize to equations in more variables? Maclaurin only noted that a similar rule holds for four equations in four unknowns. He gave no proof nor any mention of systems of more than four unknowns. But students might want to try to extend his procedure. The attempt at such extension would lead to the desirability of giving a name to the complicated sum and difference of terms referred to by both Cramer and Maclaurin and then of deriving ways of calculating with the determinant.

3. Cayley did not give formal proofs of his general results in matrix theory. He merely calculated in the low dimensional cases and 'assumed' that everything would work out for any size matrix. Cayley did notice, however, where certain analogies between matrix algebra and ordinary algebra fail. For example, when $\det A = 0$, one cannot find an inverse of A; matrix multiplication is not commutative; and the product of two matrices can be zero without either factor being zero. These are ideas which can be discussed in class and which one can easily guide the students to discover.

4. Cayley noted in the paper on matrices of 1858 that he had verified the Cayley-Hamilton theorem for a 3×3 matrix, but actually wrote out only the third degree equation that the matrix must satisfy. Students should be encouraged to make this computation themselves. The next line of Cayley's paper [3, p. 483] is also worth discussion: "I have not thought it necessary to undertake the labor of a formal proof of the theorem in the general case of a matrix of any degree." Should such a proof be given in a first course? Is a general computational proof even possible, or is the verification in the 2×2 and 3×3 cases enough to convince the students by "induction"? Cayley himself apparently felt that a proof of this theorem in general would entail a great amount of computational labor. He did not realize, however, that an appropriate use of his own algebra of matrices would significantly reduce the labor involved. Such a proof was given some twenty years later by Frobenius and could be presented after the students attempt a computational proof.

5. The calculation of the square root of a matrix by Cayley in [3, p. 484] provides the basis for a series of class exercises in matrix manipulation. For example, one first needs to show the conditions under which the entire manipulation works. (Cayley neglected this entirely; but, after all, X appears in a denominator and the solution for both X and Y involve square roots.) One could then apply the result to find possible square roots of I or of $-I$. In the latter case, one can attempt to derive a system of matrices which represent either the complex numbers or the quaternions.

6. Wessel realized that as far as his geometric representation of addition was concerned, it could as easily be done in three dimensions as in two. But when he attempted to extend his multiplication to three dimensions by introducing a third perpendicular unit, he ran into problems he could not solve. A class should attempt to define multiplication of 3-vectors and see what the difficulties are. William Rowan Hamilton some 40 years after Wessel ran into these same problems.

7. As mentioned in the text, there are many possible sources of the idea of a vector space. One example which could be discussed in class is the vector space generated over the rational number field by the solutions of the equation $x^3 - 1 = 0$. Given the three solutions $1, \omega, \omega^2$, where $\omega = -\frac{1}{2} + i\frac{\sqrt{3}}{2}$, one can show that any linear combination of these elements also satisfies an algebraic equation of degree 3 over the rationals.

Bibliography

1. Cauchy, Augustin-Louis: "Sur l'equation à l'aide de laquelle on determine les inegalités seculaires des Mouvements des Planétes," *Exercises du mathématique* 4 (1829) = *Oeuvres complètes* (2) 9, 174–195.

2. Cayley, Arthur: "Remarques sur la Notation des Fonctions Algébriques," *Journal für Mathematik* 50 (1855), 282–285 = *Collected Mathematical Papers*, vol. 2, 185–188.

3. Cayley, Arthur: "A Memoir on the Theory of Matrices," *Philosophical Transactions* 148 (1858), 17–37 = *Collected Mathematical Papers*, vol. 2, 475–496.

4. Cramer, Gabriel: *Introduction à l'Analyse des Lignes Courbes Algébriques*. Geneva, 1750.

5. Crowe, Michael J.: *A History of Vector Analysis*. South Bend: Notre Dame University Press, 1967.

6. Euler, Leonhard: *Elements of Algebra*. London: Longmans, 1840. (Reprint edition, New York: Springer-Verlag, 1984).

7. Gauss, Carl Friedrich: *Disquisitiones Arithmeticae*. Leipzig, 1801. Available in English translation by Arthur Clarke, New York: Springer-Verlag, 1986.

8. Gauss, Karl Friedrich: "Disquisitio de Elementis ellipticis Palladis ex Oppositionibus Annorum 1803, 1804, 1805, 1807, 1808, 1809." *Comm. soc. reg. scien. Gott.* 1 (1811) = *Werke*, Vol. VI, 1–24.

9. Hamilton, William Rowan: "Theory of Conjugate Functions, or Algebraic Couples, with a Preliminary and Elemntary Essay on Algebra as the Science of Pure Time," *Transactions of the Royal Irish Academy* 17 (1837), 293–422 = *Mathematical Papers*, vol. III, 3–96.

10. Hamilton, William Rowan: "On the Existence of a Symboic and Biquadratic Equation, which is Satisfied by the Symbol of Linear Operations in Quaternions," *Proceedings of the Royal Irish Academy* 8 (1864), 190–191.

11. Hawkins, Thomas: "Cauchy and the spectral theory of matrices," *Historia Mathematica* 2 (1975), 1–29.

12. Hawkins, Thomas: "Another Look at Cayley and the Theory of Matrices," *Archives Internationales d'Histoire des Sciences* 27 (1977), 82–112.

13. Heath, Thomas: *A History of Greek Mathematics*. New York: Dover, 1981.

14. Katz, Victor J.: "Who is the Jordan of Gauss-Jordan?" *Mathematics Magazine* 61 (1988), 99–100.

15. Maclaurin, Colin: *A Treatise of Algebra*. London: Millar and Nourse, 1748.

16. Maxwell, James Clerk: *A Treatise on Electricity and Magnetism*. London: Oxford University Press, 1873.

17. Newton, Isaac: *Mathematical Principles of Natural Philosophy*, (Motte's translation, revised by Cajori). Berkeley: University of California Press, 1966.

18. Peano, Giuseppe: *Calcolo geometrico secondo l'Ausdehnungslehre di H. Grassmann preceduto dalle operazioni della logica deduttiva*. Torino, 1888.

19. Sylvester, James Joseph: "Additions to the Articles 'On a New Class of Theorems' and 'On Pascal's Theorem'" *Philisophical Magazine* (3) 37 (1850), 367–370 = *Collected Mathematical Papers*, vol. 1, 145–151.

20. Vogel, Kurt: *Neun Bucher arithmetischer Technik*. Braunschweig: Vieweg, 1968.

21. Wessel, Caspar: "On the Analytical Representation of Direction," *Memoirs of the Royal Academy of Denmark* (1799); English translation in Smith, David: *A Source Book in Mathematics*, Vol. 1. New York: Dover, 1959, 55–66.

22. Weyl, Hermann: *Space-Time-Matter*. New York: Dover, 1952. (Original German edition published in 1918.)

Wessel on Vectors

Otto B. Bekken

Introduction

As a Norwegian, I find it appropriate to draw your attention to the work of a fellow countryman, Caspar Wessel. It concerns how to add and multiply directed lines in the plane and in 3-space. Wessel's solution, first submitted in 1796, provides a good introduction to our teaching of complex numbers and quaternions.

This paper is really part of my notes [3], see also [2], which were put together to help discuss

- issues from our teaching of algebra, through historical material
- the growth of ideas and their forms in algebra in a problem solving style
- with excerpts from sources
- with mathematical exercises/problems
- with discussions of teaching styles
- with modern "take-offs" on calculators/computers.

Subthemes to be handled are developments of number concepts, like irrationals and imaginaries, symbolization, and accepted proofs, or demonstrations.

My discussion of Caspar Wessel's work will be based on the contents of these booklets, i.e., [2] and [3]. He was the first to give us our geometric representation of complex numbers as it is taught today. It is often overlooked that this came out of his attempt to add and multiply directed line segments, **vectors** as we now call them. In some places, where Wessel is properly mentioned in this connection, his work is said to be vague or obscure. The only obscurity results from the fact that he wrote in Danish. Thus, he did not become generally known until his work was later published in other languages, see [15] and [10].

Who Was He?

Caspar Wessel was born 8 June 1745 in Vestby, a few miles south of Christiania, as Oslo was then called. He died 25 March 1818 in Copenhagen, see [8] and [16]. Caspar's father was a Lutheran minister of an important Norwegian family. His uncle was Peter Wessel Tordenskjold, a famous sea-captain, known from the battle of Dynekil with the Swedes. Caspar's brother, Johan Herman Wessel, became a well-known poet, who described him in these words:

> Maps he will draw,
> Reading the law!
> Industrious is he
> while I am lazy!

Caspar went to Christiania Cathedral School 1757–63, before entering Copenhagen University. Here he took his Law Degree in 1778, but already, by 1764, he had begun work as a surveyor. In 1798, he became a Royal Surveying Inspector with the Danish Academy of Sciences, but in 1805 he retired with a pension because of poor health.

Caspar Wessel's only known work on mathematics is the paper entitled "On the analytic representation of direction, an attempt applied to the problems of plane and spherical polygons."

What Was the Problem?

As early as 1679, we find Gottfried Wilhelm Leibniz writing to Christiaan Huygens:

> So far as geometry is concerned, we need still another analysis which is distinctly geometrical or linear and which will express situation [situs] directly as algebra expresses magnitude directly.... Algebra is the characteristic for undetermined numbers or magnitudes only, but it does not express situation, angles, and motion directly. Hence, it is often difficult to analyze the properties of a figure by calculation. [7, p.3]

Leibniz's attempts were not successful in creating a computational device—some generalized numbers—for working with geometrical entities in which they could be added and multiplied. For plane geometry this was first done successfully by the Norwegian surveyor, Caspar Wessel.

Wessel also tried to extend his ideas to 3-space. This was done successfully by William Rowan Hamilton on 16 October 1843 with his quaternions, which in turn led to today's vector calculus mainly developed by Josiah Willard Gibbs by 1881. This development is well described by Crowe [7].

Caspar Wessel presented his treatise to the Danish Academy of Sciences on 10 March 1797, in it he noted:

> This attempt concerns ... how straight lines should be represented to give both their length and their direction
>   (a) represented analytically and governed by algebraic rules;
>   (b) to be applied to other lines than those of equal or opposite direction.
>   I will try:
>   I. First to determine the rules for such operations;
>   II. To show through a couple of examples their application to lines on a plane;
>   III. To define direction of lines in different planes by another method;
>   IV. Look at plane and spherical polygons;
>   V. To derive by this method the well-known formulas of spherical trigonometry.
>   [14, pp. 5–7]

Possibly his aim was to develop a practical tool for his surveying, where he was forced to do trigonometric computations involving directed line segments. The imaginary numbers came up again as exactly the computation device to add and multiply directed line segments in a plane.

What Exactly Do We Want?—and How Do We Actually Do It?

In Wessel's own words:

> §1. Two lines are added when the second starts where the first ends, and when the line between the first and the last point is drawn and taken as their sum.

§4. The product of two lines of length 1 in the same plane as the positive unit and with the same starting point, should be in the same plane, with an angle of direction to the unit being the sum of the direction angles of the factors.

§5. Let +1 denote the positive unit, and let a certain perpendicular unit with the same starting point be $+\epsilon$. The direction angle of $+1 = 0°$, of $-1 = 180°$, $+\epsilon = 90°$ and of $-\epsilon = 270°$. To obtain the rule of §4, we have to multiply according to:

	1	-1	ϵ	$-\epsilon$
1	1	-1	ϵ	$-\epsilon$
-1	-1	1	$-\epsilon$	ϵ
ϵ	ϵ	$-\epsilon$	-1	1
$-\epsilon$	$-\epsilon$	ϵ	1	-1

From this we see that ϵ becomes $\sqrt{-1}$, and the product follows the usual algebraic rules.

§7. The line having direction angle v to the unit +1 is $\cos v + \epsilon \sin v$, and when multiplied with the line $\cos u + \epsilon \sin u$, the product becomes the line with direction angle $v + u$, denoted by $\cos(v + u) + \epsilon \sin(v + u)$.

§9. The general representation of a line of length r and direction angle v to the positive unit +1 is $r(\cos v + \epsilon \sin v)$. [14, pp. 9–13]

Then Wessel demonstrates that he knows very well how this relates to imaginary numbers. He explains the fractional Euler–De Moivre formula, see problem 1 in the exercises. Thus, Wessel found a new application of imaginary numbers—to the geometry of plane positions sought by Leibniz.

Earlier, imaginaries had come to be useful in algebra, first in the works of Cardano [6] and Bombelli [5], later also for Descartes, Wallis, and Viète [12]. Without really trying, Wessel had also solved another important problem of his time—*to give imaginary numbers a geometric representation*—to reconnect the meaning of general numbers to something geometric. In fact, this problem is nowhere mentioned by Wessel. The geometric representation of complex numbers taught today (Figure 1), is similar to his.

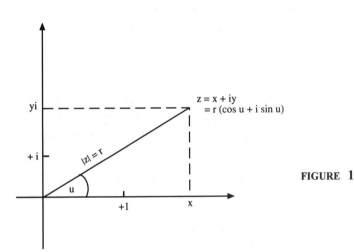

FIGURE 1

Because Wessel's article was not translated into French until 1897, his representation was rediscovered by many, among them the Swiss Jean Robert Argand in 1806 and the German Carl-Friedrich Gauss in 1831 [7]. In the preface to our edition of Wessel's work of 1896, we read Sophus Lie:

> If the work of Caspar Wessel had been appreciated earlier, he would long ago have established himself with as great a name in mathematics as his brother Johan Herman Wessel in Scandinavian literature and his uncle Peter Wessel Tordenskjold in warfare.

Can You Really Multiply Triplets?

Wessel extended his ideas to 3-space by introducing a third unit η, where $\eta^2 = -1$, and which is perpendicular to both $+1$ and $+\epsilon$. A radius from the origin to (x, y, z) could then be represented as $x + \eta y + \epsilon z$. Addition was easy, but Wessel was in fact, the first to add vectors in 3-space. He does it carefully through several steps.

His multiplication became more tricky. He defines easily

$$(x + \eta y)(a + \eta b) \qquad \text{and} \qquad (x + \epsilon z)(a + \epsilon c),$$

because this is the two-dimensional case he has already dealt with. More problematic is his §30–31:

$$(x + \eta y + \epsilon z)(a + \eta b) = (x + \eta y)(a + \eta b) + \epsilon z$$

$$(x + \eta y + \eta z)(a + \epsilon c) = (x + \epsilon z)(a + \epsilon c) + \eta y$$

He applies these operations to spherical trigonometry, but it is difficult reading. His problem really is how to do it generally, with

$$(x + \eta y + \epsilon z)(a + \eta b + \epsilon c) = ?$$

As was later realized by Hamilton in the 1840s, and proved by Hermann Hankel in 1867, this cannot be done without introducing a fourth unit μ, also satisfying $\mu^2 = -1$, and the quaternions

$$x + \eta y + \epsilon z + \mu w$$

However, Wessel came close enough to be characterized by Sophus Lie as "the forerunner of the Englishman Hamilton's celebrated quaternions."

Exercises

1. Can you fill in all details of Wessel's arguments in the following two sections?

    Consequently, whether m is positive or negative, it is always true that

    $$\cos \frac{v}{m} + \epsilon \sin \frac{v}{m} = (\cos v + \epsilon \sin v)^{1/m}.$$

    Therefore, if both m and n are integers, we have;

    $$(\cos v + \epsilon \sin v)^{n/m} = \cos \frac{n}{m} v + \epsilon \sin \frac{n}{m} v.$$

In this way we find the value of such expressions as $\sqrt[n]{b + c\sqrt{-1}}$ or $\sqrt[m]{a\,\sqrt[n]{b + c\sqrt{-1}}}$. For example, $\sqrt[3]{4\sqrt{3} + 4\sqrt{-1}}$ denotes a right line whose length is 4 and whose angle with the absolute unit is $10°$.

If m is an integer and α is equal to $360°$, then $(\cos v + \epsilon \sin v)^{1/m}$ has only the following m different values:

$$\cos\frac{v}{m} + \epsilon\sin\frac{v}{m}, \quad \cos\frac{\alpha + v}{m} + \epsilon\sin\frac{\alpha + v}{m}, \quad \cos\frac{2\alpha + v}{m} + \epsilon\sin\frac{2\alpha + v}{m}\cdots,$$

$$\cos\frac{(m - 1)\alpha + v}{m} + \epsilon\sin\frac{(m - 1)\alpha + v}{m};$$

for the numbers by which π is multiplied in the preceding series are in the arithmetical progression $1,\ 2,\ 3,\ 4, \ldots,\ m - 1$.

2. Hamilton's quaternions are quadruples

$$w = (a, b, c, d), \qquad z = (e, f, g, h),$$

multiplying according to $w * z = (A, B, C, D)$ where

$$A = ae - bf - cg - dh \qquad B = af + be + ch - dg$$

$$C = ag + ce + df - bh \qquad D = ah + bg + de - cf$$

(a) Analyze the products of the basic elements $(1, 0, 0, 0)$, $(0, 1, 0, 0)$, $(0, 0, 1, 0)$, and $(0, 0, 0, 1)$. You should find three imaginary units. Is $*$ commutative?

(b) Prove that $w^2 + 1 = 0$ if and only if $a = 0$ and $b^2 + c^2 + d^2 = 1$. Thus, $w^2 + 1 = 0$ has infinitely many quaternion solutions.
Associate vectors $\bar{u} = (u_1, u_2, u_3)$ and $\bar{v} = (v_1, v_2, v_3)$ to the quaternions $\mathbf{u} = (0, u_1, u_2, u_3)$ and $\mathbf{v} = (0, v_1, v_2, v_3)$.

(c) Prove the connections

$$\mathbf{u} * \mathbf{v} = \bar{u} \times \bar{v} - \bar{u} \cdot \bar{v}$$

$$\bar{u} \cdot \bar{v} = -\frac{\mathbf{u} * \mathbf{v} + \mathbf{v} * \mathbf{u}}{2}$$

$$\bar{u} \times \bar{v} = \frac{\mathbf{u} * \mathbf{v} - \mathbf{v} * \mathbf{u}}{2}$$

between the quaternion product and the scalar and vector product.
For each quaternion $\mathbf{z} = (e, f, g, h)$ we define $\tilde{\mathbf{z}} = (e, -f, -g, -h)$ and a mapping on the space of quaternions by:

$$T_z(\mathbf{w}) = \mathbf{z} * \mathbf{w} * \tilde{\mathbf{z}}.$$

(d) Prove that T_z is a linear operator, that $T_z \circ T_w = T_{z*w}$, that $\mathbf{R}^3$ is invariant under T_z, and that T_z represents a rotation on $\mathbf{R}^3$.

3. In retrospect we can see that Francois Viète [12] had ideas very close to Wessel's geometric representation; see [4]. Viète discusses a calculus of triangles, described in [4] by Figure 2 and

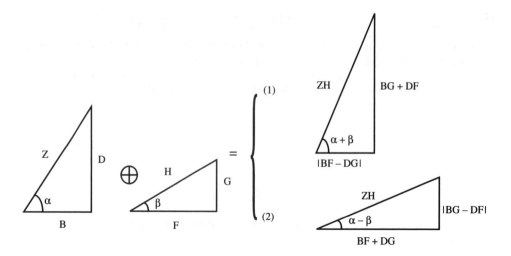

FIGURE 2

based on the algebraic identities

$$Z^2 H^2 = (B^2 + D^2)(F^2 + G^2)$$
$$= (BF - DG)^2 + (BG + DF)^2$$
$$= (BF + DG)^2 + (BG - DF)^2$$

Explain this and the connection it has to complex numbers.

Bibliography

1. Andersen, Kirsti: "Hvor kommer vektorerne fra?" *NORMAT* 33 (1985), 1–16.
2. Bekken, Otto B.: *Una Historia Breve del Algebra*. Lima: Sociedad Matematica Peruana, 1983.
3. Bekken, Otto B.: *Themes from the History of Algebra*. ADH-skrifter 1984.
4. Bashmakova, Isabela G.: "Diophantine Equations and the Evolution of Algebra," (trans. by A. Shenitzer), in *Proceedings of the ICM*. Berkeley, 1986.
5. Bombelli, Rafael: *L'algebra Opera*. Bologna, 1572 (New edition edited by E. Bortolotti, Milan: Feltrinelli, 1966.)
6. Cardano, Girolamo: *The Great Art, or The Rules of Algebra*, (trans. and ed. by T. Richard Witmer). Cambridge: MIT Press, 1968.
7. Crowe, Michael J.: *A History of Vector Analysis*. South Bend: Notre Dame University Press, 1967.
8. Jones, Phillip S.: "Caspar Wessel," *Dictionary of Scientific Biography*, vol. 14. New York: Charles Scribner's Sons, 1981.
9. Möller Pedersen, Kirsti: "Caspar Wessel og de komplekse tals repræsentation," *NORMAT* 27 (1979), 49–55.
10. Nordgaard, Martin A.: "Wessel on Complex Numbers," translation of [13], pp. 469–480; in [11], pp. 55–66.
11. Smith, David Eugene: *A Source Book in Mathematics*. New York: Dover, 1959.
12. Viète, François: *The Analytic Art*, (trans. by T. Richard Witmer). Kent, Oh: Kent State University Press, 1983.
13. Wessel, Caspar: "Om Directionens analytiske Betegning," *Nye Samling af det Kongelige Danske Videnskabernes Selskabs Skrifter* 1799, 469–518.

14. Wessel, Caspar: *Om Directionens Analytiske Betegning, med en fortale av Sophus Lie*. Kristiania: Cammermeyers Forlag, 1896.

15. Wessel, Caspar: *Essai sur la representation analytique de la direction*, (edited and translated by H. Valentiner, T.N. Thiele, and H.G. Zeuthen). Copenhagen, 1897.

16. Zeuthen, Hieronymus G.: "Caspar Wessel," *Dansk biografisk lexicon*, Vol. XVIII. København, 1904.

Karl Wilhelm Theodor Weierstrass

Who Needs Vectors?

Karen Reich

Introduction

Mathematics, as it is taught at schools, generally appears as a collection of brilliant ideas proposed by geniuses. Normally there is no place at all for disagreement; school-pupils and even students in the university have the feeling that mathematics is a complete structure, correct and beyond criticism. The paths to understanding that structure seem like a set of one-way roads, as if evolution had only built up one ideal form of mathematics. But history shows, in nearly every branch, that this view is not true. It is absolutely necessary that pupils and students learn, that as with all other kinds of evolution, mathematics also went through a selection process. Competition among several similar mathematical theories is quite common and ensuing discussions about usefulness, applicability and representation are quite normal. These discussions are able to point out the advantages of a theory and to clarify the aims. In reality, there are no one-way roads in mathematics. Vector-calculus is a good example which proves easily and clearly the above made statements. In this field, even today, there does not exist international agreement on how vectors and the operations with them are labelled. Different books use different signs and symbols, for example:

> vectors: $\mathbf{a}$, ω, $\vec{a}$
> interior product: $\vec{a} \cdot \vec{b}$, $(\vec{a}\vec{b})$
> vector product: $[\vec{a} \times \vec{b}]$, $[\vec{a}\vec{b}]$, $\vec{a} \times \vec{b}$.

History provides the reason: in the past, several mathematicians created their own theories on the basis of different aims and foundations. As a result there still are several different forms of representation in use. In vector-calculus, you can still sense the nineteenth century struggle for existence between quite different conceptions [6, pp. 183–224], which reveals itself in the symbolism employed.

The vector-calculus of today is divided into two main parts: vector-algebra and vector-analysis. Only vector-algebra is taught at school level; vector-analysis is normally reserved for universities; therefore, I am going to discuss only vector-algebra here.

Historical Outline

Vector-calculus has not one but at least two histories. One branch goes back to the *Ausdehnungslehre* of Hermann Grassmann (1809–1877) and one to the theory of quaternions of William Rowan Hamilton (1805–1865). Some examples of vectors existed even before Grassmann and Hamilton, in the seventeenth and eighteenth centuries, but these were of little importance either for Grassmann or for Hamilton. Both of them had different approaches to vector-calculus and different aims from their predecessors.

Grassmann's starting-point was the theory of general magnitudes; for these he wanted to create a calculus, which should be as universal as possible. In the first edition of his *Ausdehnungslehre* [12], he started with general forms of thinking in mathematics, by means of which he specified his theory of forms. After treating the meaning of equality and connection he defined two kinds of connection between the members, symbolized by $\cup$ and $\cap$, a synthetic connection $\cap$ with associative and commutative properties, namely $a \cap (b \cap c) = (a \cap b) \cap c$, and $a \cap b = b \cap a$, and an analytical connection $\cup$ with the property that $(a \cup b) \cap b = a$.

In the second edition of 1862 [11], Grassmann defined the extensive-magnitude as the polynomial $a_1 e_1 + a_2 e_2 + \cdots = \sum a_i e_i$, where $a_i \in \mathbf{R}$ and the e_i form what he called a system of unities. Later Grassmann described the operations of addition, subtraction, and multiplication with a real number. Then he defined a linear combination. It is important that Grassmann does not place any dimensional restrictions on his extensive magnitudes, which can have any number of dimensions. According to Grassmann, there exists a general kind of product of extensive magnitudes, given by

$$\sum a_r e_r \sum b_s e_s = \sum a_r b_s [e_r e_s].$$

At first Grassmann did not give any further specification for $[e_r e_s]$. But later he distinguished between the so-called exterior multiplication where $[e_r e_s] = -[e_s e_r]$ and the interior multiplication defined by the condition $[e_r e_r] = 1; [e_r e_s] = 0, (r \neq s)$.

Hamilton's starting point was quite different, arising from his consideration of complex numbers. He interpreted them as pairs of real numbers, and showed how computation with complex numbers can be seen as computation with ordered pairs of numbers. Following this, Hamilton decided to develop a computation with ordered triples of numbers, but in vain. In 1843, Hamilton found out that instead of triples of numbers, he had to take a system with four units $1, i, j, k$, i.e., quaternions. Multiplication no longer obeys the commutative law. The rules are: $ij = k, jk = i, ki = j, ji = -k, kj = -i, ik = -j, ii = jj = kk = -1$ [6, pp. 29–30]. These quaternions have four dimensions, one real, the scalar part, and three imaginary ones, together called the vector part. We can therefore write:

$$Q = \text{Scalar}Q + \text{Vector}Q = SQ + VQ$$

or

$$\omega = r \quad + \quad xi + yj + zk.$$

The operations of addition and subtraction pose no problem. When you multiply two vectors,

$$\alpha = xi + yj + zk \quad \text{and} \quad \alpha' = x'i + y'j + z'k,$$

you get two parts, a scalar part

$$S\alpha\alpha' = -(xx' + yy' + zz'),$$

and a vector part

$$V\alpha\alpha' = (yz' - zy')i + (zx' - xz')j + (xy' - yx')k$$

[5].

Grassmann's ideas were not adhered to immediately. It took a while for other mathematicians to become interested. In contrast, the quaternions of Hamilton were accepted almost immediately after their publication in 1853. Not only mathematicians but also physicists accepted them enthusiastically. One of the reviews read:

> The discoveries of Newton have done more for England and for the race, than has been done by whole dynasties of British monarchs; and we doubt not that in the great mathematical birth of 1853, the Quaternions of Hamilton, there is as much real promise of benefit to mankind as in any event of Victoria's reign. [6, p. 37]

Perhaps the most important influence on the future of vector-calculus was that physicists and engineers now became involved. In physics, a new theory, electrodynamics, needed to be described in a new mathematical language. James Clerk Maxwell (1831–1879) picked up the concept of quaternions, which he used in the form of three-dimensional vectors. Maxwell employed two different kinds of vectors: force vectors, which he related to a length unit, and flux vectors, which he related to a surface unit [22, p. 11].

Following Maxwell, everybody who worked in electrodynamics used vectors and contributed to vector-calculus, including Peter Guthrie Tait (1831–1901), Oliver Heaviside (1850–1925), and Josiah Willard Gibbs (1893–1903).

Gibbs is the person who reformed vector-calculus and transformed it into the theory which is nearly identical with the one used in schools today. Gibbs' main publication was *Elements of Vector Analysis* [10]. In this he gave up the idea of 4-dimensional quaternions, and adopted only the vector part $\alpha = xi + yj + zk$. Furthermore, he introduced two kinds of products, the direct product, $\alpha\beta = xx' + yy' + zz'$ (without the negative sign), and the skew product, $\alpha \times \beta = (yz' - zy')i + (zx' - xz")j + (xy' - yx')k$.

After Gibbs' work most physicists eventually supported vector-calculus as Gibbs represented it. But there were others who did not want to give up the idea of quaternions, and there were at least some mathematicians who supported Grassmann's ideas.

Who Adopted Which Conception of Vectors?

Grassman's Theory. The Grassmann-theory held to an n-dimensional representation. Its proponents were pure mathematicians. One of the most outstanding of these was Jan Arnoldus Schouten (1883–1971). But Schouten did not pick up Grassmann's extensive magnitudes. He had his own magnitudes in mind, the affinors, a special mixture of vectors and tensors. So Schouten's concept is not exactly the same as Grassmann's. Further, Schouten specified many more kinds of multiplication; in his *Foundations of Vector- and Affinor-analysis* [30], he defined nineteen different kinds of multiplication, concluding with "and so on."

Quaternions. In 1893 Pieter Molenbroek, Professor of Mathematics in Leiden, published his *Application of the Quaternions in Geometry* [24]. In the introduction, Molenbroek informed the reader that quaternions should be preferred in solving difficult problems. The resulting methods would be more elegant. Molenbroek rewrote elementary geometry and differential geometry by means of quaternions. The first chapter is dedicated to problems of trigonometry, the second to points, planes, straight lines, and spheres, the third to conic sections and special surfaces, the fourth to the general theory of surfaces.

In 1907, the "International Association for Promoting the Study of Quaternions and Allied Systems of Mathematics" was organized. Its founder was the Japanese mathematician Kimura, who had studied in the U.S., and one of its presidents was the Irish astronomer, Sir Robert Ball (1873–1942). But this association did not last much beyond the first world war.

Vectors in Gibbs' Style. This kind of representation was commonly adopted by most of the physicists and engineers, especially by people who worked in electrodynamics. One of these was August Föppl (1854–1924), and his book *Introduction to Maxwell's Theory of Electricity* [7] became a very successful manual which was republished several times.

Varia. There were several authors who developed their own type of vector-calculus and tried to promote it. One of them was the crystallographer, Woldemar Voigt (1850–1919), who worked with so-called polar vectors and axial vectors [32], similar to Maxwell's force vectors and flux vectors.

The Situation in Teaching. The same discussion was carried on in teaching. A small selection of examples illustrates the situation. In 1896/7 Hermann Wiener (1857–1939) gave a lecture in the Technische Hochschule Darmstadt on Grassmann's *Ausdehnungslehre* [18, p. 226]. In the summer semester of 1902, Eugen Jahnke (1861–1921) began teaching a course on "Vectors and their Application to Problems of Mechanics according to Grassmann" in Berlin, which lasted to the winter semester 1904/05 and was published afterwards [15]. Föppl, of course, used the Gibbs version of vectors in his famous lectures on technical mechanics [8].

The Encyklopädie der mathematischen Wissenschaften

An interesting insight into the state of vector-calculus at the turn of the century can be found in the great *Encyklopädie* (1898–1935). One might think that there would be only one article on vector-calculus, but that is wrong; there existed no less than three major articles on the subject!

In his article "The Geometrical Foundation of Mechanics of a Rigid Body," which was finished in February, 1902 [31], Heinrich Emil Timerding (1873–1945) mentioned that a vector is defined by the direction from one point to another point, i.e., the difference of the two points. Timerding quoted Hamilton and mentioned that Hamilton's i, j, k were identical with Grassmann's e_1, e_2, e_3. In some detail Timerding also explained Grassmann's definitions of Plangrösse and Ebenengrösse which correspond to the vector parallelogram defined by vectors (x_1, y_1, z_1) and (x_2, y_2, z_2) and the so-called supplementary vector, whose components are $L = y_1 z_2 - z_1 y_2$, $M = z_1 x_2 - x_1 z_2$, $N = x_1 y_2 - y_1 x_2$. Timerding's vectors are only three dimensional.

Max Abraham (1875–1922) authored the article "Mechanics of the deformable bodies" [1], which was completed in February, 1901. Abraham, a colleague of August Föppl, took care of the later editions of Föppl's above-mentioned manual on electrodynamics. Although referring to Grassmann, Abraham adopted Hamilton's quaternions in the Gibbs-version. Like Voigt he employed polar vectors $A = \alpha A_x + \beta A_y + \gamma A_z$, and axial vectors $P_x = A_y B_z - A_z B_y$, $P_y = A_z B_x - A_x B_z$, and $P_z = A_x B_y - A_y B_x$. Abraham's vectors are also only three dimensional.

Within the chapter "Maxwell's electromagnetic theory," Hendrik Antoon Lorentz (1853–1928) presented a third article on vector calculus [20], which was finished in June, 1903. Lorentz discussed scalar products and vector products and, in quoting Maxwell, symbolized them as $(\alpha.\beta) = |\alpha||\beta|\cos(\alpha\beta)$ and $[\alpha \times \beta] = (A_y B_z - A_z B_y, A_z B_x - A_x B_z, A_x B_y - A_y B_x)$, where $\alpha = (A_x, A_y, A_z)$ and $\beta = (B_x, B_y, B_z)$.

Experiments in Standardizing Vector-Calculus

In the beginning of the twentieth century, mathematicians, physicists and engineers tried to find a system of standardization in vector-calculus. In 1903, Ludwig Prandtl (1875–1953) emphasized that due to the popularization of vector-calculus, one should aim for a unification. Prandtl made some suggestions which were published in the *Jahresbericht der Deutschen Mathmematiker Vereinigung* [26]. In the same year, on September 24, Prandtl gave a report on this problem on occasion of the "Naturforscherversammlung" in Kassel. The discussion there consisted of comments made by outstanding physicists and mathematicians: Arnold Sommerfeld (1868–1951), Arthur von Oettingen (1836–1920), Georg Hamel (1877–1954), Rudolf Mehmke (1857–1944), Ernst Zermelo (1871–1953), Felix Klein (1849–1925), Heinrich Burkhardt (1861–1914), Ludwig Boltzmann (1844–1906) and Eugen Jahnke [27]. The American mathematician Alexander Macfarlane (1851–1913) answered Prandtl in the form of a memoir [21].

Because so many scientists of different nations were involved, many ideas of defining and symbolizing vectors and the operations with them were created. In 1904, R. Mehmke from Stuttgart compared the American version of vector-calculus with the so-called German-Italian one [23]. Prandtl continued the debate by adding a so-called "physical direction" [28]. In 1908, Franz Jung (1872–1957) distinguished three major versions of vector-calculus: The German-Italian one, the geometric version, and the physical version [16].

In March of 1907, Cesare Burali-Forti (1861–1931) and Roberto Marcolongo (1862–1943) published a first note on the unification problem [4] and declared that this issue would be a main subject for the next International Mathematical Congress which was scheduled to take place in Rome from April 5 to 12, 1908. At this congress, Marcolongo spoke within section III, "Mechanics, mathematical physics, geodesy," on the unification of vectorial notation. Afterwards, this section formulated the following resolution:

> Section III (Mechanics), after an exchange of views in which the importance of a unification of vectorial notation was recognized, proposes to the Congress the appointment of an International Commission to study this question. [14, p. 65]

This international commission met again in 1912 during the next International Mathematical Congress in Cambridge. There the commission had to declare that they could not accomplish the task, and asked for an extension; but the next international congress, which was to be in Stockholm in 1916, did not take place because of the war [9].

In the meantime, Burali-Forti and Marcolongo published four other notes on the subject. In note IV they proposed a new system which was also published in French in *L'Enseignement mathématique* [5]. In this journal several other mathematicians discussed these propositions and expressed their own opinions: Gaston Combébiac, Timerding, Klein, Edwin Wilson (1879–1964), Giuseppe Peano (1858–1932), Emmanuel Carvallo (1856–1945) and Eugen Jahnke. In 1907/8 Felix Klein delivered his lectures on "Elementarmathematik vom höheren Standpunkte aus" in which he compared these unification problems in vector-calculus with those which had come up earlier in the field of electro-engineering. Here a unification took place in 1881, which had resulted in the adoption of the Volt, Ohm and Ampère as the main units [17, vol. 1, p. 158]. However, this unification came about only because it was forced by economic interests. According to Klein, there seemed to be no equivalent situation in vector-analysis, where there was no direct pressure to act, so that every mathematician was able to use the system he preferred. This statement is repeated with the same words in the later editions of Klein's lectures.

In 1921 another attempt at unification emerged. Again, it was hoped that a commission would help resolve this matter; this time it was the so-called "Ausschufür Einheiten und Formelgröen" (AEF), which presented a new suggestion for vectorial symbolism [2]. These propositions were discussed in 1922 by L. Prandtl, Lothar Schrutka (1881–1945) and C. Böhm; J.A. Schouten and Jean Spielrein added even new propositions [3]. In fact, a real unification never happened, but nevertheless the participants seem to have understood each other.

Discussion about the Usefulness of Vector-Calculus

The controversies concerning the standardizing of vector-calculus were accompanied by discussions on its usefulness: is it only a new and convenient form of writing, or is it also able to produce new results? Behind this issue stands the question as to whether a good formalism, in itself, is of any advantage or not [19].

If one looks at how vector-calculus is taught in schools today, one obtains a similar feeling: it is used only as a kind of language for writing. Pupils have difficulties in understanding what kind of other advantages vector-calculus might offer. But there remain opportunities for teachers to point out what a good formalism is able to accomplish. They can make it clear what kind of results can only be achieved by means of vector-calculus. Here the work of Felix Klein remains of great importance, as we shall see.

Klein was not too enthusiastic about vector-calculus simply as a language. He had difficulties in understanding why these methods became so popular and had been adopted all over the world. His lectures "Elementarmathematik vom höheren Standpunkte aus" (Elementary mathematics from an advanced point of view) were especially directed to teachers of mathematics to show them new developments through an easy but encouraging point of view. Here Klein explains the adoption and application of vectors by the observation that there are many people who just adore formal analogies [17, vol. 2, p. 117]. He also mentioned that vector-calculus is occasionally even used in schools.

In 1912, Rudolf Rothe (1873–1942) pointed out [29] that from the beginning onwards, enthusiasm and reservation accompanied the development of vector-calculus. However, according to Rothe, in mechanics and in electrodynamics as well as in other physical fields, vectors had been accepted. So they have finally achieved their "civil right" (Bürgerrecht), an expression used by Gauss in favor of the adoption of the complex numbers. The vectorial method is elegant and short and that is the reason why vector-calculus had generally come into use. But it is only a method, a kind of short hand, without supplying any deeper information. It does not bring anything new and is not able to do so. Rothe concluded with the observation that it also should be possible to admire only the beautiful vectorial leaves and blossoms on the mathematical tree and to forget about the roots.

Appendix 1: Klein's introduction to vector-calculus as an introduction of vectors in school

Klein [17, p. 21] starts with a straight line with the endpoints x_1 and x_2, which are located on the x-axis. You can write its length as a determinant:

$$x_1 - x_2 = \frac{1}{1} \begin{vmatrix} x_1 & 1 \\ x_2 & 1 \end{vmatrix}.$$

In analogy to this expression, a triangle in the x-y-plane with the coordinates of the vertices being (x_1, y_1), (x_2, y_2), and (x_3, y_3) has the area

$$A = \frac{1}{1 \cdot 2} \begin{vmatrix} x_1 & y_1 & 1 \\ x_2 & y_2 & 1 \\ x_3 & y_3 & 1 \end{vmatrix}.$$

Similarly, a tetrahedron in space with vertices (x_1, y_1, z_1), (x_2, y_2, z_2), (x_3, y_3, z_3), (x_4, y_4, z_4) has volume

$$V = \frac{1}{1 \cdot 2 \cdot 3} \begin{vmatrix} x_1 & y_1 & z_1 & 1 \\ x_2 & y_2 & z_2 & 1 \\ x_3 & y_3 & z_3 & 1 \\ x_4 & y_4 & z_4 & 1 \end{vmatrix}.$$

Later [17, pp. 29–30], Klein describes Grassmann's principles for the plane and for 3-space. He starts with three points and the determinant

$$\begin{vmatrix} x_1 & y_1 & 1 \\ x_2 & y_2 & 1 \\ x_3 & y_3 & 1 \end{vmatrix}.$$

Out of this determinant you can form the matrices

$$\begin{vmatrix} x_1 & y_1 & 1 \\ x_2 & y_2 & 1 \end{vmatrix} \qquad \text{and} \qquad |x_1 \; y_1 \; 1|,$$

which are achieved by omitting one or two rows of the determinant.

The first matrix leads to $\mathcal{Y} = y_1 - y_2$, $\mathcal{X} = x_1 - x_2$, and, by omitting the third column of the determinant, to $\mathcal{N} = x_1 y_2 - x_2 y_1$. So you get the system $\mathcal{X}$, $\mathcal{Y}$, $\mathcal{N}$. This triple defines the straight line between the first two points (Figure 1), because you can write the equation

$$\begin{vmatrix} x & y & 1 \\ x_1 & y_1 & 1 \\ x_2 & y_2 & 1 \end{vmatrix} = 0 \qquad \text{as} \qquad \mathcal{Y}x - \mathcal{X}y + \mathcal{N} = 0.$$

This straight line is determined by the proportion $\mathcal{X} : \mathcal{Y} : \mathcal{N}$. Such an oriented straight line between the points is called a vector.

In the case of 3-space, Klein starts with 4 points and the following matrices:

$$(x_1 \; y_1 \; z_1 \; 1), \quad \begin{pmatrix} x_1 & y_1 & z_1 \\ x_2 & y_2 & z_2 \end{pmatrix}, \quad \begin{pmatrix} x_1 & y_1 & z_1 & 1 \\ x_2 & y_2 & z_2 & 1 \\ x_3 & y_3 & z_3 & 1 \end{pmatrix}, \quad \begin{pmatrix} x_1 & y_1 & z_1 & 1 \\ x_2 & y_2 & z_2 & 1 \\ x_3 & y_3 & z_3 & 1 \\ x_4 & y_4 & z_4 & 1 \end{pmatrix}.$$

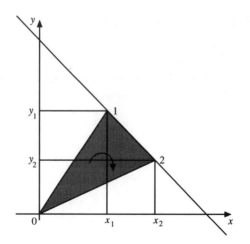

FIGURE 1

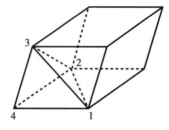

FIGURE 2

The first matrix gives the coordinates of one point. The determinant of the last matrix gives the volume of the tetrahedron. In German this volume is also called *Spatprodukt*. The word *Spat* = spar has its origin in the language of mining, because of the crystal-like shape of the content. The second matrix produces six determinants $\mathcal{X}$, $\mathcal{Y}$, $\mathcal{Z}$, $\mathcal{L}$, $\mathcal{M}$, $\mathcal{N}$, which, as above, define the line between the points:

$$\mathcal{X} = x_1 - x_2; \quad \mathcal{Y} = y_1 - y_2; \quad \mathcal{Z} = z_1 - z_2;$$

$$\mathcal{L} = y_1 z_2 - y_2 z_1; \quad \mathcal{M} = z_1 x_2 - z_2 x_1; \quad \mathcal{N} = x_1 y_2 - x_2 y_1$$

The third matrix produces four determinants of the third order:

$$L = \begin{vmatrix} y_1 & z_1 & 1 \\ y_2 & z_2 & 1 \\ y_3 & z_3 & 1 \end{vmatrix}, \quad M = \begin{vmatrix} z_1 & x_1 & 1 \\ z_2 & x_2 & 1 \\ z_3 & x_3 & 1 \end{vmatrix}, \quad N = \begin{vmatrix} x_1 & y_1 & 1 \\ x_2 & y_2 & 1 \\ x_3 & y_3 & 1 \end{vmatrix}, \quad P = -\begin{vmatrix} x_1 & y_1 & z_1 \\ x_2 & y_2 & z_2 \\ x_3 & y_3 & z_3 \end{vmatrix},$$

which define the plane triangle through the three points (x_1, y_1, z_1), (x_2, y_2, z_2), (x_3, y_3, z_3). Afterwards, Klein applied these magnitudes in mechanics. So Klein's vectors have a little bit more geometrical meaning than the common arrow-vectors and show at least the close relationship between vectors and determinants, i.e., matrices.

Appendix 2: Holors

The symbolism of vector-calculus had not been standardized, neither by the beginning of the twentieth century, or later. Several forms of expression are still used. This means that not only are historians confronted with the different kinds of writing, but also everybody who gets involved with vectors. So neither Grassmann's extensive magnitudes nor Hamilton's quaternions are really dead.

Grassmann's *Ausdehnungslehre* especially seems to be a source even in today's mathematics; it is still able to stimulate new ideas. For example, Parry Moon from MIT and Domina E. Spencer from the University of Connecticut recently produced the idea of magnitudes called holors [25]. Their book is dedicated to Dirk J. Struik; in the preface one reads:

> The word holor indicates a mathematical entity that is made up of one or more independent quantities. Examples of holors are complex numbers, vectors, matrices, tensors and other hypernumbers. The very fruitful idea of representing a collection of quantities by a single symbol appeared originally in 1673, when John Wallis represented a complex number in 2-space. More complicated holors were developed by Hamilton (1843), Grassmann (1844), Gibbs (1881), Ricci (1884), Heaviside (1892), Levi-Civitá (1901), Einstein (1916), and many others.

A holor is defined as a mathematical entity built up of one or more independent merates, where merates refer to the numbers $v^1, v^2, \dots$. The word merate should remind you of the Greek word *meros* = part.

Bibliography

1. Abraham, Max: "Geometrische Grundbegriffe," *Encyklopädie der mathematischen Wissenschaften*, vol. IV, 3. Leipzig: Teubner, 1901–1908, 3–47.
2. Anonymous: "Einführung einer einheitlichen Vektorschreibweise," *ZAMM* 1 (1921), 421–422.
3. Anonymous: "Zur Einführung einer einheitlichen Vektorschreibweise," *ZAMM* 2 (1922), 161–164.
4. Burali-Forti, Cesare and Marcolongo, Roberto: "Per l'unificazione delle notazioni vettoriali," *Rendiconti del Circolo matematico di Palermo* 23 (1907), 324–328; 24 (1907), 65–80, 318–322; 25 (1908), 352–375; 26 (1908), 369–377.
5. Burali-Forti, Cesare and Marcolongo, Roberto: "Notations rationelles pour le système vectoriel minimum," *L'Enseignement mathématique* 11 (1909), 41–45.
6. Crowe, Michael: *A History of Vector Analysis: The Evolution of the Idea of a Vectorial System*. South Bend: Notre Dame University Press, 1967.
7. Föppl, August: *Einführung in die Maxwellsche Theorie der Elektrizität*. Leipzig: Teubner, 1894.
8. Föppl, August: *Vorlesungen über technische Mechanik*. Leipzig: Teubner, 1898–1910.
9. Fehr, Henri: "Compte rendu du Congres de Cambridge," *L'Enseignement mathématique* 14 (1912), 301–306, 365–391, 441–447.
10. Gibbs, Josiah Willard: *Elements of Vector-Analysis*. New Haven: Yale University, 1881.
11. Grassmann, Hermann: *Die Ausdehnungslehre. Vollständig und in strenger Form bearbeitet*. Berlin: T. C. F. Enslin, 1862.
12. Grassmann, Hermann: *Die lineale Ausdehnungslehre, ein neuer Zweig der Mathematik*. Leipzig: Wigand, 1844.
13. Hamilton, William R.: "On quaternions or a new System of Imaginaries in Algebra," *Philosophical Magazine* 29 (1846), 26–31 = *The Mathematical Papers*, Vol. 3, 236–239.
14. Internationaler Mathematiker-Kongress. *Jahresbericht der Deutschen Mathematiker Vereinigung* 17 (1908), 2. part.

15. Jahnke, Eugen: *Vorlesungen über die Vektorrechnung. Mit Anwendungen auf Geometrie, Mechanik und mathematische Physik.* Leipzig: Teubner, 1905.

16. Jung, Franz: "Einige vektoranalytische Bezeichnungs-und Benennungsfragen," *Jahresbericht der Deutschen Mathematiker Vereinigung* 17 (1908), 383–390.

17. Klein, Felix: *Elementary Mathematics from an Advanced Standpoint: Geometry* (E. R. Hedrick and C. A. Noble, trans.). New York: Macmillan, 1939.

18. Klemm, Friedrich: "Die Rolle der Mathematik in der Technik des 19. Jahrhunderts," *Kulturgeschichte der Naturwissenschaften und ter Technik.* Munich: Deutshes Museum, 1979, vol. 1, 219–229.

19. Knobloch, Eberhard: "Symbolik und Formalismus im mathematischen Denken des 19, und beginnenden 20. Jahrhunderts," *Mathematical Perspectives* (Joseph Dauben, ed.). New York: Academic Press, 1981, 139–165.

20. Lorentz, Hendrik Antoon: "Maxwells Elektromagnetische Theorie," *Encyklopädie der mathematischen Wissenschaften*, vol. V, 2. Leipzig: Teubner, 1904–1922, 63–144.

21. Macfarlane, Alexander: "The notation and fundamental principles of vector analysis," *Jahresbericht der Deutschen Mathematiker Vereinigung* 13 (1904), 228–233.

22. Maxwell, James Clerk: *A Treatise on Electricity and Magnetism.* Oxford: Clarendon Press, 1873.

23. Mehmke, Rudolf: "Vergleich zwischen der Vektoranalysis amerikanischer Richtung und derjenigen deutsch-italienischer Richtung," *Jahresbericht der Deutschen Mathematiker Vereinigung* 13 (1904), 217–228.

24. Molenbroek, Pieter: *Anwendung der Quaternionen auf die Geometrie.* Leiden: Brill, 1899.

25. Moon, Parry and Spencer, Domina Eberle: *Theory of Holors. A Generalization of Tensors.* Cambridge: Cambridge University Press, 1986.

26. Prandtl, Ludwig: "Grundsätze für eine einheitliche Schreibung der Vektorrechnung im technischen Unterricht," *Jahresbericht der Deutschen Mathematiker Vereinigung* 12 (1903), 404–405.

27. Prandtl, Ludwig: "Über einheitliche Schreibweise der Vektorrechnung im technischen und physikalischen Unterricht," *Verhandlungen der Gesellschaft deutscher Naturforscher und Ärzte*, 2 Teil, 1. Hälfte (1903), 31.

28. Prandtl, Ludwig: "Über die physikalische Richtung in der Vektoranalysis," *Jahresbericht der Deutschen Mathematiker Vereinigung* 13 (1904), 436–449.

29. Rothe, Rudolf: "Anwendung der Vektor-Analysis auf Differential-geometrie," *Jahresbericht der Deutschen Mathematiker Vereinigung* 21 (1912), 249–274.

30. Schouten, Jan Arnoldus: *Grundlagen der Vektor- und Affinoranalysis.* Leipzig: Teubner, 1914.

31. Timerding, Heinrich Emil: "Geometrische Grundlegung der Mechanik eines starren Körpers," *Encyklopädie der mathematischen Wissenschaften*, vol. IV, 1. Leipzig: Teubner, 1901–1908, 125–189.

32. Voigt, Woldemar: *Die fundamentalen physikalischen Eigenschaften der Krystalle in elementarer Darstellung.* Leipzig: Teubner, 1898.

The Teaching of Abstract Algebra:
An Historical Perspective

Israel Kleiner

What follows is a description of a somewhat nontraditional way to teach abstract algebra—nontraditional in its approach and in its content. While courses in abstract algebra, if textbooks are any guide, usually begin with the theory and then give some applications, I begin with "problems" and try to show how attempts to solve the problems give rise to the abstract theory. This is, of course, the historical sequence of events.

My approach, however, is genetic rather than strictly historical. The intent, as Otto Toeplitz (1881–1940) described it in *The Calculus: A Genetic Approach* (1963), is to look at the historical origins of an idea in order to find the best way to motivate it—to consider the context in which the originator of the idea was working in order to find the "burning problem" which he was striving to solve. The psychological motivation is just as important as the mathematical one.

In addition to serving as a motivational device, history provides the opportunity to raise a number of general issues about the nature of mathematics:
(a) The genesis of mathematical concepts and theories.
(b) The nature and role of axiomatic systems.
(c) The role of intuition versus that of logic in the creation of mathematics.
(d) The meaning of number.
(e) The nature of proof.
(f) Mathematics as a living and evolving subject.

Here are some of the "problems" I have dealt with—some are standard, others less so—and the ideas in abstract algebra they give rise to.

1. Why is $(-1)(-1) = 1$? This problem leads to the concepts of ring, integral domain, and order, and to some elementary results about them.
2. Find all solutions in integers of $x^2 + 2 = y^3$. This problem leads to the concepts of unique factorization domain, euclidean domain, and other ideas of commutative algebra.
3. Solve $x^3 = 15x + 4$. This problem leads to the introduction of complex numbers into mathematics, and to the investigation of other interesting topics.
4. Are there numbers beyond the complex numbers? This question leads to the consideration of what we mean by "numbers," the introduction of quaternions and octonions (Cayley numbers), and, more generally, to notions of noncommutative algebra.
5. Solve in integers: $n = x^2 + y^2 + z^2 + w^2$. This problem leads to integral quaternions, noncommutative factorization, one-sided ideals.
6. Trisect a $60°$ angle using straightedge and compass. This problem leads to the concept of a (number) field and to some elementary theory of field extensions.
7. Show that $a^{p-1} \equiv 1 \pmod{p}$, p a prime and $(a, p) = 1$. This problem leads to the notion of a cyclic group and to some elementary properties of finite abelian groups.

8. What is geometry? This question leads to the concept of an (infinite) group of transformations.

9. Is a square more symmetric than an equilateral triangle? This question, too, leads to the notion of a (finite) group of transformations.

10. Solve "by radicals" $x^5 - 6x + 3 = 0$. This problem leads to the concepts of field and group of permutations, and to Galois theory.

It should be noted that we begin the study of abstract algebra with rings, both commutative and noncommutative, continue with fields, and culminate with groups. Although this is not the standard order, it introduces students gradually, through numerical examples, to the abstractions of algebra.

Most of the above problems can be dealt with in a first course in abstract algebra, but some (e.g., 10) are more suitable for a second course. In fact, there is enough material here for at least two courses. The beautiful Galois theory is a nice culmination of many of the ideas in these notes.

We now discuss the problems in some detail.

Why is $(-1)(-1) = 1$?

This is an instance of the general problem of the logical justification of the laws of operation with negative numbers. It became a pressing problem, for both pedagogical and professional reasons, at Cambridge University around 1830. (This is one example where didactic considerations have been a driving force in the development of new mathematics. Others can be found in the works of Cauchy (1789–1857), Weierstrass (1815–1897), and Dedekind (1831–1916).) In fact, the very existence of negative numbers came into question. Peacock (1791–1858) and others set themselves the task of resolving this problem by codifying the laws of operation with numbers. This was perhaps the earliest instance of axiomatics in algebra. Some of the key issues of "abstract algebra" that emerge here are:

(i) Manipulation of symbols apart from interpretation—so-called symbolical algebra.

(ii) The choice of laws to be obeyed by the operations on symbols.

It is interesting to note in this connection that while in past centuries numbers were an important object of study and symbols were the language in which to express relationships among numbers, with Peacock symbols acquire for the first time "a life of their own" and become an object of study in their own right. In fact, later in the nineteenth century it is arithmetic which becomes the language of a large part of mathematics (e.g., analysis). The roles of arithmetic and algebra would thus seem to have become reversed.

We discuss some of these matters, and focus on the following (see [5], [22], [29] for details):

(a) Reasons why the problem of the negative numbers became an important issue at the time.

(b) Some proposed solutions of this problem, with special attention given to Peacock's solution, embodied in his "principle of the permanence of equivalent forms."

(c) Implications of the symbolical approach for subsequent developments in algebra; for example, the works of De Morgan (1806–1871), Hamilton (1805–1865), Boole (1815–1864), and Cayley (1821–1895).

Pointing out some of the limitations in Peacock's development, we next take a more modern, Hilbertian approach to the problem of negative numbers. In spirit, the modern approach is very close to Peacock's. Just as Hilbert "defined" (characterized) the real numbers axiomatically as a complete ordered field (see [43]), so we characterize the integers as an ordered integral domain in which the positive elements are well-ordered—undoubtedly a twentieth century development. Once this is done we can prove such "laws" as $(-1)(-1) = 1$ (and more generally $(-a)(-b) = ab$), $a \cdot 0 = 0$, and others (see [70], [89]). In fact, Fraenkel (1891–1965) proved such laws in 1914 for arbitrary rings (see [14]).

The following are some issues which we discuss in this context:

 (i) How can we prove a law such as $(-1)(-1) = 1$? This question leads to the concept of an axiom. We cannot prove everything.

 (ii) What axioms should we set down in order to give a description of (to define, to characterize) the integers? This question enables us to introduce the concepts of ring, integral domain, and ordered structure.

(iii) How do we know when we have enough axioms? E.g., how do Z and Z_n differ as rings? This question permits us to introduce the concept of completeness of a set of axioms.

(iv) What does it mean to characterize the integers? This question sets the stage for the introduction of the concept of isomorphism. We shall have characterized the integers by means of a set of axioms when any two systems satisfying these axioms are isomorphic. For example, the axioms for an ordered integral domain do not characterize the integers; this is so because the rationals, say, are also an ordered integral domain, and—as can be readily shown—the integers and rationals are not isomorphic.

 (v) Could we have used fewer axioms to characterize the integers? In fact, the commutativity of addition can be derived from the other axioms for an integral domain. Here we confront the concept of independence of a set of axioms.

(vi) Are we at liberty to pick and choose axioms as we please? This leads us to the concept of consistency of a set of axioms and, more broadly, to the question of "freedom of choice" in mathematics. Raising these questions in such a "natural" setting makes this problem especially interesting.

Find all solutions in integers of $x^2 + 2 = y^3$

Every problem deserves (and gets) some historical background. The study of diophantine equations goes back at least to Diophantus (c. A.D. 250) and, in one form or another, to the Babylonians (c. 1600 B.C.). Beginning in the nineteenth century it has inspired the development of algebraic number theory and thus of concepts such as ring, field, and ideal. Diophantine equations have also played an important role in the evolution of "classical" (pre-nineteenth-century) algebra (see [1]). The equation $x^2 + 2 = y^3$ was discussed by Fermat (1601–1665), while $x^2 + k = y^3$ for general k (sometimes called Bachet's equation) has not been solved to date; see [57].

It is easy to guess solutions of $x^2 + 2 = y^3$; e.g., $x = \pm 5$, $y = 3$. The problem is, of course, to find all solutions. To get quickly at the conceptual heart of the matter, we begin with the equation $x^2 + y^2 = z^2$ and factor it, getting $(x + yi)(x - yi) = z^2$, an equation in Gaussian integers. We invoke here Hadamard's powerful principle, namely that the shortest path between two truths in the real domain passes through the complex domain. The property of the ordinary integers that if a product of two relatively prime integers is a square then each is a square carries over to the Gaussian integers, yielding $x + yi = (a + bi)^2$. Thus $x + yi = (a^2 - b^2) + 2abi$

and hence $x = a^2 - b^2$, $y = 2ab$, from which we get $z = a^2 + b^2$. It is easy to verify that, conversely, for arbitrary $a, b \in Z$, the above equations yield a solution of $x^2 + y^2 = z^2$. Thus we obtain the familiar formula for Pythagorean triples which appears in Euclid's Elements (c. 300 B.C.) but was apparently known to the Babylonians over a millennium earlier.

The study of divisibility in the ring of Gaussian integers and, in particular, the characterization of its primes, readily yields the solution of Fermat's problem: When is an integer a sum of two squares? See [77], [83]. (An interesting proof of Fermat's two-squares theorem using some elementary group theory and Minkowski's theorem in the "geometry of numbers" is given in [86].) The answer to Fermat's problem leads, in turn, to the following result: If Z_p is the field of integers mod p, let $Z_p(i) = \{a + bi : a, b \in Z_p\}$—the "complexification" of Z_p. Then Z_p is a field, with addition and multiplication the same as for complex numbers, if and only if $p \equiv 3 \pmod 4$. These are precisely the primes which cannot be written as sums of two squares.

Coming back to $x^2 + 2 = y^3$ and proceeding as in the case of $x^2 + y^2 = z^2$, we get $(x + \sqrt{2}i)(x - \sqrt{2}i) = y^3$, hence $x + \sqrt{2}i = (a + b\sqrt{2}i)^3$, arguing now in the domain of "integers" $D = \{a + b\sqrt{2}i : a, b \in Z\}$. Cubing, equating coefficients, and doing some routine algebra, we find that $x = \pm 5$, $y = 3$ are, in fact, the only solutions of $x^2 + 2 = y^3$.

There is, of course, considerable work to be done to justify the above steps, both for $x^2 + y^2 = z^2$ and $x^2 + 2 = y^3$. This leads naturally to the concepts of unique factorization domain and Euclidean domain. See [41], [45], [57], [83] for details.

Should we bring in principal ideal domains at this stage to prove that a Euclidean domain is a unique factorization domain? Introducing them in this context may be an unnecessary detour. However, they can be brought in "naturally" via the greatest common divisor, which is a useful device in showing that a Euclidean domain is a unique factorization domain. The needed result about greatest common divisors is that in a Euclidean domain D the greatest common divisor of a and b can be written in the form $ax_0 + by_0$, x_0, $y_0 \in D$. To show this we consider $\{ax + by : x, y \in D\}$. This set is an ideal of D; proving the above result about the greatest common divisor is, essentially, proving that a Euclidean domain is a principal ideal domain.

The initial impetus for algebraic number theory came from Fermat's conjecture about $x^p + y^p = z^p$ (and from higher reciprocity laws). The case $p = 3$ can be settled by using essentially the same arguments as for the two diophantine equations considered above: $x^3 + y^3 = (x+y)(x+\omega y)(x+\omega^2 y)$, where ω is a primitive cube root of 1, and we thus consider the problem in the domain of "integers" $D = \{a + b\omega : a, b \in Z\}$, which is a unique factorization domain. The case of arbitrary p is handled analogously: $x^p + y^p = (x + y)(x + \omega y) \cdots (x + \omega^{p-1}y)$, ω a primitive pth root of 1. For a "proof" see [46].

The "proof," of course, breaks down (at $p = 23$), as Lamé (1795–1870) found (in 1847) to his great consternation. The problem is that the domain of cyclotomic "integers" $D = \{a_0 + a_1\omega + \cdots + a_{p-2}\omega^{p-2} : a_j \in Z\}$ is not a unique factorization domain for general p. This observation was the start of the fruitful attempts, by Kummer (1810–1893), Kronecker (1823–1891), and especially Dedekind, to restore unique factorization to such, and eventually more general, domains of "integers." The concepts of field, ring, and ideal were introduced by Dedekind in the 1870s in this setting. (See [2], [9], [11], [26], [41], [45], [73], [77], [86] for details.) The resulting algebraic number theory is rich and deep, but some of it can be discussed at this level; for example, quadratic fields and representation of integers by quadratic forms. See [46], [57], [73], [77], [78].

The following are some interesting "projects" related to the above topic.

(a) The polynomial $x^2 + x + 41$ produces primes for all nonnegative integers $x \leq 40$. Using unique factorization in quadratic number fields one can prove the following more general result, first observed by Euler (1707–1783): The polynomial $x^2 + x + q$ produces primes for all nonnegative integers $x \leq q - 2$ if and only if the ring of integers of the quadratic field $Q(\sqrt{1 - 4q})$ is a unique factorization domain. Since $1 - 4q < 0$ and the complex quadratic fields which are unique factorization domains are known (see [83]), it follows that the above result holds exactly when $q = 3,\ 5,\ 11,\ 17,\ 41$. See [53] for details.

(b) It is always instructive to see the interrelation of different areas of mathematics. With a knowledge of some analysis (power-series expansions) and some commutative ring theory (prime and maximal ideals) one can prove the following result: Every trigonometric identity is a consequence of the identity $\sin^2 x + \cos^2 x = 1$. See [67].

(c) Here is another link between algebra and analysis. It is "easy" to do nonstandard calculus (à la Robinson) once the field R^* of hyperreal numbers is available. Using a few elementary concepts and results in commutative ring theory having to do mainly, as in (b), with prime and maximal ideals, one can readily construct R^*—it turns out to be the quotient ring of the ring of sequences of real numbers modulo a properly chosen maximal ideal. See [60].

Solve $x^3 = 15x + 4$

The problem of the algebraic solution of the cubic may seem out of place in a course such as this—it is "classical" rather than "modern" algebra. I find, however, that it is an interesting topic to present, especially to secondary-school teachers. It was perhaps the first major achievement in algebra since the Babylonians, who 3000 years earlier had solved the quadratic. It brings into focus the problems which beset algebra at the time—lack of proper notation and symbolism, and a poor understanding of the number system and of manipulation of numbers. This gives students a sense of what algebra was like before there was abstract algebra. Moreover, the problem of the cubic has a "modern algebra" postscript—the "irreducible case of the cubic." Although such a cubic has 3 real roots, these cannot be given in terms of real radicals. The proof uses the considerable sophistication of modern field theory; see [43] and problem 10.

Another point of contact with abstract algebra: In 1847 Cauchy defined the complex numbers, which grew out of the solution of the cubic, as congruence classes of real polynomials modulo $x^2 + 1$. (The concept of isomorphism cries out for definition here.) His model for this construction was Gauss's congruence classes of integers modulo a prime p. Cauchy's idea, in turn, was extended by Kronecker in the 1880s, in his theory of algebraic number fields, to general polynomial rings, and evolved into the notion of quotient ring. Other instances of quotient structures around this time are: The notion of a quotient group, and the notion—due to Cantor (1845–1918)—of the real numbers as, essentially, the quotient ring (without the use of this terminology) of the ring of Cauchy sequences modulo the ideal of null sequences.

Most students think, if they give the issue any thought, that complex numbers arose in response to the desire to solve $x^2 + 1 = 0$. This is not the case. It was, in fact, the solution of the cubic which gave rise to complex numbers. The complex numbers are an interesting case study of the genesis, evolution, and acceptance of a "mathematical system" (a story which I have pursued in some detail in connection with this problem—see [20]). An interesting plot of this story is the controversy between Leibniz (1646–1716) and Jean Bernoulli (1667–1748), subsequently resolved by Euler, about the logarithms of negative and complex numbers; see [23] for details. The story of complex numbers raises interesting historical/philosophical questions such as:

(a) The meaning of "number" in mathematics.

(b) The roles of "physical" need versus intellectual curiosity as motivating factor in the development of mathematics.

(c) The role of heredity versus environment in the creation of mathematics.

(d) The nature of proof in mathematics.

See [20] for details.

The problem of the cubic also provides a useful perspective on the first problem (why is $(-1)(-1) = 1$?), and a nice entry point into the next problem (are there numbers beyond the complex numbers?) and the last one (solution of 5th-degree equations).

Are there numbers beyond the complex numbers?

The answer depends on what we mean by "numbers." A brief history of the evolution of the various number systems with an indication of gains and losses of various properties at each transition stage (e.g., from rationals to reals) helps set the scene for the introduction of quaternions and octonions (Cayley numbers).

The starting point for Hamilton's work on quaternions was his 1835 definition of complex numbers as ordered pairs of reals (see [7]). He found it unsatisfactory to view the complex numbers in the form $a + bi$—he (apparently) felt that addition of bi to a was like addition of apples to oranges. Gauss (1777–1855) and Cauchy, too, were led to their respective representations of complex numbers by misgivings about the nature of $\sqrt{-1}$. These are instances of aesthetic and philosophical considerations as causes in the evolution of mathematical ideas. One might explore similar examples.

Van der Waerden [33] gives a "blow by blow" account of Hamilton's invention of quaternions in 1843. It is a rare glimpse of the creative process in mathematics at work, and guiding students through it is well worthwhile. Hamilton's motivation (the "why" rather than the "how") for the introduction of the quaternions is, however, not as easy to reconstruct. His own retrospective view of them, given in 1855, is:

> The quaternion [was] born, as a curious offspring of a quaternion of parents, say of geometry, algebra, metaphysics, and poetry.... I have never been able to give a clearer statement of their nature and their aim than I have done in two lines of a sonnet addressed to Sir John Herschel: "And how the one of Time, of Space the Three. Might in the Chain of Symbols girdled be." [19, p. 232]

The metaphysical connection, via ideas of Kant (1724–1804), was important in all of Hamilton's work in algebra (see [15] for a full account). It is a further example of extramathematical influences on a mathematician's work. Cantor is another case in point.

Hamilton's work—the first example of a noncommutative number system—was a watershed, a "revolution in arithmetic which is entirely similar to the one which Lobachevsky affected in geometry," according to Poincaré. As all revolutions, however, it was first viewed with suspicion and mistrust. John Graves' comments (in 1844) are telling:

There is still something in this system [of quaternions] which gravels me. I have not yet any clear view as to the extent to which we are at liberty arbitrarily to create imaginaries, and to endow them with supernatural properties. [19, p. 233]

Most mathematicians, however, quickly came around to Hamilton's point of view. Hamilton let the genie escape from the bottle—his work soon led to the exploration of diverse "number systems" whose properties differed in various ways from those of the real and complex numbers. Among those were the octonions (Cayley, Graves (1806–1870), 1844), triple algebras (De Morgan, 1844), exterior algebras (Grassmann (1809–1877), 1844), biquaternions (Hamilton, 1853), group algebras (Cayley, 1854), and matrices (Cayley, 1855/1858). See [19], [34] for details.

A new subject—noncommutative algebra (ring theory)—began to take shape. Among its first results was the classification of finite-dimensional associative division algebras over R. It turns out that there are only three such algebras: The real numbers, the complex numbers, and the quaternions. The proof of this theorem, given independently by Frobenius (1849–1917) in 1878 and C. S. Peirce (1839–1914) in 1881, is accessible to college juniors and seniors. (See [40], [61], [64], [65] for details.) A related result, established recently (1950s) but accessible, is that the only finite-dimensional alternative division algebra over R, that is, nonassociative, algebras for which $(aa)b = a(ab)$ and $a(bb) = (ab)b$, is the octonions. (See [40], [65] for details.)

Incidentally, it is a relatively easy exercise to show, once we know what we want to show, that triples of real numbers cannot form a division algebra. [68] Hamilton worked for ten years attempting to define multiplication of triples so as to turn them into a division algebra! The benefits of hindsight.

Here is another "easy" exercise: Show that $x^2 + 1 = 0$ has infinitely many roots in the domain of quaternions; $bi + \sqrt{1 - b^2}j$, $b \in [0, 1]$, will do. What about a polynomial of degree n having at most n roots? And are the quaternions algebraically closed?

Another result established (by Dedekind and Weierstrass in the 1860s but published about twenty years later) in the formative stages of the theory of associative algebras was a characterization of finite-dimensional commutative algebras without nilpotent elements: They are finite direct sums of copies of a field (R or C in the original proofs). This result, too, is within reach of upper-level students; see [49], [64], [69].

A further entry-point into the subject of noncommutative algebra is the beautiful theorem of Wedderburn, proved in 1905, that every finite division ring is a field. Again, the proof is accessible to the enterprising undergraduate; see [49], [61]. Division rings, moreover, connect with coordinatization of projective planes, and Wedderburn's theorem provides the only known proof that in a finite projective plane Desargues' theorem implies Pappus's theorem; see [44].

Solve in integers: $n = x^2 + y^2 + z^2 + w^2$

This is Lagrange's four-squares theorem, namely that every positive integer is a sum of at most four squares. There are many elementary, number-theoretic proofs of this result (see e.g., [59], [72]); but note that Euler tried for 20 years to prove the theorem, without succeeding. (See [8], [30], [36] for a history of the problem.) We want to focus, however, on a nice, conceptual, algebraic proof along the lines of the solutions of the Diophantine equations discussed in problem 2. The following steps give an outline of the proof (see [59], [61] for details):

(a) Write the equation $n = x^2 + y^2 + z^2 + w^2$ in the form

$$n = (x + yi + zj + wk)(x - yi - zj - wk).$$

This introduces the division ring H of quaternions.

(b) Note that it suffices to prove the theorem for n a prime: For any $\alpha = a + bi + cj + dk$, define the norm $N(\alpha) = a^2 + b^2 + c^2 + d^2$. It is readily seen that $N(\alpha\beta) = N(\alpha)N(\beta)$, from which (b) follows.

(c) Define the ring of "integral quaternions" $R = \{(a + bi + cj + dk)/2 : a,\ b,\ c,\ d \in Z$ all of the same parity$\}$. (Note the analogy with the definition of "integers" in the quadratic field $Q(\sqrt{d})$, where $d \equiv 1 \pmod{4}$, as $\{(a + b\sqrt{d})/2 : a,\ b \in Z, a,\ b$ both of the same parity$\}$.)

(d) Show that R is a (noncommutative) left (and right) Euclidean domain, hence a left principal ideal domain. The notion of one-sided ideal is fundamental here.

(e) Prove that an element α is prime in R if and only if its norm $N(\alpha)$ is prime in Z. Thus no integer prime remains prime in R. Hence if $p \in Z$ is prime, then $p = \alpha\beta$ nontrivially for some $\alpha,\ \beta \in R$. Hence $N(p) = N(\alpha)N(\beta)$, or $p^2 = N(\alpha)N(\beta)$. Thus $p = N(\alpha) = N(\beta)$ ($N(\alpha) = 1$ makes the factorization $p = \alpha\beta$ trivial), and hence p is a sum of four squares.

There is another very nice algebraic proof of this theorem. It is a consequence of factorization of certain 2×2 matrices with entries from the ring of Gaussian integers. See [82].

Lagrange's theorem is the starting point of deep researches in number theory on the "Waring problem" which asks if every integer can be written as a sum of a fixed number of kth powers. See [59].

Trisect a 60° angle with straightedge and compass

The previous problems required for their solution concepts and result mainly from ring theory, both commutative and noncommutative. The solution of this problem gives rise, in a natural way, to the concept of (number) field.

In investigating the constructibility (which we will take to mean, in the classical sense, with straightedge and compass) of geometric entities or—equivalently—of the numbers that describe them, it is quickly noted that if a and b are constructible (line segments or real numbers) then so are $a + b$, $a - b$, $a \cdot b$, and if $b \neq 0$, a/b. In particular, the field Q of rational numbers is constructible (if anything is constructible). Moreover, if $d \in Q$ (i.e., d is constructible) then $\sqrt{d}$ is constructible; e.g., it is obvious that $\sqrt{2}$, the diagonal of a unit square, is constructible. But then, by the above remarks, the field $Q(\sqrt{d}) = \{(a_1 + a_2\sqrt{d})/(a_3 + a_4\sqrt{d}) : a_i \in Q\}$ is constructible. (Such number fields will have also been encountered in the study of algebraic numbers in problem 2, although there it is rings of integers of number fields which are the main objects of study.) One quickly notes that $Q(\sqrt{d}) = \{a + b\sqrt{d} : a,\ b \in Q\}$ (rationalize the denominator). It follows readily that the only constructible numbers are those obtained by a repetition of the above process, namely adjunction of square roots. A few elementary results about field extensions yield the proof of the impossibility of trisecting a 60° angle with straightedge and compass.

The first to deal with the problem of the impossibility of construction with straightedge and compass was Wantzel (1814–1848), in 1837. (See [35].) Although Wantzel does not introduce the language of fields, the essentials of the ideas we use today to resolve these construction problems are there: The reduction of the problem to the solution of equations, the notion of an irreducible (polynomial) equation, the concept of a rational function of a given number of elements, and the conditions for constructibility given in terms of iteration of the solution of quadratic equations. We bypass these early gropings and get to the heart of the matter—field extensions. This illustrates the difference between the historical and the genetic approaches.

This problem appears on my list because I think that it is a good one with which to begin the study of fields. One ought to proceed with a study of field extensions (even in a first course) to obtain a characterization of finite fields. It is a modest investment yielding a considerable payoff.

Again some history. The concept of field came into being through a confluence of sources:

(a) Classical number theory. Here we encounter the field of integers modulo p in Gauss's study (in 1801) of congruence, and the field with p^n elements constructed by Galois (1811–1832) in 1830 as solutions of the congruence $F(x) \equiv 0 \pmod{p}$, where $F(x)$ is an irreducible polynomial of degree n.

(b) Algebraic number theory. Here are introduced, beginning in the 1830s, fields of algebraic numbers (finite extensions of Q)—by Jacobi (1804–1851), Eisenstein (1823–1852), Dirichlet (1805–1859), Dedekind, et al. In 1871 Dedekind gives a definition of a "number field" as a subset of C closed under the four algebraic operations. Kronecker, in his 1882 study of algebraic number theory, views an algebraic number field as a quotient field (our terminology) of $Q[x]$ modulo an irreducible polynomial; cf. (problem 3) Cauchy's similar definition of C.

(c) Geometry. In algebraic geometry Kronecker introduced in 1882 the notion of "domain of rationality"—the field of rational functions in arbitrary quantities R', R'', R''', ... This notion, as we noted, was adumbrated by Wantzel in his treatment of geometric constructions with straightedge and compass.

(d) Classical algebra. We are referring to the problem of solution of polynomial equations. In Galois' work (1831, but not published until 1846) the notion of "formal adjunction" of the roots of a polynomial equation to the (field of) coefficients is clearly present. The concept of solvability by radicals is defined in terms of a sequence of fields, essentially as it is now. (Wantzel's ideas on constructibility, although much less explicit, also bear on the question of adjunction.)

(e) Symbolical algebra. We noted in problem 1 that members of the school of symbolical algebra (Peacock, De Morgan, Gregory (1813–1844), et al,), in attempts to justify the laws of operation with numbers, listed various sets of postulates which numbers were to satisfy. These came close to the axioms for a field. Hamilton, too, in his definition of complex numbers and quaternions, stated the laws which these systems were to obey. He was the first to single out the associative law as a property worthy of consideration. (The commutative and distributive laws were discussed by Servois (1767–1847) in 1814.)

The first abstract definition of a field was given in 1893 by Weber (1842–1913), in connection with his rather abstract treatment of Galois theory. E. H. Moore (1862–1932) characterized finite fields in a paper presented at the International Congress of Mathematicians in Chicago in 1893. Steinitz (1871–1928) founded abstract field theory in a groundbreaking paper in 1910. See [16], [25], [26], [27], [28], [31], [35] for historical details.

Back to the technical comments. Here are several "morsels" that can be discussed with a minimal knowledge of field theory:

(i) A one-page proof that an arbitrary angle cannot be p-sected, p an odd prime, is given in [52].

(ii) If a is constructible from Q then $(Q(a) : Q) = 2^n$. A nice example to show that the converse is not true is given in [63]. By the way, a necessary and sufficient condition that a be constructible from Q is that $(\overline{Q(a)} : Q) = 2^n$, where $\overline{Q(a)}$ is the normal closure of $Q(a)$. See [88].

(iii) A finite extension of Q is algebraic but not conversely. An explicit counterexample is $Q(\sqrt{2}, \sqrt{3}, \sqrt{5}, \ldots, \sqrt{p}, \ldots)$. See [80].

(iv) An interesting application of the properties of cyclotomic polynomials to a special case of Dirichlet's theorem on primes in arithmetic progression is given in [56, Appendix B], where it is shown that for any positive integer a, there are infinitely many primes of the form $na + 1$, $n = 1, 2, \ldots$.

(v) In a lighter vein, one can use the field of four elements to determine the possible final positions of the last peg in the game of "solitaire." See [85, Chapter 6].

Show that $a^{p-1} \equiv 1 \pmod{p}$

This problem may seem contrived; however, there are good historical reasons for dealing with it. In studying congruences in the *Disquisitiones Arithmeticae* (1801) Gauss shows that the nonzero integers modulo a prime p are all powers of a single element or, in our terminology, that the group of such integers is cyclic. Given any nonzero integer a, he defines its order modulo p (without using that terminology) as the smallest positive integer n such that $a^n \equiv 1 \pmod{p}$. He then shows that the order of an element is a divisor of $p - 1$ and uses this result to prove Fermat's Little Theorem: $a^{p-1} \equiv 1 \pmod{p}$. This proof was one of the early examples of the use of group-theoretic ideas to prove number-theoretic results. (Implicit group-theoretic thinking in number theory is already present in Euler; see [36], [38].)

Lagrange's theorem for cyclic groups, that is, that the order of an element divides the order of the group, can be exploited to solve other number-theoretic problems. For example, the group of units of Z_n is intimately related to the period of the decimal expansion of $1/n$. See [45], [59], [74].

Number-theoretic problems were, as we indicated earlier (problem 2), instrumental in introducing concepts in commutative ring theory. As this problem shows, they have also played a role in the evolution of (finite, abelian) group theory. See [13], [18], [38] for details.

What is geometry?

We are referring here to Klein's Erlangen Programme of 1872 in which he described a geometry as the study of the invariants of a group of transformations of a "space." Some of the groups (and the corresponding geometries) that appear in this work are the projective group, the group of isometries, and the group of similarities. See [17], [21], [38], [39].

The nineteenth century witnessed an explosive growth in geometry, both in scope and in depth. New geometries emerged: Projective geometry, non-Euclidean geometries, differential geometry, algebraic geometry, n-dimensional geometry, and Grassmann's geometry of extension. Various geometric methods competed for supremacy: The synthetic versus the analytic, the metric versus the projective. At mid-century a major problem had arisen, namely the classification of the relations and inner connections among the different geometries and geometric methods. It is against this background that Klein's Erlangen Programme must be viewed. See [38], [39].

The intent here is not to discuss in detail the background or content of the Erlangen Programme, but rather to indicate an important source in the evolution of the concept of an abstract group. Just as number theory was the source of (finite) abelian groups, and classical algebra the source of permutation groups, so geometry was the source of transformation groups—infinite groups for the most part, in contrast to the finite groups arising from the other two sources. (Analysis was another source of transformation groups; see [18], [75].) These concrete theories of groups gave rise in the latter part of the nineteenth century to the abstract group concept (see [18], [38]), and that is the message I am trying to convey. Modern elaborations of groups in geometry are given in [48], [66], [71], [75], [87], [90].

Is a square more symmetric than an equilateral triangle?

Symmetry is all-pervasive in nature (see [37], [47], [79], [81]). Its mathematical analysis is very recent, originating in the mid-nineteenth century (see [3], [37], [39]). It gave rise to the concept of finite transformation group—yet another type of group.

The intent of this discussion, as of problems 7, 8, and 10, is to indicate that the abstract group concept originated from examples of concrete groups which arose in distinct and important areas of mathematics. The purpose, then, is to motivate the study of abstract group theory.

The (group of) symmetries of a set A with given structure relative to a subset (or a property) consists of the one-to-one and onto structure-preserving mappings of A (the "automorphisms" of A) under which the given subset (or property) remains invariant. In this sense the concept of symmetry is shared by problems 8, 9, and 10. Thus the group of a geometry (problem 8) is the group of symmetries of a "space" S relative to some given property of the space; the symmetries of a geometric figure (e.g., in the plane—problem 9) are the isometries of Euclidean space which leave the figure unchanged; and the symmetries of a polynomial equation (problem 10) consist of the permutations (automorphisms) of the roots of the equation which leave invariant all relations among the roots over the field of coefficients of the equation. This is, in fact, Galois' definition of the "Galois group" of the equation; in modern terms—the group of automorphisms of the splitting field of the equation leaving the field of coefficients element-wise invariant.

With some elementary notions of groups defined by generators and relations, one can obtain the classification of finite symmetry groups of the Euclidean plane. See [55].

Solve "by radicals" $x^5 - 6x + 3 = 0$

Lagrange's 1770 work on the solvability of equations was the first implicit use of (permutation) groups in mathematics. It was continued by Ruffini (1765–1822) and Abel (1802–1829) and achieved its crowning glory with Galois (see [12], [16], [18], [28], [32], [34] for details). The resulting Galois theory is, in its modern incarnation, a grand symphony on two major themes, groups and fields, and two minor themes, rings and vector spaces. Thus, the four major concepts of "modern" algebra—group, ring, field, and vector space—all come to bear on the solution of the "classical" problem of the solvability of polynomial equations by radicals. Galois theory would thus seem to be a "must" topic in a (second) course in abstract algebra. Moreover, there are nice "applications" of the theory, aside from solvability of equations by radicals:

(a) An essentially algebraic proof of the Fundamental Theorem of Algebra. See [84].

(b) The resolution of the problem of the "irreducible case" of the cubic, namely the nonexistence of a solution by real radicals. See [43] and problem 3.

(c) A classification of the regular polygons constructible by straightedge and compass. See [84].

(d) A characterization of finite simple (primitive) extensions of a field. See [62].

(e) A proof of the irrationality of expressions such as $\sqrt[4]{3} + \sqrt[5]{4} + \sqrt[6]{72}$. See [76].

One of the problems in presenting Galois theory to students is that the proofs of many of its theorems are fairly sophisticated, and the payoff, although considerable, is rather long in coming. In any case, Lagrange's ideas of the systematic treatment of third and fourth degree equations utilizing the notion of (a group of) permutations are easy to convey (see 32], [41], [50]) and should precede the formal treatment of Galois theory. For a computational proof of the insolvability of the general quintic, using ideas of Abel which contain very little group theory, see [42].

Conclusion

Teaching abstract algebra is a challenging task. For students, the subject is often their first encounter with abstract mathematics. I find that the introduction of historical elements helps motivate the learning of abstractions, enables one to raise important mathematical issues, lends a humanizing element to the subject, and often illuminates the array of new concepts being introduced.

Bibliography

A. Historical

1. Basmakova, I. G,: "Diophantine equations and the evolution of algebra," (A. Shenitzer and H. Grant, trans.), *American Mathematical Society Translations* (2) 147 (1990), 85–100.

2. Bourbaki, N.: *Elements d'Histoire des Mathématiques*. Paris: Hermann, 1969.

3. Burn, R. P.: *Groups: A Path to Geometry*. Cambridge: Cambridge University Press, 1985. (See the "Historical Note" at the end of each chapter.)

4. Burton, D. M.: *The History of Mathematics: An Introduction*. Boston: Allyn and Bacon, 1985.

5. Clock, D. A.: *A New British Concept of Algebra: 1825–1850*. Ph.D. Dissertation, University of Wisconsin, 1965. (University Microfilms, no. 64-10221, 1985.)

6. Crossley, J. N.: *The Emergence of Number*. Singapore: World Scientific Pub., 1987 (orig. 1980).

7. Crowe, M. J.: *A History of Vector Analysis*. South Bend: University of Notre Dame Press, 1967.

8. Dickson, L. E.: *History of the Theory of Numbers*, 3 vols. New York: Chelsea, 1971 (orig. 1919-1923).

9. Edwards, H. M.: *Fermat's Last Theorem: A Genetic Introduction to Algebraic Number Theory*. New York: Springer-Verlag, 1977.

10. Edwards, H. M.: "Fermat's Last Theorem," *Scientific American* 239:4 (October 1978), 104–122.

11. Edwards, H. M.: "The genesis of ideal theory," *Archive for History of Exact Sciences* 23 (1980), 321-378.

12. Edwards, H. M.: *Galois Theory*. New York: Springer-Verlag, 1984.

13. Feit, W.: "Theory of finite groups in the twentieth century," in D. Tarwater, *et al.*, *American Mathematical Heritage: Algebra and Applied Mathematics*. Lubbock: Texas Tech. University Press, 1981, 37–60.

14. Fraenkel, A.: "Über die Teiler der Null und die Zerlegung von Ringen," *Journal für die Reine und Angewandte Mathematik* 145 (1914), 139–176.

15. Hankins, T. L.: *Sir William Rowan Hamilton*. Baltimore: The Johns Hopkins University Press, 1980.

16. Kiernan, B. M.: "The development of Galois theory from Lagrange to Artin," *Archive for History of Exact Sciences* 8 (1971/72), 40–154.

17. Klein, F.: "A comparative review of recent researches in geometry," [The Erlangen Programme, in German, with a review by H. S. M. Coxeter] *Mathematical Intelligencer* 0 (1977), 22–30. (An English translation is available in *New York Mathematical Society Bulletin* 2 (1893), 215–249.)

18. Kleiner, I.: "The evolution of group theory: A brief survey," *Mathematics Magazine* 59 (1986), 195–215.

19. Kleiner, I.: "A sketch of the evolution of (noncommutative) ring theory," *L'Enseignment Mathématique* 33 (1987), 227–267.

20. Kleiner, I.: "Thinking the unthinkable: The story of complex numbers (with a moral)," *Mathematics Teacher* 81 (1988), 583–592.

21. Kline, M.: *Mathematical Thought from Ancient to Modern Times*. New York: Oxford University Press, 1972.

22. Koppelman. E.: "The calculus of operations and the rise of abstract algebra," *Archive for History of Exact Sciences* 8 (1971/72), 155–242.

23. Leapfrogs: *Imaginary Logarithms*. Leapfrogs (England), 1978.

24. Leapfrogs: *Complex Numbers*, Leapfrogs (England), 1980.

25. Lichtenberg, D. R.: *The Emergence of Structure in Algebra*. Ph.D. Dissertation, University of Wisconsin, 1966. (University Microfilms, no. 66-13,426.)

26. Merzbach, U.: *Quantity to Structure: Development of Modern Algebra Concepts from Leibniz to Dedekind*. Ph.D. Dissertation, Harvard University, 1964.

27. Moore, E. H.: "A doubly-infinite system of simple groups," *Proceedings of the International Congress of Mathematicians*. Cambridge, 1893.

28. Novy, L.: *Origins of Modern Algebra*, Leiden: Noordhoff, 1973.

29. Pycior, H.: "George Peacock and the British origins of symbolical algebra," *Historia Mathematica* 8 (1981), 23–45.

30. Scharlau, W. and Opolka, H.: *From Fermat to Minkowski: Lectures on the Theory of Numbers and Its Historical Development*. New York: Springer-Verlag, 1985.

31. Steinitz, E.: "Algebraische Theorie der Körper," *Journal für Mathematik* 137 (1910), 167–308.

32. Tignol, J. P.: *Galois Theory of Algebraic Equations*. New York: John Wiley, 1988.

33. Van der Waerden, B. L.: "Hamilton's discovery of quaternions," *Mathematics Magazine* 49 (1976), 227–234.

34. Van der Waerden, B. L.: *A History of Algebra*. New York: Springer-Verlag, 1985.

35. Wantzel, P. L.: "Recherches sur les moyens de reconnaitre si un Problème de Géométrie peut se résoudre avec la règle et le compas," *Journal de Mathématique Pures et Applique* 2 (1837), 366–372.

36. Weil, A.: *Number Theory: An Approach Through History*. Boston: Birkhäuser, 1984.

37. Weyl, H.: *Symmetry*. Princeton: Princeton University Press, 1952.

38. Wussing, H.: *The Genesis of the Abstract Group Concept* (A. Shenitzer, trans.). Cambridge: MIT Press, 1984.

39. Yaglom, I. M.: *Felix Klein and Sophus Lie: Evolution of the Idea of Symmetry in the Nineteenth Century* (S. Sossinsky, trans.). Boston: Birkhäuser, 1988.

B. Technical

40. Albert, A. A. (ed.): *Studies in Modern Algebra*. Washington: Mathematical Association of America, 1963.

41. Allenby, R.: *Rings, Fields and Groups*. London: Edward Arnold Publ., 1983.
42. Ayoub, R. G.: "On the nonsolvability of the general polynomial," *American Mathematical Monthly* 89 (1982), 397–401.
43. Birkhoff, G. and MacLane, S.: *A Survey of Modern Algebra*, 3rd ed. New York: Macmillan, 1977 (orig. 1941).
44. Blumenthal, L. M.: *A Modern View of Geometry*. San Francisco: W. H. Freeman, 1961.
45. Bolker, E. D.: *Elementary Number Theory: An Algebraic Approach*. New York: W. A. Benjamin, 1970.
46. Borevich, Z. I and Shafarevich, I. R.: *Number Theory*. New York: Academic Press, 1966.
47. Budden, F. J.: *The Fascination of Groups*. Cambridge: Cambridge University Press, 1972.
48. Burn, R. P.: *Groups: A Path to Geometry*. Cambridge: Cambridge University Press, 1985.
49. Burton, D. M.: *A First Course in Rings and Ideals*. Reading, Ma.: Addison-Wesley, 1970.
50. Clark, A.: *Elements of Abstract Algebra*. New York: Dover, 1984 (orig. 1971).
51. Dobbs, D. and Hanks, R.: *A Modern Course on the Theory of Equations*. Passaic, NJ: Polynomial Publishing House, 1980.
52. Duncan, D. and Barnier, W.: "On trisection, quintisection, . . ., etc.," *American Mathematical Monthly* 89 (1982), 693.
53. Fendel, D.: "Prime-producing polynomials and principal ideal domains," *Mathematics Magazine* 58 (1985), 204–210.
54. Gaal, L.: *Classical Galois Theory with Examples*. New York: Chelsea, 1973.
55. Gallian, J. A.: *Contemporary Abstract Algebra*. Lexington: Heath, 1986.
56. Goldstein, L. J.: *Abstract Algebra: A First Course*. Englewood Cliffs, NJ: Prentice-Hall, 1973.
57. Grosswald, E.: *Topics from the Theory of Numbers*, 2nd ed. Boston: Birkhäuser, 1984 (orig. 1966).
58. Hadlock, C. R.: *Field Theory and Its Classical Problems*. Washington: Mathematical Association of America, 1978.
59. Hardy, G. H. and Wright, E. M.: *An Introduction to the Theory of Numbers*. Oxford: Oxford University Press, 1938.
60. Hatcher, W. S.: "Calculus is Algebra," *American Mathematical Monthly* 89 (1982), 362–370.
61. Herstein, I. N.: *Topics in Algebra*. New York: Blaisdell, 1964.
62. Jacobson, N.: *Basic Algebra*, Vol. 1. San Francisco: W. H. Freeman, 1974.
63. Kalmanson, K.: "A familiar construction criterion," *American Mathematical Monthly* 79 (1972), 277–278.
64. Kantor, I. L. and Solodovnikov, A. S.: *Hypercomplex Numbers* (A. Shenitzer, trans.). New York: Springer-Verlag, 1989.
65. Kurosh, A. G.: *Lectures on General Algebra*. New York: Chelsea, 1963.
66. Lyndon, R. C.: *Groups and Geometry*. Cambridge: Cambridge University Press, 1985.
67. Magid, A. R.: "Trigonometric identities," *Mathematics Magazine* 47 (1974), 226–227.
68. May, K. O.: "The impossibility of a division algebra of vectors in three-dimensional space," *American Mathematical Monthly* 73 (1966), 289–291.
69. McCoy, N. H.: *Rings and Ideals*. Washington: Mathematical Association of America, 1948.
70. McCoy, N. H.: *Introduction to Modern Algebra*. Boston: Allyn and Bacon, 1968.
71. Nikulin, V. V. and Shafarevich, I. R.: *Geometries and Groups*. New York: Springer-Verlag, 1987.
72. Niven, I. and Zuckerman, H. S.: *An Introduction to the Theory of Numbers*. New York: John Wiley, 1960.
73. Pollard, H. and Diamond, H. G.: *The Theory of Algebraic Numbers*. Washington: Mathematical Association of America, 1975.
74. Rademacher, H.: *Lectures on Elementary Number Theory*. New York: Blaisdell, 1964.
75. Rademacher, H.: *Higher Mathematics from an Elementary Point of View*. Boston: Birkhäuser, 1983.
76. Richards, I.: "An application of Galois theory to elementary arithmetic," *Advances in Mathematics* 13 (1974), 268–273.
77. Richman, F.: *Number Theory: An Introduction to Algebra*. Pacific Grove, Ca.: Brooks/Cole, 1971.
78. Robinson, A.: *Numbers and Ideals*. San Francisco: Holden-Day, 1965.
79. Rosen, J.: *Symmetry Discovered*. Cambridge: Cambridge University Press, 1975.

80. Roth, R. L.: "On extensions of Q by square roots," *American Mathematical Monthly* 78 (1971), 392–393.
81. Shubnikov, A. V. and Koptsik, V. A.: *Symmetry in Science and Art* (G. D. Archard, trans.). New York: Plenum Press, 1974.
82. Small, C.: "A simple proof of the four-squares theorem," *American Mathematical Monthly* 89 (1982), 59–61.
83. Stark, H.: *An Introduction to Number Theory*. Cambridge: MIT Press, 1978 (orig. 1970).
84. Stewart, I.: *Galois Theory*. London: Chapman and Hall, 1973.
85. Stewart, I.: *Concepts of Modern Mathematics*. Baltimore: Penguin, 1975.
86. Stewart, I. and Tall, D.: *Algebraic Number Theory*. London: Chapman and Hall, 1979.
87. Tuller, A.: *A Modern Introduction to Geometries*. Princeton: Van-Nostrand, 1967.
88. Van der Waerden, B. L.: *Modern Algebra*. Vol. 1. New York: Frederick Ungar, 1953.
89. Weiss, E.: *A First Course in Algebra and Number Theory*. New York: Academic Press, 1971.
90. Yale, P. B.: *Geometry and Symmetry*. San Francisco: Holden-Day, 1968.

Georg Friedrich Bernhard Riemann

Toward the Definition of an Abstract Ring

David M. Burton and Donovan H. Van Osdol

A definite characteristic of the mathematics of the last century has been its tendency towards abstraction. More and more, mathematicians have studied the relations between abstract entities defined in an arbitrary manner, restricted only in that these definitions must not lead to a contradiction. (Needless to say, those definitions which endure were first suggested by analogies to "real" objects.) As a course in the university curriculum, what is known as "abstract algebra" develops the properties of a number of abstractly defined systems, but too often the material is presented on a purely formal basis without reference to its historical origins. This led Jacques Barzun to complain in his *Teacher in America*:

> There is no sense of history behind the teaching [of algebra], so the feeling is given that the whole system dropped down ready-made from the skies, to be used only by born jugglers. [1, p. 82]

Of the major branches of what we call "modern" algebra, only the abstract theory of rings and ideals is entirely a product of the twentieth century. Indeed, almost the whole of ring theory as it is studied and taught today is the work of mathematicians of the last seventy years. This article gives a brief sketch of the early evolution of this engaging and useful concept. It derives from the firm conviction that a true appreciation of any branch of mathematics is impossible without some acquaintance with the history of that subject.

Unique Factorization for Algebraic Integers

The abstract theory was, needless to say, preceded by investigations of a more specialized nature. It was in the course of studying the ring of integers of an algebraic number field that Richard Dedekind (1831–1916) first introduced in 1871 the notion of an ideal. His goal was to do for an algebraic number field what Ernst Kummer (1810–1893) had done in the particular case of the cyclotomic integers built up from a prime root of unity. Kummer, in 1847, had saved the property of unique factorization into primes for cyclotomic integers by inventing the concept of "ideal prime factor." Dedekind's problem was how to define ideal prime factors of integers in an algebraic number field. In Supplement 10 to the second edition of Dirichlet's *Number Theory* (1871), Dedekind's new conception was to identify an ideal number with the set of all algebraic integers that it divides. In an adaptation of Kummer's terminology, he called such a set an "ideal." (One might view this Supplement as the birthplace of the set-theoretic approach to defining basic mathematical concepts.) This replaces the problem of defining an ideal number by that of defining ideals. The precise formulation of a Dedekind ideal is a "system" [set] I

of algebraic integers such that if α, $\beta \in I$, then $\alpha \pm \beta \in I$ and also $\mu\alpha \in I$ where μ is any algebraic integer.

In Dedekind's theory, a nonzero ideal I is prime if it has the property that $\alpha\beta \in I$ implies that either $\alpha \in I$ or $\beta \in I$. This definition permits him to establish the ideal factorization theorem in algebraic number fields: every ideal has a unique representation as a product of prime ideals. Thus, when prime ideals are used in place of prime numbers we obtain a natural generalization of the number theory of the ordinary integers.

The Study of Polynomials

The modern theory of rings had a double origin in the nineteenth century, the second source being the work of David Hilbert (1862–1945), Edmund Lasker (1868–1941) and F. S. Macaulay (1862–1927) on polynomial rings. In the second half of the nineteenth century, the ring $C[x_1, \ldots, x_n]$ of polynomials in n variables, where C is the complex number field, was studied in considerable detail by methods which depended on the special nature of the ring. In a monumental piece of work, "On the Theory of Algebraic Forms" (1890) [8], Hilbert proved what is now known as the Hilbert Basis Theorem: Every ideal I in $C[x_1, \ldots x_n]$ has a finite set of generators; that is, there exists a finite set $\{f_1, \ldots, f_m\}$ of polynomials in I such that any polynomial g in I can be expressed as $g = u_1 f_1 + \cdots + u_m f_m$, where the coefficients are again polynomials (not necessarily in I). The term "ring" itself, or *zahlring* (number ring), seems to have made its first appearance in Hilbert's 1897 paper "The Theory of Algebraic Number Fields" [10]. Previously, Dedekind had used the word *einordnung* (an order) to describe a similar system of algebraic numbers.

Shortly after the turn of the present century, Lasker [12], the world chess champion, established a ring-theoretic analogue of the Fundamental Theorem of Arithmetic for polynomial rings. Namely, in 1905 he showed that each ideal in $C[x_1, \ldots, x_n]$ admits a representation as an intersection of primary ideals. (An ideal I is primary if $\alpha\beta \in I$ implies that either $\alpha \in I$ or $\beta^k \in I$ for some positive integer k.) Then Macaulay [13] in 1913 gave an algorithmic process for determining the primary representation of an ideal of such a polynomial ring, given a finite set of generators for $C[x_1, \ldots, x_n]$.

Although there were a growing variety of rings by the end of the nineteenth century, the essential notions concerned in the concept of a ring (that is, the properties of its laws of composition and not the nature of the elements which make up the ring) only slowly came to be realized. A necessary prelude seemed to be an axiomatic foundation for the theory of commutative fields. The abstract definition of a field as we know it today was given in 1893 by H. Weber in an article on Galois theory [19]. Then, in 1903, L. E. Dickson (1874–1954) and E. V. Huntington (1874–1952) separately presented an explicit set of postulates together with a proof of their independence. Both papers appeared in the same issue of the *Transactions of the American Mathematical Society* under essentially the same title: "Definitions of a Field by Independent Postulates" [4], and "Definitions of a Field by Sets of Independent Postulates" [11], respectively! This early contribution to the modern postulation method (outside geometry) is noteworthy for the impetus it imparted to abstract algebra in the United States.

Fraenkel's Work on Rings

A forerunner in the abstract treatment of ring theory was the 1914 paper "On the Divisors of Zero and the Decomposition of Rings" [5] by Adolf Fraenkel (1891–1965). (Fraenkel was later to favor his middle name over his first name, and thus to publish his results as Abraham Fraenkel.) This article gives the first axiomatic characterization of the notion of a ring, although the definition is not the one in use today. Fraenkel's stated goal was to move away from studying only fields, so as to provide a comprehensive theory applicable to the integers modulo n, the p-adic numbers of Hensel, and the systems of "higher complex numbers" (hypercomplex number systems) studied by Weierstrass, Schwarz, Dedekind, and others.

Fraenkel's reference to hypercomplex number systems indicates another path taken en route to the abstract ring concept. These are systems whose "numbers" can be expressed in the form $x_1e_1 + x_2e_2 + \cdots + x_ne_n$, where the x_i are real numbers; the e_i are the fundamental units of the system. Addition of such numbers affords no difficulty and is defined by

$$\sum_{i=1}^{n} x_ie_i + \sum_{i=1}^{n} y_ie_i = \sum_{i=1}^{n}(x_i + y_i)e_i.$$

Multiplication is completely determined by the n^3 real numbers a_{ijk} which appear in the products

$$e_je_k = \sum_{i=1}^{n} a_{ijk}e_i \qquad (j, k = 1, 2, \ldots, n),$$

since the general definition of multiplication is

$$\left(\sum_{i=1}^{n} x_ie_i\right)\left(\sum_{i=1}^{n} y_ie_i\right) = \sum_{i,j=1}^{n} x_iy_je_ie_j.$$

The feature to notice is that multiplication will not be commutative or associative unless the numbers a_{ijk} are subject to certain restrictions. The conditions (due to Dedekind [3] in 1885):

$$a_{ijk} = a_{ikj}, \qquad \text{corresponding to} \qquad e_je_k = e_ke_j,$$

and

$$\sum_{i=1}^{n} a_{irs}a_{kit} = \sum_{i=1}^{n} a_{ist}a_{kri} \qquad \text{corresponding to} \qquad (e_re_s)e_t = e_r(e_se_t)$$

will make the system into a commutative ring. But as Karl Weierstrass (1815–1897) observed in 1884 [20], when $n > 2$ it is not possible for this ring to be a field since divisors of zero always exist.

The field of complex numbers is an obvious illustration of a hypercomplex number system. Here, there are two fundamental units $e_1 = 1$ and $e_2 = i$ obeying the familiar rules

$$1^2 = 1, \quad 1i = i1 = i, \quad i^2 = -1.$$

When there are four fundamental units $e_1 = 1$, $e_2 = i$, $e_3 = j$, $e_4 = k$ with the multiplication table

	1	i	j	k
1	1	i	j	k
i	i	-1	k	$-j$
j	j	$-k$	-1	i
k	k	j	$-i$	-1

then the resulting hypercomplex number system is the ring of real quaternions. This ring, described by William Hamilton (1805–1865) in 1843, provided the first example of a system in which all the axioms for a field are satisfied with one notable exception: multiplication need not be commutative [7]; nowadays such a system is called a skew field or division ring. The discovery of the quaternions soon inspired the manufacture of other hypercomplex number systems in which the most common laws of arithmetic did not necessarily apply. For instance, in 1845, Arthur Cayley (1821–1895) concocted his octonions or Cayley numbers, a system involving eight fundamental units where neither the commutative nor the associative law for multiplication holds [2].

Prompted as much by the subsequent studies of these systems as by the work of Dedekind and Hilbert on ideals, Fraenkel proposed an axiomatic declaration of exactly what constituted a ring. His formulation aimed to be general enough to subsume the structure of both commutative and noncommutative rings. He asserted that one should study the consequences of general algebraic laws, and use special properties of a given ring only when they are necessary.

According to Fraenkel, a ring is a set R subject to ten conditions. The first four, r1–r4, simply specify that with respect to a given operation "+" (called addition), R is a group. The commutativity of addition is not assumed. The remaining six conditions are:

r5. There is given an operation $*$ called multiplication;
r6. Multiplication is an associative operation;
r7. R contains a right identity element 1 for multiplication;
r8. Regular elements of R (see below) have multiplicative inverses;
r9. For any $a, b \in R$ there exist regular elements $\alpha_{a,b}$ and $\beta_{a,b}$ such that $a * b = \alpha_{a,b} * b * a$ and $a * b = b * a * \beta_{a,b}$;
r10. Multiplication satisfies the left and right distributive laws over addition.

Fraenkel divides the elements of R into two parts: the divisors of zero (those elements a for which there is a nonzero b with $a * b = 0$) and the regular elements. The effect of r5–r8 is then to force the set of regular elements of R to be a group with respect to multiplication. Notice that the system of integers does not satisfy Fraenkel's definition.

He points out that r1–r7 and r10 are identical to conditions assumed by Dickson and Loewy in their work on fields. The conditions r5–r8, says Fraenkel, allow R (with respect to multiplication) to form a group "to the extent possible, namely with the exception of the divisors of zero." He showed that the element 1 is actually a two-sided identity for multiplication by use of condition r9. That implies that $a = a * 1 = 1 * a * \beta_{a,1}$ for some $\beta_{a,1}$ in R; hence, upon multiplying on the left by 1,

$$1 * a = 1^2 * a * \beta_{a,1} = 1 * a * \beta_{a,1} = a$$

for any a in R. Fraenkel's argument for ensuring that addition is commutative is equally interesting. For arbitrary elements a and b in R, he appeals to the distributive laws twice; first, to obtain

$$(a + b) * (1 + 1) = (a + b) * 1 + (a + b) * 1 = a + b + a + b,$$

and thereafter to arrive at

$$(a + b) * (1 + 1) = a * (1 + 1) + b * (1 + 1) = a + a + b + b.$$

Equating these expressions and using the cancellation law yields the commutativity of addition.

The existence of divisors of zero, and the impossibility of division by arbitrary elements, lead to difficulties. As an example, Fraenkel points out that a polynomial of degree n with coefficients from a ring may well have more that n roots—even infinitely many. Hensel had been able to deal with some of these problems for the p-adic numbers by reducing them to the consideration of finitely many associated fields. Fraenkel's motivation was to see to what extent such a result could be proved for his abstractly given rings.

He defines a prime divisor of zero to be a divisor of zero in R which, aside from regular elements, has no proper divisors. A ring R is defined to be decomposable if it satisfies "condition Z":

> If a, b, $c \in R$ and $a * b$ is divisible by c, then c can be written in at least one way as $c = c_1 * c_2$ so that a is divisible by c_1 and b is divisible by c_2.

For Fraenkel, the analog of a field was to be what he called a simple ring. A ring R is simple if all of its prime divisors of zero are equivalent to each other (that is, if any two prime divisors of zero divide each other). Notice that this is not the current use of the word "simple" in ring theory. His fundamental theorem (Satz 7, page 172) is then: each decomposable ring R possessing only n essentially different prime divisors of zero can be written in only one way as a direct sum of n simple rings.

One problem Fraenkel faced was that he knew he wanted to be able to consider certain elements as equal even when they were not. But he was unable to take this conceptual leap and simply define a new set whose elements were the equivalence classes determined by the new "equality" relation. Fraenkel explicitly states that the relation "=" is to be a reflexive, symmetric, transitive relation, and that any operations defined are to have the property that equally defined elements can replace each other in making calculations. That is, he assumes that = is a congruence relation. But he does not study the relationship between congruence relations and ideals as Sono later did, nor does he consider quotient rings *per se*.

Six years later in 1920, in a paper entitled "On Simple Extensions of Decomposable Rings" [6], Fraenkel moved further in the direction of the modern definition of a ring. For him, a (commutative) ring is now a "system" R with two operations, addition and multiplication, satisfying the following five conditions:

1. The associative law for addition.
2. The associative law for multiplication.
3. The commutative law for multiplication.
4. The distributive law of multiplication over addition.
5. For any a and b in R, there exist unique elements x and y satisfying the equations $a + x = b$ and $y + a = b$.

The commutative law for addition still does not appear as an axiom, but is derived from an additional assumption concerning regular elements, namely:

6. R contains at least one regular element; and if a is a regular element then the equation $a * x = b$ has a solution for any element b in R.

This last requirement implies the existence of a multiplicative identity e (from which as above the commutativity of addition can then be obtained), as follows. Take a to be a regular element and let e satisfy $a * e = a$. For any element r in R, solve the equation $a * x = r$; then

$$r * e = (a * x) * e = a * (x * e) = a * (e * x) = (a * e) * x = a * x = r,$$

so that e is an identity for multiplication.

Fraenkel's goal in this paper was to see to what extent some results of Steinitz in 1910 ("Algebraic Theory of Fields" [10]) could be generalized to his decomposable or simple rings. The difficulties which he encounters are of a "different type and essentially deeper" than those encountered in the theory of fields. The main tool which he uses is a version of the Euclidean algorithm.

For a decomposable ring R, Fraenkel is interested in simple (in the current usage) extensions $R(\alpha)$, where α may be either algebraic or transcendental over R. First of all, do such extensions even exist? For transcendental α, the answer is "yes." He first constructs $R(X)$ when R is a simple ring with essentially unique prime divisor of zero p, as follows. Let $R(X)$ consist of all formal quotients with coefficients from R:

$$\frac{a_0 + a_1 X + a_2 X^2 + \cdots + a_m X^m}{b_0 + b_1 X + b_2 X^2 + \cdots + b_n X^n},$$

subject only to the condition that not all of the b_i should be divisible by p, and that therefore at least one of them should be regular. Fraenkel calls polynomials of this latter type primitive functions; the largest k such that b_k is regular is called the order of the polynomial. Here "X should in general be considered as a mere symbol, for which occasionally an element of R may be substituted." Of course, two such quotients are to be considered equal under the same conditions that two rational numbers are considered equal. In short, Fraenkel localized $R[X]$ at the multiplicatively closed set of all polynomials satisfying the specified condition. He proves that $R(X)$ is itself simple whenever R is and that the prime divisors of zero in $R(X)$ are all equivalent to those for R. For general decomposable R (with finitely many inequivalent prime divisors of zero) he then defines $R(X)$ to be the direct sum of the $R_i(X)$, where R is the direct sum of the simple rings R_i (as above).

For α algebraic, things proceed differently. A polynomial $f(X)$ in $R[X]$ is called arithmetically divisible by a polynomial $g(X)$ in $R[X]$ if there exists a polynomial $h(X)$ in $R[X]$ with $f(X) = g(X)h(X)$; $f(X)$ is irreducible if its only arithmetic divisors are polynomials of order zero and polynomials with the same order as $f(X)$. Then α is algebraic over R if there is an irreducible primitive $f(X)$ in $R[X]$ with $f(\alpha) = 0$. In this case, Fraenkel defines $R(\alpha)$ to be $R[X]$ modulo the ideal generated by $f(X)$—but not, of course, in those terms. Rather, it is $R[X]$, but with equality defined by considering $g(X)$ and $h(X)$ to be equal if their difference is arithmetically divisible by $f(X)$. He proves that $R(\alpha)$ is a ring, but in general is not decomposable—even if R is simple. With more restrictive hypotheses, he is able to get some results concerning the decomposability of rings of the form $R(\alpha)$ with α algebraic over R.

The version of the Euclidean algorithm which Fraenkel proves is: if R is a simple ring, and $f(X)$, $g(X)$ are polynomials in $R[X]$ with $g(X)$ primitive, then there exist unique polynomials $G(X)$ and $h(X)$ in $R[X]$ so that $f(X) = g(X)G(X) + h(X)$ and the degree of $h(X)$ is less than the order of $g(X)$. He proves this as a corollary of his Fundamental Theorem: if R is a simple ring and $f(X)$ is a primitive polynomial of order n in $R[X]$, then there is a primitive polynomial $g(X)$ in $R[X]$ of order zero such that $f(X)g(X)$ has both degree and order equal to n. The proof requires six pages to reach its conclusion.

The Work of Masazo Sono

The currently used definition of a (commutative) ring seems first to have appeared in 1917 in a paper written in English and entitled "On Congruences" [16] by the Japanese mathematician Masazo Sono. Sono seems to be relatively unknown in the West, but his legacy is great. According to Professor Kakutani, one of Sono's students was Akizuki, who in turn was the teacher of both Hironaka and Matsumura!

Sono defines an abstract ring by using nine postulates, "after consideration of the points which are essential in rings." The axioms look very familiar:

1. If $a, b \in R$, then $a + b \in R$.
2. If $a, b \in R$, then $a + b = b + a$ (commutative law).
3. If $a, b, c \in R$, then $(a + b) + c = a + (b + c)$ (associative law).
4. There exists $z \in R$ such that $z + b = b$ for every $b \in R$.
5. Corresponding to every $a \in R$, there exists in R another x such that $a + x = z$, where z is the element referred to in 4.
6. If $a, b \in R$, then $a * b \in R$.
7. If $a, b \in R$, then $a * b = b * a$ (commutative law).
8. If $a, b, c \in R$, then $(a * b) * c = a * (b * c)$ (associative law).
9. If $a, b, c \in R$, then $a * (b + c) = (a * b) + (a * c)$ (distributive law).

Sono goes on to say that "this set of postulates is part of a set by which Prof. Dickson has defined a field abstractly; hence, there is no necessity of again showing the independence of the postulates." Sono is referring to Dickson's 1903 article: "Definitions of a Field by Independent Postulates." [4]

Since the first five of his postulates guarantee that every ring is a commutative group with respect to addition, Sono notes that the additive identity and additive inverses are unique. Thereafter the element z in the definition is denoted by the symbol 0 and the element x corresponding to a by the symbol $-a$. Sono then presents the now familiar proof that $a * 0 = 0$:

$$a * 0 + a * b = a * (0 + b) = a * b = 0 + a * b.$$

Given that commutative ring theory was just getting established, the approach taken by Sono is surprisingly modern. He starts with an equivalence relation, or "congruence" that is compatible with addition and multiplication; that is, if $a \equiv a'$ and $b \equiv b'$, then $a + b \equiv a' + b'$ and $a * b \equiv a' * b'$. As Sono puts it, "Suppose the definition of equality is changed so that R remains a ring." Then the set M of elements which are congruent to zero form an ideal of R and two elements congruent to one another belong to the same coset of M. It is then shown that the cosets of M themselves form a ring under the usual operations

$$(a + M) + (b + M) = (a + b) + M$$
$$(a + M) * (b + M) = (a * b) + M$$

Conversely, any ideal M of R gives rise to a congruence on R compatible with the ring operations; simply take $a \equiv b \pmod{M}$ whenever $a - b$ lies in M. Thus, "ideals must be naturally introduced as moduli in congruences."

Sono then "adds a few theorems on rings." Among them is the now standard result that if B is an ideal of R which contains another ideal A, then B/A is an ideal of the quotient ring R/A. Also among them is the so-called Third Isomorphism Theorem, which states that if A and B are two ideals of the ring R, then the quotient rings $(A + B)/B$ and $A/(A \cap B)$ are isomorphic. Notice in this regard that it is important that Sono's rings are not required to possess multiplicative identities; otherwise these quotients are not even rings! One of Sono's main contributions is his characterization of simple rings R (that is, rings having no ideals other than R and $\{0\}$). To be more specific, he establishes that a simple ring R is either a field or one which consists of p elements, (p a prime): $0, v, 2v, \ldots, (p-1)v$, where $pv = 0$ but $kv \neq 0$ for $0 < k < p$ and $v^2 = 0$.

Emmy Noether

The mathematician who best advanced the abstract point of view in ring theory was Emmy Noether (1882–1935). Indeed, the abstract theory of rings is frequently said to have been inaugurated by her 1921 paper "Ideal Theory in Rings" [14]. Its elegant, axiomatic treatment of the material was novel at the time. As Kaplansky has aptly written, "The importance of this paper is so great that it is surely not much of an exaggeration to call her the mother of modern algebra."

Noether was apparently unaware of Sono's 1917 article. Owing to its relatively obscure source and the fact that it was published during the Great War, this is not entirely surprising. In framing her concept of an abstract ring, Noether cites both Fraenkel's 1914 paper and the 1920 article as references; she remarks that the four different decompositions of a prime ideal of which she speaks in her paper coalesce in the presence of Fraenkel's added hypotheses. Noether's and other definitions of abstract rings given at the time applied to commutative rings only. She employs the same five axioms that appeared in Fraenkel's second paper, adding a sixth postulate which explicitly calls for the commutativity of addition. Thus, the modern definition of a (commutative) ring was firmly in place by 1921, and thereafter began to take hold. It is interesting to note that the subject was in such an early stage of development that Noether felt compelled to give a proof (in a footnote) that a commutative ring with identity has a unique multiplicative identity.

In "Ideal Theory" [14] Noether generalized Lasker's ideal theory to rings satisfying the conclusion of Hilbert's basis theorem. Specifically, she proved that each ideal in a ring is finitely generated if and only if the ascending chain condition on ideals holds. (A ring satisfies the ascending chain condition provided every sequence $I_1, I_2, I_3, \ldots$ of ideals with $I_1 \subset I_2 \subset I_3 \subset \cdots$ has only a finite number of terms.) Noether demonstrated that in rings with the ascending chain condition an arbitrary ideal can be represented as a finite intersection of primary ideals. It is appropriate that rings with the ascending chain condition are now usually called "Noetherian."

In a subsequent (1927) paper, "Abstract Study of Ideal Theory in Algebraic Number- and Function-fields" [15], Noether undertook to do for an abstract ring what Dedekind had done for any ring of algebraic numbers. The opening sentence indicates that the purpose of the paper is to provide an abstract characterization of those rings whose ideal theory agrees with that of the ring of algebraic integers of an algebraic number field. She gave five axioms for what are now called Dedekind rings; and then proved that, in any ring in which these axioms are satisfied, each proper ideal is uniquely expressible as a finite product of prime ideals. The long study of the decomposition of ideals started in the previous century was thus completed with Noether's ring-theoretic version of the Fundamental Theorem of Arithmetic.

Conclusion

The revolutionary idea of working abstractly with rings and their ideals—first put forth by Fraenkel and Sono and later advanced by Noether's two landmark papers—essentially created the subject of what is now called Commutative Algebra. Within four years, van der Waerden's famous *Modern Algebra* [18] made all of these beautiful ideas (and more) readily available to a new generation of algebraists. The struggle to arrive at the "correct" setting for the study of factorization and other equally important algebraic ideas provides an interesting case study in how the science of mathematics is advanced over the course of time. Certainly the subject of ring theory has not "dropped down ready-made from the skies," and we owe it to our students to make them aware of this fact.

For Further Consideration

1. Suppose that $=$ is a relation on a set X which satisfies the following two conditions:
   (a) it is reflexive;
   (b) for all $a, b, c \in X$, if $a = b$ and $a = c$, then $b = c$.
   Prove that $=$ is an equivalence relation on X.

2. Suppose that R is a ring; prove that if $a * b = 0$ then $b * a = 0$.

3. If 0 is the only divisor of zero then r9 is superfluous.

4. Divisors of zero do not have multiplicative inverses.

5. $\{x : x \text{ is a regular element}\} = \{x : x \text{ is invertible}\}$.

6. In Z_{360} (the ring of integers modulo 360), $25 \sim 5$ and $7 \sim 1$, but 4 does not divide 2. (Here $a \sim b$ if each divides the other; a divides b if there exists c with $b = a * c$.)

7. If $\alpha * a * \alpha' = \beta * b * \beta'$ with $\alpha, \alpha', \beta, \beta'$ regular then $a \sim b$. The converse is false.

8. To show that r9 is independent of the other defining conditions for a ring, Fraenkel considers a "quaternion" system over Z_8. Let R consist of all $a_0 + 2a_1i + 2a_2j + 2a_3k$ where all of the $a_i \in Z_8$. For purposes of calculation, suppose that $ai = ia$ for all $a \in Z_8$ and similarly for j and k. Notice that i, j, k are not themselves in R, but suppose that they multiply according to the following table:

	i	j	k
i	2	k	$2j$
j	$2k$	2	i
k	j	$2i$	2

Notice that R has 512 elements. Prove that $a_0 + 2a_1i + 2a_2j + 2a_3k$ is a divisor of zero if and only if a_0 is divisible by 2. Otherwise, its inverse is

$$b_0 + 2b_0^3(2a_2a_3 - a_0a_1)i + 2b_0^3(2a_3a_1 - a_0a_2)j + 2b_0^3(2a_1a_2 - a_0a_3)k,$$

where $a_0b_0 = 1$ in Z_8. Using $2i$ and $2j$, show that r9 does not hold in this ring.

9. Give an example of a polynomial which has more roots than its degree.

10. In Z_{360}, the prime divisors of zero are of three types: those divisible by 2, but by no higher power of 2 nor by 3 nor by 5 (2, 14, 22, 26, 34,...); those divisible by 3, but by no higher power of 3 nor by 2 nor by 5 (3, 21, 33, 39, 51,...); and those divisible by arbitrary powers of 5, but not by 2 nor by 3.

11. Show that $Z_{360} = Z_8 \oplus Z_9 \oplus Z_5$ is an example of Fraenkel's fundamental theorem (here we identify Z_8 with $\{0, 45, 90, 135, 180, 225, 270, 315\}$; Z_9 with $\{0, 40, 80, 120, 160, 200, 240, 280, 320\}$; and Z_5 with $\{0, 72, 144, 216, 288\}$).

12. Prove that if R is a simple ring with prime divisor of zero p, if a_i is the coefficient of $a_0 + a_1X + a_2X^2 + \cdots + a_mX^m$ in $R[X]$ with largest subscript among all a_k which are regular, and if b_j has the same property for $b_0 + b_1X + b_2X^2 + \cdots + b_nX^n$ in $R[X]$, then the product of these two polynomials has the property that the coefficient of X^{i+j} is itself regular. (In fact, the order of a product is the sum of the orders of its factors.) Thus the set of polynomials which Fraenkel uses as denominators for $R(X)$ is a multiplicatively closed set.

13. In $Z_8[X]$, $X^2 + 3$ is a primitive irreducible function. Denote one of its roots by $\sqrt{5}$; then $1 + \sqrt{5}$, $1 + 7\sqrt{5}$, and 2 are prime divisors of zero in the ring $Z_8(\sqrt{5})$. Compute $(1 + \sqrt{5}) \times (1 + 7\sqrt{5})$ and conclude that unique factorization does not hold in $Z_8(\sqrt{5})$.

14. Let p be a prime integer. Prove that $X/(X + p)$ satisfies over Z_{p^2} the polynomial $Y^2 - 2Y + 1$.

15. In a finite commutative ring R with identity, every nonzero element is either a zero-divisor or invertible; hence R is a ring in the sense of Fraenkel.

Bibliography

1. Barzun, Jacques: *Teacher in America*. Boston: Little, Brown, 1945.
2. Cayley, Arthur: "On Jacobi's Elliptic Functions, in Reply to Rev. Brice Brownin and on Quaternions," *Philosophical Magazine and Journal of Science* 3 (1845), 210–213.
3. Dedekind, Richard: "Zur Theorie der aus n Haupteinheiten gebildeten complexen Grössen," *Göttingen Nachrichten* (1885), 141–159.
4. Dickson, Leonard: "Definitions of a Field by Independent Postulates," *Transactions of the American Mathematical Society* 4 (1903), 13–20.
5. Fraenkel, Adolf: "Über die Teiler der Null und die Zerlegung von Ringen," *Journal für die Reine und Angewandte Mathematik* 145 (1914), 139–176.
6. Fraenkel, Adolf: "Über einfache Erweiterungen zerlegbarer Ringe," *Journal für die Reine und Angewandte Mathematik* 151 (1920), 121–166.
7. Hamilton, William: "On a New Species of Imaginery Quantities Connected with the Theory of Quaternions," *Proceedings of the Royal Irish Academy* 2 (1844), 424–434.
8. Hilbert, David: "Uber die Theorie der Algebraischen Formen," *Mathematische Annalen* 36 (1890), 473–534.
9. Hilbert, David: "Über die Vollen Invariantensysteme," *Mathematische Annalen* 42 (1893), 313–373.
10. Hilbert, David: "Die Theorie der algebraischen Zahlkörper," *Jahresbericht der Deutschen Mathematikervereinigung* 4 (1897), 175–546.
11. Huntington, Edward V.: "Definitions of a Field by Sets of Independent Postulates," *Transactions of the American Mathematical Society* 4 (1903), 31–37.
12. Lasker, Edmund: "Zur Theorie der Modulen und Ideale," *Mathematische Annalen* 60 (1905), 20–116.
13. Macaulay, Francis: "On the Resolution of a given Modular System into Primary Systems," *Mathematische Annalen* 74 (1913), 66–121.
14. Noether, Emmy: "Idealtheorie in Ringbereichen," *Mathematische Annalen* 83 (1921), 24–66.
15. Noether, Emmy: "Abstrakter Aufbau der Idealtheorie in algebraischen Zahl- und Funktionenkörpern," *Mathematische Annalen* 96 (1927), 203–226.
16. Sono, Masazo: "On Congruences," *Memoirs of the College of Science of Kyoto* 2 (1917), 203–226.
17. Steinitz, Ernst: "Algebraische Theorie der Körper," *Journal für die Reine und Angewandte Mathematik*, 151 (1910), 167–309.
18. van der Waerden, B. L.: *Moderne Algebra*, vol. II. Berlin: Springer-Verlag, 1931.
19. Weber, Heinrich: "Beweis des Satz, das jede eigentlich-primitive quadratische Form unendliche viele Primzahlen darzustellen fähig ist," *Mathematische Annalen* 43 (1893), 521–549.
20. Weierstrass, Karl: "Zur Theorie der aus n Haupteinheiten gebildeten complexen Grössen," *Göttingen Nachrichten* (1884), 395–414.

Emmy Noether

In Hilbert's Shadow

Notes Toward a Redefinition of Introductory Group Theory

Anthony D. Gardiner

> Nothing is more difficult than to begin.
> And what is more fundamental than a good beginning.
> —Euripides

The Basic Problem

In his lovely article "What groups mean in mathematics and what they should mean in mathematical education," Hans Freudenthal starts out from a child's drawing and leads on to a sequence of unusual examples from other branches of mathematics and science.

> In the examples I have displayed, experts will have recognized a common feature—
> each of them shows in its particular way how groups arise and are used to study
> regularities in nature and in mathematics. Why did I refrain from using the word
> groups? If systematics is pursued, one starts by defining a group, continues by proving
> a few general theorems about the group concept, then develops the general principles
> according to which groups can be applied, and finally arrives at some applications
> of groups according to these principles, provided of course that sufficient time is left
> for this minor concern. Yet mathematics develops systematically only in an 'objective
> mind.' In the individual it takes the path from the particular case to the general
> principle, from the concrete to the more abstract, and so it happens in history too.
> Groups and group theory methods preceded the conscious organization of this complex
> of investigations in terms of the explicit group concept by at least half a century. This
> is a common way in mathematics. In order to organize a field of knowledge you first
> have to acquire knowledge about it by exploring it. [11, pp. 106–107]

What form then should this exploration take? Freudenthal fails to stress that it must take place in a familiar context that is sufficiently rich to embody the essential ideas of the new field and, for those interested in using a historical approach to introduce group theory, there's the rub. At that stage students' understanding of mathematical topics such as the solution of equations, nineteenth century geometry, or differential equations is very limited.

So how should one begin? Whatever approach one chooses, Freudenthal insists

> Fundamental definitions do not arise at the start but at the end of the exploration,
> because in order to define a thing you must know what it is and what it is good for.
> [11, p. 107]

In real life this step from preliminary exploration to final definition can be both painful and messy, as Samuel Butler observed in his *Notebooks*:

Every new idea has something of the pain and peril of childbirth about it. [5]

Unfortunately neither Freudenthal, nor anyone else, has shown us how this transition—from exploration to formal presentation—can be achieved. The struggle to give birth, whether to infants or ideas, is generally felt to be a private matter. This is one reason why popular texts find it so tempting to skip the messy exploration. Most algebra texts betray little by way of self-doubt when it comes to deciding how the subject is best organized for the beginner.

A strong characteristic of the mathematics of the last hundred years or so has been its tendency to go abstract. To a non-mathematician this statement may seem ludicrous; after all, he could say, mathematics is already totally abstract. ... however, in the framework of mathematics we can have diverse levels of abstraction.

This characteristic will be well illustrated in what we are about to do in this chapter. We have just studied something about permutations of a set of objects. Looking at the properties enjoyed by permutations we make a long jump, extracting what final properties seem essential to that set of permutations to define a formal algebraic system called a group. ... We begin immediately with the definition of a group ...

Definition: A non-empty set G is said to be a **group** if in G there is defined an operation $\circ$ such that ... [14, pp. 129–130]

The chapter in this text which precedes this sudden leap into abstraction has several nice features, in that it includes a very clear and elementary discussion of cyclic decomposition, even and odd permutations and their application to combinatorial problems, riffle shuffles, and the Josephus problem. However, though these features may help students get used to permutation notation, it is hard to see how any of it can be said to convince students that the properties about to be extracted in the definition of a group are precisely "the formal properties [which] seem essential in that set up." Permutations which, up to this point, have been studied as individual entities cannot suddenly be conceived as ensembles. Moreover, in the absence of any explicit discussion of the role of associativity, or the need for inverses, the permutation setup is more likely to camouflage rather than reveal their central importance.

"Complete axiomatization is the obituary of an idea." [source unknown] What we are faced with in this situation is an attempt to motivate which fails to address a beginner's main difficulties. Moreover, one suspects that even this attempt was included only because the book is designed to serve a course for non-math majors—I know of nothing like it in any text written by either author for those majoring in mathematics.

It is easy to criticize this familiar cold-blooded approach, but it is too widespread to be simply shrugged off. One has to assume that some authors have tried to do things differently but have felt obliged, for one reason or another, to revert to the no-nonsense axiomatic approach. A good example is the recent text by Allenby [1] in which one can see the author trying to motivate—using Lagrange's *Réflexions sur la Résolution Algébrique des Equations*—before scurrying back to a none too inspired version of traditional systematics.

What drives grown men and women who love their subject to such things? Is there really no alternative? The question is all the more interesting since most of these authors certainly admire Herman Weyl's marvelous essay, *Symmetry*, yet find his clearly laid out scheme hard to implement.

The course these lectures will take is as follows. First I will discuss bilateral symmetry in some detail and its role in art as well as organic and inorganic nature. Then we shall generalize this concept gradually, ... first staying within the confines of geometry, but then going beyond these limits through the process of mathematical abstraction along a road that will finally lead us to a mathematical idea of great generality, the Platonic ideal as it were behind all the special appearances and applications of symmetry. To a certain degree this scheme is typical of all theoretical knowledge. We begin with some general but vague principle (symmetry in the first sense), then find an important case where we can give that notion a concrete precise meaning (bilateral symmetry), and from that case we gradually rise again to generality, guided more by mathematical construction and abstraction than by the mirages of philosophy, and if we are lucky we end up with an idea no less universal than the one from which we started. Gone may be much of its emotional appeal, but it has the same or even greater unifying power in the realm of thought and is exact instead of vague. [27, pp. 5–6]

A common superficial reaction is merely to assert that textbook authors should follow this admirable plan. However, Weyl is not trying to teach introductory group theory. Those who do try to follow his plan find it exceedingly difficult to mesh their chosen motivation with their eventual approach to group theory itself—a stage which Weyl never reaches. This leaves unanswered the question as to whether it is possible for a textbook to bring out this transition from vague beginnings to full generality via a precise special case.

Mathematicians who do not like the idea of discussing their vague beginnings in public often forget that, though they themselves have outgrown the need for such vague beginnings, their students have not.

There is nothing wrong with vague questions; it is the combination of vague questions and vague answers that is bad. [16]

The whole of science is nothing more than a refinement of everyday thinking. [8]

Group Theory at School Level

In the late 1960s there were numerous advocates of the benefits of teaching the elements of group theory to high school students. Here the basic problems are more marked and more clearly visible. Some authors evidently preferred a no-nonsense abstract approach.

PREFACE

The role of the notion of 'group' as a fundamental element of mathematics being universally recognized, it is desirable that the elements of the theory of groups should be taught in secondary school, not as a supplementary chapter, but as a central element in the construction of the mathematical edifice. .... The author hopes that a great part of the material of this book will be integrated into the secondary curriculum.

1. INTRODUCTION

In its most elementary parts, mathematics makes systematic use of binary operations. ... An internal binary operation defined on the set E is a function from $E \times E$ to E. .... If the operation $*$ is an internal operation defined on E, we say that the set E is provided with the operation $*$, and we will call $E, *$ an algebraic structure. [21, pp. vii, 1]

Other authors such as Zoltan Dienes envisage a more humane approach.

Symbol manipulation in mathematics is all too often utterly meaningless simply because there is no corresponding transformation of images. [7]

This is all too true. However Dienes gives one no convincing evidence that he, any more
than those who would criticize, has any real idea how to generate the "images" relevant to
introductory group theory. Indeed what follows suggests that he has allowed himself to be
carried away by his own sales talk.

What is being suggested is simply that the type of game which the research mathematician plays with his apparently austere toys should be translated to the level of
secondary school children, thereby giving them a taste and so almost certainly a *liking* [my italics] for this most exciting and most human of all games: building up and
playing with mathematical structure. [7, p. 103]

Such naiveté would be excusable, perhaps even admirable, in an adolescent. But one has
to point out that Dienes was not alone, and that he was taken seriously. (Some of his earlier
proposals for curricular change had been imaginative and timely, with the possible result that
people were too uncritical of his subsequent ideas.)

The preliminary motivations which appeared in print and which were meant to cultivate
this taste, even liking, for group theory among high school students never made contact with
elementary group theory itself. And the attempts to dabble in theory at this level never got far
enough to make the exercise convincing. The exam questions for school leavers which were
until recently set in the UK to assess the topic caused embarrassment even to those who set the
questions and were thoroughly disliked by many teachers.

Alternatives

Could we have done better? Or does the apparently simple facade of elementary group theory,
with its many beautiful links with symmetry and ornament, and with science and nature, conceal
unsuspected and unavoidable subtleties which are way beyond high school students, and which
remain difficult even for undergraduates? The modern abstract treatment certainly seems both
simple and elegant to the initiated. But as in science, it is easy to forget what this simplicity
ultimately derives from.

The simplicities of natural laws arise through the complexities of the language we
used for their expression. [29]

In the realm of algebra, history has another significant tale to tell. The use of a letter
to represent a fixed but unknown number dates from Greek times. However, the use
of a letter or letters to stand for a whole class of numbers was not conceived until
the late sixteenth century. ... Why was the use of letters for general coefficients so
long delayed? The answer would seem to be that this device constitutes a higher
level of abstraction in mathematics, a level farther removed from intuition. It is more
difficult to think about $ax^2 + bx + c = 0$ than about $3x^2 + 5x + 6 = 0$. Yet to reason
deductively about the algebraic procedures for significant general expressions became
possible only after these general coefficients were introduced. [19, pp. 36–37]

If the use of symbols to denote known classes of familiar numbers presents difficulties,
what is the student to make of symbols that stand for nothing in particular?

For Hilbert the words point, line, and plane, were so pregnant with meaning that he had to struggle to insist that they must be formally treated as meaningless terms (table, chair, mug) which obey certain laws. For us the problem is the other way round. When we calculate, or ask our students to calculate, with hypothetical elements a, b, or with subgroups H, K, in a hypothetical group G, we have achieved at a stroke what Hilbert found so hard. For our students these symbols are utterly devoid of meaning! Yet it is meaning which guides the mathematician's manipulations, even if these manipulations must subsequently be justified in term of the formal rules.

> Reason alone cannot create truth. The attempt to construct a theory of things by the play of empty formulas is as vain a pretence as that of the weaver who would produce linen without putting thread in his shuttle. [22]

> The whole modernist argument rests essentially on the assumption: By making the implicit mechanisms, or techniques, of thought conscious and explicit [via axioms] one makes these techniques easier. ... [However] knowing the theory of Freudian slips will not necessarily prevent you from making one.... The real problem which confronts mathematics teaching is not that of rigor, but the problem of the development of 'meaning,' of the existence of mathematical objects. ... One has not, I believe, extracted from Hilbert's axiomatics the true lesson to be found there; it is this: One accedes to rigor only by eliminating meaning; absolute rigor is only possible in and by, such destitution of meaning. But if one must choose between rigor and meaning, I shall unhesitatingly choose the latter. It is this choice one has always made in mathematics. [25, pp. 197, 198, 202]

That Hilbert himself understood the primacy of meaning is suggested by the quotation from Kant with which he opens his *Foundations of Geometry*.

> *All human knowledge thus begins with intuitions* [my italics], proceeds thence to concepts and ends with ideas. [15, p. 1]

Lesser mortals have missed the point that such a treatment is, in some sense, the final stage. We have been paying the price ever since. The author of the Foreword to the 1971 Open Court translation of the *Foundations of Geometry* ended thus:

> Hilbert was a giant and we are fortunate to be able to live in his shadow. [15, p. iii]

The word "shadow" here is deliciously double-edged (even if the word "fortunate" is perhaps not quite so fortunate). René Thom would clearly like us to escape from some aspects of this shadow.

> Pedagogy must strive to recreate (according to Haeckel's law of recapitulation—ontogenesis recapitulates phylogenesis) the fundamental experiences which, from the dawn of historic time, have given rise to mathematical entities. Of course this is not easy for one must forget all the cultural elaboration (of which axiomatics is the last) which have been deposited on these mathematical objects, in order to restore their original freshness. One must forget culture in order to return to nature. The modernist tendency represents, on the contrary, an increase of culture to the detriment of nature. ... The mathematical community has in these last years allowed itself to be led astray by declarations and ill-considered promises. There has been talk of a 'revolution in mathematics' and assertions that, thanks to new syllabuses and new

methods, the most average pupil would be able to complete his secondary studies in mathematics. It is time to put a stop to these utterances which border on deception. No miracle is possible and one can only hope to ameliorate the situation step by step, and by small local improvements. [25, pp. 206, 208]

Thom, like some of the other critics of unmotivated axiomatic formalism we have quoted, was responding mainly to the premature formalization of familiar mathematical objects from traditional high school mathematics. These were used and understood (in some nontrivial sense) by mathematicians for many years before they were eventually axiomatized. For such systems the axiomatization tells us very little that is new—at least on an elementary level; it merely systematizes "facts" that were previously familiar for other reasons.

The same is by no means true for groups. Though many interesting results were discovered before the triumph of the abstract approach (in the 1890s and thereafter), it was in some sense impossible to think properly about groups before this. For example, the historical emphasis on permutation groups focused early attention on many questions which, from a strictly group theoretical viewpoint, are either wrongly posed, or of secondary importance (such as the idea that each group comes with a single well-defined degree). Moreover, when teaching introductory group theory, there is no simple "nature" to which one could return (to use Thom's language): the mathematical experiences needed in order to introduce group theory "concretely" are no longer (and perhaps never were) part of students' backgrounds.

Some writers are prepared to go out on a limb and hint at possible ways of getting round this dilemma.

> After so many examples of what groups mean in mathematics and beyond, it is time to single out the factor common to all of them: Groups are important because they arise from structures as systems of automorphisms of those structures. ... The automorphisms of a structure form a group with composition as group operation. ... The way of introducing guarantees that the thing defined is a group; rather than by an algorithmic verification, this result is obtained in one conceptual blow, and this is a great advantage. Preferring conceptual to algorithmic approaches is one of the most conspicuous features of what is really modern in modern mathematics.... It is not good mathematics to trust algorithms better than insight. [11, pp. 109–110]

However, in practice, group theory is almost universally taught and learnt—or not learnt as the case may be—from the inside, by immersing students in an abstract approach.

> The learner can master the theory of groups only by doing much practice work in connection with his reading and frequent re-reading of the text. This will not only familiarize him with the theorems and give him a needed facility in applying them, but will also render him an especially important service in helping him to understand the relative importance of the principal theorems. ... Chapter 1 is introductory in character. The learner is led to some of the principal elementary ideas of group theory and is given the opportunity of becoming familiar with them by using them in the analysis of several notions which are important in the later development. [6, pp. vi–vii]

Is this internal approach really the best motivation one can offer the prospective student? If so, then as Dorothy Parker is said to have remarked (on being challenged to produce a one sentence aphorism incorporating the word "horticulture"):

You can lead a 'horticulture,' but you can't make her think.

Group theorists, however, are more or less unanimous in declaring (through their silence) that there is no obvious alternative to adopting, right from the outset, a cold-blooded axiomatic approach to their particular culture. The best of them are prepared to travel the rugged path with you, and to pace the journey to assist the beginner as much possible. But at the end of the day the novice is offered no alternative but to learn to think in the required way from the inside.

> What were isolated and separate insights before, now begin to fit into a unified, if not yet final, pattern. I have set myself the task of making this pattern apparent to the reader. ... While any question concerning a single object (e.g., a single finite group) may be answered in a finite number of steps, it is the goal of research to divide the infinity of objects under investigation into classes of types with similar structure. The idea of O. Hölder for solving this problem was later made a general principle for investigation in algebra by E. Noether. We are referring to the consistent application of the concept of a homomorphic mapping. With such mappings one views the object, so to speak, through the wrong end of a telescope. These mappings, applied to finite groups, give rise to the concepts of normal subgroup and factor group. Repeated application of the process of diminution yields the composition series, whose factor groups are the finite simple groups. These are accordingly the bricks of which every finite group is built. [30, p. v]

Many authors feel mildly embarrassed about their strictly internalist treatment, and so begin with an apology of a motivation—perhaps a brief look at the symmetries of a square as in [2], or a page on the elementary operations of arithmetic as in [1] and [6]. Other authors give their examples after introducing the initial definitions and basic ideas. In either case, the examples given are usually untypical of the groups to be studied in the sequel. For example, the initial examples are often abelian, and may well be infinite even when the rest of the book is exclusively finite. As illustrations of why one should study groups abstractly, the early examples are generally unconvincing; and they are rarely of any use to the student in helping him or her to make sense of the particular line of development which is followed in the rest of the book. Too often, the examples play no role at all in the subsequent discussion and are never referred to again. I know of only two books which try to moderate the extremes of this internalist approach. The first, as one might expect, is that of Andreas Speiser. [23] His attempt to address larger issues (for example, by presenting group theory in the context of other historical developments in algebra, and by giving a fully motivated complete analysis of two-dimensional crystallographic groups) makes his book particularly interesting, but it does not really make it any easier, since the approach he adopts in the strictly group theoretic parts of the book is fairly uncompromising. Thus the cultural and historical sections are adjuncts which do not really affect the basic approach. (It is interesting to note that this lovely book does not appear to have ever been translated into English.)

The other book [4] is quite a different matter. *Groups: a Path to Geometry* is not a conventional text, but consists of a sequence of exercises, based on the "Do-It-Yourself" principle that

> One must learn by doing the thing; though you think you know it, you have not certainty till you try. [Sophocles]

Those who fear to experiment with their hands will never know anything. [22]

The hand is the cutting edge of the mind. [Jacob Bronowski]

This book contains a first course in group theory, pursued with conventional rigor. ... the book consists of over 800 problems [and not much else!] ... Mathematics is something we do rather than something we learn and, all too often, lectures give the opposite impression. ... at the outset, the groups under discussion are groups of transformations. This is faithful to the historical origins of the theory. It provides the one context in which the proof of the associative law is immediate, and it makes the study of sets with only a single defined operation obviously worthwhile. For Galois (1830), Jordan (1870), and even in Klein's *Lectures on the Icosahedron* (1884), groups were defined by the one axiom of closure. The other axioms were implicit in the context of their discussions—finite groups of transformations. Our work on abstract groups starts in Chapter 6. [4, p. xi]

Burn's approach is mainly guided by two principles. First, he wants to get the student to do the work and so avoid the stultifying effect which lectures can have on students' independence. Next, he wants to use groups to elucidate certain geometrical examples (and to use the study of these geometrical examples to consolidate work on groups). The appeal to history is incidental: the works of Lagrange, Cauchy, Abel, Galois, and even Jordan do not play any real part—though Klein clearly does. And though Burn claims that the book contains a first course in group theory, one has to point out that it contains no Sylow theorems, nothing on normalisers, nothing at all on composition series, no mention of soluble groups, and just one mention of simplicity (of A_5). He does not say whether such topics were excluded deliberately because they were felt to be unsuitable in a first course (with the geometry being presumably used to reinforce the basic ideas before moving on), or whether they were omitted because he was determined to include the geometrical topics and this forced out those harder group theoretical ideas which play little role in the geometry. The judgement is nevertheless an interesting one. Though many group theorists would object, one suspects that Burn's strategy is to try to interest the students and to consolidate basic ideas first so that they want to take a second course. This practice, in the long run, may well produce just as many students who understand the more advanced topics as the usual uncompromisingly give-it-to-them straight first course, which makes considerable demands on ordinary students without providing much by way of a pay-off.

There can be nothing more destructive of true education than to spend long hours in the acquirement of ideas and methods which lead nowhere. It is fatal to all intellectual vitality. [28]

But suppose one sees the Jordan-Hölder and Sylow theorems as the *sine qua non* of a first course in group theory. What then? Are we stuck with the standard treatment?

Any didactic introduction to a field of knowledge passes through a period during which purely dogmatic teaching is dominant. An intellect is prepared for a given field; it is received into a self-contained world and, as it were, initiated. If the initiation has been disseminated for generations ... it will become so self-evident that the person will completely forget he has ever been initiated, because he will never meet anyone who has not been similarly processed. [9, p. 54]

Of course, it is always possible that some uninitiated outsider might blunder in and try to do things differently, unaware of the unstated conventions and taboos. Bob Burn is in some ways an outsider. He was trained as a geometer, and has spent most of his professional life training teachers. Another outsider who saw the need for something different was Frank Budden—a gifted and unusual teacher in a boys' grammar school.

This is not a textbook on group theory, and it does not pretend either to ideal balance and emphasis, or to a universally acceptable sequence of development and arrangement, nor does it make a fetish of mathematical rigour. ... The volume aims to interest, to enlighten and to transport the reader, rather than to provide him with the strict discipline of a mathematical education.

It takes 545 pages to cover what would be completed in most textbooks in one to two hundred pages. But that is precisely its *raison d'être*—to be expansive, to examine in detail with care and thoroughness, to pause—to savour the delights of the countryside in a leisurely stroll with ample time to study the wild life, rather than to plunge from definition to theorem to corollary to next theorem in the feverish haste of a cross-country run. [3, p. xi]

As one reads these words, one can almost hear the mathematicians closing ranks ready to counterattack! The book is not perfect. What book is? But it is a brave attempt to supplement the standard treatises and should have been welcomed as a valuable source of ideas and readings for all kinds of students.

So how did the group theorists respond to Budden's implied criticism of the standard approach to group theory? *Mathematical Reviews* despatched their copy to a renowned hardliner, with predictable results.

The *Fascination of Groups* is a book for laymen. It is a somewhat peculiar book ... groups, subgroups, cosets and normal subgroups are not defined until page 75, page 214, page 333 and page 382 respectively. Not surprisingly only the rudiments of group theory are presented by the end of the book. ... For the audience of teachers, university students and senior pupils in school for whom the book is expressly written, it is questionable whether this is the appropriate book. ... The reviewer's most serious criticism of the book is that it gives an often misleading and occasionally dead wrong impression of mathematics. ... For all these reasons the book must be judged unsuitable for an audience with any mathematical bent. [10]

What is one to say? No comparison is made with other books in the same genre. No alternative, more suitable books are suggested. There is not even any sign that the reviewer understood the author's motives in wanting to make groups appear "fascinating." Instead of trying to do better, the initiates simply ignored the upstart and went about their business. The book was never reprinted and is now out of print. I know of no text by a professional mathematician, and only one popularization, which acknowledges the existence of Budden's remarkable book.

A mathematician is a blind man in a dark room looking for a black hat which isn't there. (Charles Darwin, quoted in [13])

In place of Budden's many truly fascinating examples we are left with—nothing at all.

It is as if the subject were taught in a room with mirrors on the walls rather than windows to the outside world. (Hans Freudenthal, [11])

Is there then, as John von Neumann hinted, perhaps no other way?

Young man, in mathematics you don't understand things, you just get used to them.
(J. von Neumann, quoted in [13])

The optimum system of a science, the ultimate organization of its principles, is completely incomprehensible to the novice. ... Every didactic introduction is, therefore,
literally a "leading into" or a gentle constraint. ... The initiation into any thought
style, which also includes the introduction to science, is epistemologically analogous
to the initiations we know from ethnology and the history of civilization. Their effect
is not merely formal. The Holy Ghost as it were descends upon the novice, who will
now be able to see what has hitherto been invisible to him. Such is the result of the
assimilation of a thought style. [9, p. 104]

All of which reminds one of D'Alembert's dictum

Allez en avant, et la foi vous viendra. (Go forth, and faith will follow you.)

What can we offer those whom faith and the Holy Ghost, as it were, passes over? In any
didactic introduction to group theory for novices which is based on "the ultimate organization
of its principles," it is not so much previously familiar mathematical objects which motivate,
instantiate, and justify the new theory, as the objects one constructs within the theory itself as
a result of one's initiation. This makes progress especially hard for those who get stuck at an
early stage.

The boundary line between that which is thought and that which is taken to exist is
too narrowly drawn. Thinking must be accorded a certain power to create objects,
and objects must be construed as originating in thinking. [9]

While this facility comes naturally to the mathematician, it does nothing to solve the
problem of how to retain the student's attention and diligence during the period before they
reach the state of assimilating the group-theoretical "thought style." And how can we during
this interregnum convince them that we are not as intellectually naked as Jonathan Swift's
architect?

There was a most ingenious architect who had contrived a new method for building
houses by beginning at the roof and working downwards to the foundation. [24]

There are those who would argue that one only has to use sufficiently many examples and
all will be well.

Why study examples? ... For the mathematician doing his research, examples are all
but indispensable to his work. To begin with, the direction of his research is guided
by a thorough examination of all the pertinent examples he can get his hands on.
Only after these examples are analyzed does he attempt to formulate their common
properties into some sort of theorem, and then attempt to prove the theorem. For the
student of group theory, regardless of his level, examples help to clarify and justify
the definitions and theorems of the subject he is studying. ... This book was written
with the dual purposes in mind of providing examples that serve to illustrate various
group-theoretical concepts, and providing counterexamples. [26, p. ix]

Whether or not Weinstein's description of how research mathematicians really work is
correct, there is a serious difficulty for the student: the examples are presented in the language
of abstract group theory, and so presuppose a considerable degree of fluency in the reader. It
is difficult to see how it could be otherwise (though the author makes things much worse than

they need be: the first definition, on page 2, is of a central pushout system!). The author even admits that his material is not aimed at real beginners.

> This book is intended for anyone *who has taken* [my italics] an introductory course in group theory. [26]

Conclusions

Our search for good examples of genuine introductions to group theory has borne little fruit. What does this tell us about Ernst Haeckel's Law of Epigenesis?

> Ontogenesis is a brief and rapid recapitulation of phylogenesis, determined by the physiological functions of heredity (generation) and adaptation (maintenance). [12]

The significance of this "law" for mathematics has been stressed by Felix Klein, by George Pólya, and (as we saw above) by René Thom. What would a treatment based on Haeckel's law look like? Need it be even loosely historical? Thom's comments on school mathematics suggest that it may be more important to respect human psychology than human history *per se*. Others feel that history has a more important role to play here.

> An essential obstacle to the spreading of such a natural and truly scientific method of instruction is the lack of historical knowledge which so often makes itself felt. [18]

> There is not much doubt that the difficulties the great mathematicians encountered are precisely the stumbling blocks that students experience, and that no attempt to smother these difficulties with logical verbiage can succeed. [19, p. 40]

We obviously cannot ask students to repeat all the fumbling of their historical predecessors. So selection is necessary and important.

> One can compress history and avoid many of the wasted efforts and pitfalls, but one cannot eliminate it. [19, p. 41]

To the initiated the modern approach to group theory has a naturalness and an elegance which they would like students in a first course to appreciate. At present, they are largely unsuccessful, but are reluctant to even contemplate replacing it by something more palatable (and possibly more effective) but less elegant. When challenged that a highly effective method of his own "lacked elegance," Ludwig Boltzmann is said to have replied, "Elegance should be left to shoemakers and tailors." In life it is easier to score points off others than it is to win the argument. Boltzmann was so hurt by those who consistently dismissed his work that he eventually killed himself.

Historical approaches to group theory can easily fall into the trap requiring students to learn not one subject, but two or three. They know no geometry, so before using even the simplest geometrical motivation one has to spend time working in two and three dimensions. They may well have never heard of the Platonic solids, let alone know any of their properties. They will never have solved cubic or quartic equations, or have reflected on how they solve quadratics. In general, their background historical knowledge is likely to be exceedingly thin. It is thus all too easy to end up by making the subject harder instead of making it more accessible.

I believe one can devise a treatment which respects both the students' intellectual preparedness and the essence of modern group theory by drawing selectively on historical material (starting with symmmetric functions and the solution of equations, exploring Cauchy-type problems on "the number of values a function can take," Gauss's congruence arithmetic and the composition of forms, permutation groups—including a thorough study of S_n for $n < 6$, isometries in two dimensions, and applications to the solution of equations). But given my previous remarks, the reader will rightly remain skeptical until the evidence is available.

> It is very difficult, if not impossible to give an accurate historical account of a scientific discipline. Many developing strands of thought intersect and interact with one another. . . . It is as if we wanted to record in writing the natural course of and excited conversation among several people all speaking simultaneously amongst themselves, and each clamoring to make himself heard, yet which nevertheless permitted a consensus to crystallize. [9, p. 15]

What Paul Fong and company would make of such a treatment is a question I would rather not think about just yet, though at the end of the day it is perhaps the only thing that matters. There is no point getting things right if one is then simply ignored by those who matter most.

Bibliography

1. Allenby, R. B. J. T.: *Rings, Fields and Groups*. London: E. J. Arnold, 1983.
2. Birkhoff, G. and MacLane S.: *A Survey of Modern Algebra*. New York: Macmillan, 1941.
3. Budden, F. R.: *The Fascination of Groups*. Cambridge: Cambridge University Press, 1972.
4. Burn, R. P.: *Groups: A Path to Geometry*. Cambridge: Cambridge University Press, 1985.
5. Butler, S: *The Notebooks of Samuel Butler*. London: A. C. Fifield, 1918.
6. Carmichael, R. D.: *Introduction to the Theory of Groups*. New York: Dover, 1937.
7. Dienes, Z. P.: *The Power of Mathematics*. London: Hutchinson, 1964.
8. Einstein, A.: *Out of my Later Years*. New York: Philosophical Library, 1950.
9. Fleck, L.: *Genesis and Development of a Scientific Fact*, (F. Bradley and T. Trenn, trans.). Chicago: University of Chicago Press, 1979.
10. Fong, P.: *Mathematical Reviews* 58 (entry 28147), 1979.
11. Freudenthal H.: "What groups mean in mathematics and what they should mean in mathematical education," in A. G. Howson, (ed.): *Developments in Mathematical Education*. Cambridge: Cambridge University Press, 1973, 101–114..
12. Haeckel, E.: *General Morphology of Organisms*. Berlin, 1866.
13. Henle, J. M.: *An Outline of Set Theory*. New York: Springer-Verlag, 1986.
14. Herstein, I. N. and Kaplansky, I.: *Matters Mathematical*. New York: Chelsea, 1978.
15. Hilbert, D.: *Foundations of Geometry*. Chicago: Open Court, 1971.
16. Kac, M., Rota, G.-C. and Schwarz, J.: *Discrete Thoughts*. Boston: Birkhäuser, 1986.
17. Kant, I.: *Critique of Pure Reason*. London: Macmillan, 1933.
18. Klein, F.: *Elementary Mathematics from an Advanced Standpoint*. New York: Dover, 1945.
19. Kline, M.: *Why Johnny Can't Add*. New York: St. Martin's Press, 1973.
20. Ledermann, W.: *Introduction to the Theory of Finite Groups*. Edinburgh: Oliver and Boyd, 1964.
21. Papy, G.: *Groups*, (M. Warner, trans.). New York: St. Martin's Press, 1964.
22. Sarton, G.: *The Life of Science*. New York: Henry Schuman, 1948.
23. Speiser, A.: *Die Theorie der Gruppen von endlicher Ordnung*. Berlin: Springer-Verlag, 1927.
24. Swift, J.: *Gulliver's Travels: a Voyage to Balnibari*. New York: Houghton Mifflin, 1960.
25. Thom, R.: "Modern mathematics: Does it exist?" in A. G. Howson, (ed.): *Developments in Mathematical Education*. Cambridge: Cambridge University Press, 1973, 194–209.
26. Weinstein, M.: *Examples of Groups*. Passaic, N.J.: Polygonal Publishing House, 1977.
27. Weyl, H.: *Symmetry*. Princeton: Princeton University Press, 1952.
28. Whitehead, A. N.: *Essays in Science and Philosophy: Mathematics and Liberal Education*. New York: Philosophical Library, 1948.

29. Wigner, E. P.: *Symmetries and Reflections*. Bloomington: Indiana University Press, 1967.
30. Zassenhaus, H.: *The Theory of Groups*. New York: Chelsea, 1947.

David Hilbert

An Episode in the History of Celestial Mechanics and its Utility in the Teaching of Applied Mathematics

Eric J. Aiton

The teaching of applied mathematics to late teenagers is quite a hard task. It is sometimes done by the mathematics teacher, sometimes only by the physics teacher. In whatever context students receive their first introduction to the subjects of statics and dynamics, they encounter serious difficulties in their attempts to grasp the first principles and their applications.

There are notoriously difficult concepts to be mastered. For example, teachers stress the "fictitious" nature of centrifugal force, while students reflecting on their own experience have a strong impression of a real force. How to make sense of such a paradoxical claim? Here, it seems to me, it is necessary to explain to students something of the historical context in which the concepts arose, and the choices which were made by past mathematicians in their efforts to reduce the complexity of experience to something that can be mathematically handled.

I describe in this paper an episode in the history of celestial mechanics, as a contribution towards making available historical material which can be shared with students. How might it be used in the classroom? There are, on the face of it, two main approaches: either to incorporate it into the teaching of the applied mathematics itself, or to regard it as complementary to a systematic exposition of the mathematics. My preference is for the second alternative. At this level, some knowledge of the subject matter is an essential prerequisite for understanding the history. To incorporate it into the initial teaching of the mathematics can be done, but would take very much longer, an important factor when there are so many time constraints on the curriculum. It is better to require the students to assume some matters on trust while imparting a preliminary grasp of the mathematics, and to use history subsequently to clarify the resolution of some of the problems, and also to place the mathematics within its human and cultural context.

Besides making a contribution to bridging the gap between what C. P. Snow called the two cultures, the historical material enhances the discussion of three broad areas:
- the concepts and principles of dynamics
- the role and evaluation of mathematical models
- the ethics and effectiveness of the exchange of ideas between mathematicians and scientists

The first two of these areas are straightforward, but the third may seem somewhat mysterious. Communication or the lack thereof is an important dimension of mathematical activity, however, as we shall see in this case study, it transpires that communication was not the primary purpose of one of the exchanges of ideas, and was ineffective in some of the others.

To conclude this introduction, I outline the problem under discussion. Uniform motion in a circle is only the simplest case of motion in a curve under the action of a central force. In order to develop the dynamics of motion in a curve, mathematicians of the seventeenth century represented the continuous force by a series of infinitesimal impulses, and the curve by a series of infinitesimal straight lines—in other words, a polygon. This is still a valid analysis, and in effect is what we do today. But what was not always appreciated by mathematicians of the time is that the sides of the polygon must be second order infinitesimals—an infinity of such straight lines going to make up a (first order) infinitesimal arc. The curve of motion is to be considered as consisting, in turn, of a series of arcs of this kind, along each of which the force is regarded as constant during a (first order) infinitesimal time.

Such problems of different orders of infinitesimals were confusing enough, but there were more. Two versions of the infinitesimal arc model were used in the seventeenth century: in one the infinitesimal arcs were parabolic and in the other they were circular. In this paper I explore the clarification of the principles underlying such topics that can be provided by a study of the ways in which Newton, Leibniz, and Varignon struggled to resolve the problems. For this purpose, I draw upon the conclusions of a number of historical researches undertaken over the course of several years, detailed accounts of which may be found in the references.

As I have stressed, this story is fundamentally about communication. As teachers, we have a duty to communicate effectively, and to teach our students to do likewise. It is worth raising with students the question of whether mathematicians and scientists generally should impose such a responsibility upon themselves, striving always towards clarity and comprehensibility while resisting any temptation to score debating points at the expense of the truth.

Centrifugal Force

In an early notebook, known as the Waste Book, Newton wrote:

all bodies moved circularly have an endeavor from the centre about which they move. [10, p. 147]

As late as 1681, in describing the orbit of a comet, he wrote of

the *vis centrifuga* overpow'ring the attraction and forcing the comet, notwithstanding the attraction, to begin to recede from the sun. [4, p. 103]

In the case of a body rotating in a circle on the end of a string (ideally outside a gravitational field, or shall we say on a frictionless horizontal table), there is only one real force, namely the tension in the string. And in the case of the comet, the only real force is the attraction. Yet Newton, up to 1681 at least, believed in the existence of a real centrifugal force in addition to the centripetal attraction, so that the motion of a body was the result of a contest between these two forces.

There is no mention of centrifugal force in Newton's *Principia* (1687) but in one of his attacks on Leibniz written in 1711, Newton says that

centrifugal force is always equal and opposite to the force of gravity by the third law of motion. [1, p. 32]

This is not in agreement with what he wrote in 1681, for the centrifugal force and the attraction were then regarded as unequal. Quite clearly, at that time they were also regarded as acting on the same body, namely the comet. However, action and reaction do not act on the same body. If they did, they would cancel each other out and the body would remain at rest or move in a straight line in accordance with the first law of motion. Of course, they act on different bodies. For example, the string pulls the body to which it is attached and the reaction is the pull of the body on the string. The sun attracts the comet and the comet attracts the sun. Evidently

Newton had changed his whole conception of centrifugal force, though without disclaiming his earlier notion, and he supposed that his definitive conception was the only true one. The question arises whether the earlier concept can be interpreted meaningfully. Considered as an endeavor of the circulating body, or a force acting on the body itself, it does not exist. But if we consider a reference frame fixed in the body and rotating with it, the body will appear to have an endeavor to recede from the centre. This of course is a fictitious force reflecting the acceleration for the reference frame. There will then be a contest between the attraction and the centrifugal force engendered by the rotation to move the body along the rotating radius vector. Leibniz in fact described the motion of a planet or comet in this way and deduced the radial acceleration, expressed in modern notation as

$$\ddot{r} = \frac{a}{r^3} - \frac{b}{r^2},$$

where r is the distance from the centre of attraction and a and b are constants [6, p. 157].

Newton objected that, since centrifugal force was equal and opposite to the attraction by his third law of motion, Leibniz's reasoning implied that $\ddot{r} = 0$ [12, p. 117]. Leibniz's formula in fact gives a correct measure of the radial acceleration and is a notable contribution which his contemporaries failed to appreciate.

Quite clearly, Newton and Leibniz are using the term centrifugal force in different senses. Newton should have recognized this, as he had for nearly twenty years himself understood the term in Leibniz's sense. His intention, however, was not to understand Leibniz but to denigrate his work.

The historical considerations we have introduced so far should help to clarify some of the difficulties surrounding the concept of centrifugal force and the meaning of the law of action and reaction. Also they should help the student to recognize the importance of grasping the exact meaning of concepts and laws.

Infinitesimals

If we examine the demonstrations of circle squarers, we find that somewhere they equate an arc and a chord. This is where they go wrong. It was Cusanus (Nikolaus von Kues) (1401–1464) who first stated the conditions under which an arc and a chord could be validly equated. In his *De perfectione mathematica* (1458), he declared his intention to complete mathematics out of the coincidence of opposites; that is, the curved and the straight [11, p. 161]. When he knew the relationship between the arc and the corresponding chord, he would have the means of rectification of curves and of quadratures. Cusanus recognized that opposites coincide in the infinite (that is, the infinitely great or infinitely small) and indeed the coincidence of the infinitesimal chord and arc is the basis of the calculus.

It should be remembered that there is an irreducible problem concerning the magnitude of an infinitesimal, so that a completely rigorous treatment of dynamical problems is probably unattainable. The infinitesimal is less than any assignable continuum. Zeno considered the problem in antiquity and, Cantor notwithstanding, it remains unsolved today. For the dynamical problems relate to the real world, whereas our models belong to the ideal world of mathematics.

Newton

In proposition 1 of Book I, of the *Principia*, Newton demonstrated that for any centripetal force, the areas swept out in equal times are equal. If no force acted, the planet would traverse AB, Bc (Figure 1) in equal successive infinitesimal intervals. Triangles SAB, SBc are equal in area.

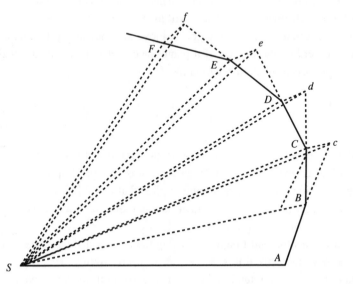

FIGURE 1

Now suppose that the force acts instantaneously at B, deflecting the planet into BC so that $BC = Bc$. The centripetal force has drawn the planet towards S, but in the small time, Cc may be taken parallel to SB. (BC is the resultant motion for Bc and cC.) Triangles SBc, SBC are equal in area, being on the same base SB and between the same parallels SB and Cc. As triangles SAB, SBc are equal in area, it follows that triangles SAB, SBC are equal in area. Hence in successive equal small times, the areas swept out are equal.

Now let the number of triangles be augmented and their breadth diminished *ad infinitum*; their ultimate perimeter ADF will be a curved line. Therefore the centripetal force by which the body is drawn perpetually from the tangent to this curve will act continuously and the area swept out will be proportional to the time.

Note that ADF is supposed to be a finite arc, so that Newton represents the curve by a polygon with first-order infinitesimal sides. When Newton applies this result, however, he does so for an infinitesimal arc, as if the curve were represented by a polygon with second-order infinitesimal sides.

In proposition 6 of Book I of the *Principia*, Newton considers the general case of a body moving in a curve under the action of a centripetal force.

The curve PQ (Figure 2) is an infinitesimal arc and PR is the tangent at P. Newton takes the deflection RQ to be proportional to the force (by law II) and to the square of the time (by Lemma X). The deflection QR is described with uniformly accelerated motion. In effect, Newton has justified by his Lemma X the application of Galileo's law of fall to the infinitesimal

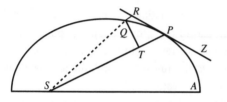

FIGURE 2

deflection QR. This means that, in the infinitesimal time, the force can be regarded as constant. Thus QR is proportional to $f\,dt^2$, where f represents the force and dt the element of time. Also QR is proportional to $f \cdot SP^2 \cdot QT^2$. It should be noted that Kepler's area law has only been applied to an infinitesimal arc.

In proposition 11, Newton deduces the result for an ellipse with force directed towards the focus. Here geometry implies that QT^2/QR is equal to the latus-rectum. Hence $f \cdot SP^2$ is constant and f is proportional to $1/SP^2$.

The Parabolic Arc Approximation

The analysis which follows is based on a review of D.T. Whiteside [13]. Suppose the body traverses the arc AB (Figure 3) in time t. Divide t into n parts. Represent the continuous force by a series of impulses acting at the beginning of each part of time. Taking force = f, each impulse equals $f \cdot \frac{t}{n}$. This creates a speed $f \cdot \frac{t}{n}$ in unit mass. So the distance traversed in the first part of time is $f \cdot (\frac{t}{n})^2$.

In the second part of time, the speed created by the first impulse is conserved, while the second impulse creates an equal speed. Hence the distance traversed in the second part of time is $2f \cdot (\frac{t}{n})^2$. In the whole time t, the total distance

$$BT = (1 + 2 + 3 + \cdots + n)f\left(\frac{t}{n}\right)^2 = \frac{1}{2}n(n+1)f\left(\frac{t}{n}\right)^2,$$

which tends to $\frac{1}{2}ft^2$ as n tends to infinity.

In adding the velocities we have assumed that they are all parallel to AS. Also we have assumed that f is constant. These assumptions are only justified if AB is taken to be an infinitesimal arc. In this case, the time t is an infinitesimal and the arc AB is approximated by a parabola. Hence we have represented the curve by a series of first-order infinitesimal

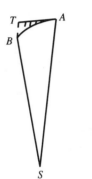

FIGURE 3

parabolic arcs. Each of the arcs (according to our construction) is represented by a polygon with second-order infinitesimal sides. For in each part of time, the inertial motion and the speed created by the impulse are taken as constant, so that the resultant motion is rectilinear. We can say that representing a curve by a series of first-order infinitesimal parabolic arcs is equivalent to representing it by a polygon with second-order infinitesimal sides.

For a more detailed description of the models used by Newton and Leibniz, see [7].

A Comedy of Errors

The problem of the correct representation of a curve by a polygon was first debated in correspondence between Leibniz and Varignon in 1704, an account of which may be found in [2]. Leibniz had contributed his own paper on planetary motion to the *Acta Eruditorum* in the autumn of 1688. According to his own account, it was written after he had seen a review of Newton's *Principia* but before he had seen the *Principia* itself. This claim is now known to be false, for Domenico Bertoloni Meli has found manuscripts demonstrating the influence of Newton's *Principia* on the development of the theory proposed by Leibniz in his essay [8]. As Leibniz had been too busy to pursue these researches himself he had set Varignon to work on the many-body problem, a daunting task for which Varignon was ill equipped.

Varignon had been investigating the relationship between the centripetal force required to move a body and the weight of the body. In other words, he was trying to measure the attraction in terms of terrestrial gravity. As a corollary from his result, he deduced that, for a body moving in a circle, the force was twice the value obtained by Huygens. He consulted Leibniz and Johann Bernoulli. Neither was of any real help to him, for instead of attempting to locate the error in his reasoning, they supplied him with new demonstrations of their own. Leibniz had in fact also made a mistake, without recognizing it; he had found the centrifugal force to be half the value obtained by Huygens. Varignon solved the problem for himself while out walking, and he understood the point at issue perfectly, though he was never able to convince Leibniz of the correctness of his analysis.

Varignon divided the circle into a polygon with infinitesimal sides such as EA, AG (Figure 4). He supposed the centrifugal force to act instantaneously at the vertices. He then regarded the motion in the side AG (identified with the arc AG) to be compounded of inertial uniform motion AF and uniformly accelerated motion FG (because force produces acceleration). Leibniz had

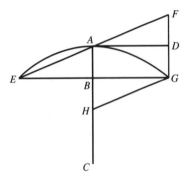

FIGURE 4

taken the motion in AG to be compounded of inertial uniform motion in the tangent AD and a uniform motion in DG (supposing the force to act with an instantaneous impulse at A).

Leibniz immediately accepted Varignon's suggestion that the "tangent" should be regarded as the prolongation of the chord. Thus he now took the motion in the arc AG to be compounded of uniform motion in the "tangent" AF and uniform motion in FG. This gave the correct result, so that Leibniz was able to correct his error.

Varignon's flash of inspiration during his walk was the recognition that the approximation of the curve by a polygon (that is, with first-order infinitesimal sides) did not represent reality. The circle must be divided into true arcs. The motion in the arc AG is then compounded of a uniform motion in the tangent AD and a uniformly accelerated motion in DG. He was never able to convince Leibniz, who told him that it was better not to introduce the acceleration into the elements when there was no need.

How can we explain an incorrect model giving a correct result? The difference between the arc and the chord (say in the middle) is of the same order of magnitude as the effect of the force we are trying to measure. Such an error is therefore logically unacceptable in the model. The first-order infinitesimal polygon model, however, does bring the body to the right place at the vertices of the polygon (though not in between) and since the calculation is based only on how the body arrives at these positions, it does give correct results.

Another explanation is as follows. Only a continuous function can be differentiated and therefore you need a continuous velocity function from which to derive the force of acceleration. This means that, in first-order infinitesimal time, the body must be supposed to move in a smooth curve.

In Newton's demonstration of proposition 1, the chords are taken to be the tangents of the limiting curve, just as in Varignon's original calculation and in Leibniz's corrected one. Only Varignon's corrected version is strictly correct. That is, the curve must be represented by a polygon with second-order infinitesimal sides, or equivalently by a series of infinitesimal parabolic arcs having common tangents at the joins.

Leibniz

Leibniz based his planetary theory on Kepler's analysis of planetary motion. [5] According to Kepler, the planet was coaxed around the sun by an immaterial vortex, while a magnetic attraction and repulsion action alternately varied the distance so that the planet moved in an ellipse. According to Leibniz, the planet was carried around by a harmonic vortex (that is, a vortex rotating in accordance with Kepler's area law), while the distance from the sun was varied by the joint action of the centrifugal force engendered by the circulation and the gravity towards the sun.

Leibniz resolves the orbital motion in two ways:

(1) M_2M_3 (Figure 5) is composed of M_2L (the inertial motion at M_2) and the new impression of gravity $LM_3(M_2G)$. The point G is missing from the diagram. It is the point on OM_2 where M_3G is drawn parallel to LM_2.

(2) M_2M_3 is composed of D_2M_3 (strictly T_2M_3, the circulation) and M_2D_2 (strictly M_2T_2, the radial velocity already acquired).

The first resolution is hypothetical and represents what would happen in the absence of the harmonic vortex, when gravity would act alone. It enables him to find the gravity LM_3.

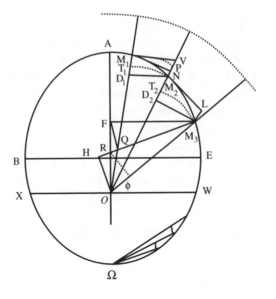

FIGURE 5

The second resolution represents reality. The planet is moved trans-radially by the harmonic vortex (in circular arcs) and simultaneously along the rotating radius vector by the joint action of the centrifugal force (engendered by the circulation) and gravity.

Leibniz finds that

$$ddr = (OM_1 - OM_2) - (OM_2 - OM_3) = 2D_2T_2 - M_2G.$$

After making the correction inspired by Varignon, he interpreted this equation as

$$ddr = \text{centrifugal force} - \text{gravity}.$$

In the case of the ellipse with gravity directed towards a focus, this takes the form

$$ddr = \left(\frac{a}{r^3} - \frac{b}{r^2} \right) dt^2,$$

where a and b are constants.

For the calculation of the centrifugal force Leibniz originally used the model involving deflection from the tangent. His error arose because he regarded the deflection as produced by a uniform motion instead of a uniformly accelerated motion. For the representation of the orbit, however, he used the first-order infinitesimal model. His diagram should be interpreted as shown in Figure 6.

The line M_2L of Leibniz's diagram (Figure 5) is not the tangent at M_2 but the prolongation of the chord M_1M_2 as shown in Figure 6. What is the evidence for this?

(1) In a letter to Huygens [3, p. 113], he describes M_2L as the prolongation of the chord.
(2) He states that the triangles M_1NM_2, M_3D_2G are congruent. This is untrue if M_2L is taken to be the tangent. It is true if M_2L is taken to be the prolongation of the chord.
(3) There is a manuscript in which M_2L is described as equal and parallel to M_1M_2 [5, p. 217].

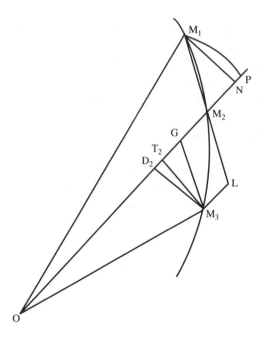

FIGURE 6

(4) There is a manuscript describing and illustrating with a diagram the deflection from the chord $M_1 M_2$ into the chord $M_2 M_3$ resulting from the instantaneous impulse at M_2. The same manuscript shows the calculation of the centrifugal force using the deflection from the tangent. [5, p. 216]

Notwithstanding the fundamental error of Leibniz's model, it gives the correct result, because it brings the body in succession to the points M_1, M_2, M_3, even though at intervening points it does not represent the true position of the body.

Second Differentials

In the course of the priority dispute on the invention of the calculus, both sides accused their opponents of not understanding second-differentials. When John Bernoulli in June 1713 informed Leibniz of the contents of the *Commercium epistolicum*, the Royal Society's report on the dispute, he also informed him that his nephew Nikolaus had found an error in *Principia*, Book II, proposition 10, where Newton had identified the terms of the binomial expansion of $(x + o)^n$ with the differentials. [6, p. 338] There is indeed an error which Newton corrected in the second edition. It is just a coincidence that, if Newton's error is supposed to be false identification of the binomial terms with differentials, then when that is put right, the correct result is obtained. However, Newton's error lies elsewhere, as was first brought to light by Lagrange.

Leibniz had been badly treated in the *Commercium epistolicum*, and encouraged by Bernoulli, repeated the criticism concerning higher differentials in his own response known as the *Charta Volans*. Leibniz in turn was accused of not understanding second and higher differentials by John Keill, who obtained all his arguments from Newton. Believing that the criticisms came from Keill himself and not Newton, Leibniz refused the entreaties of his friends to reply, since he wished to have no dealings with a man like Keill.

One of Newton's criticisms was that Leibniz made an error involving second differentials in asserting that the triangles M_1NM_2, M_3D_2G (Figure 5) were congruent [1, p. 37]. He had, Keill claimed, confused the chord and the tangent. The criticism is based on a misunderstanding of the model used by Leibniz to represent the curve and the action of the forces. According to the interpretation we have described, and which is well documented, there is no error.

Another Newton criticism was that Leibniz made an error involving second differentials in asserting that $NP = D_2T_2$ [1, p. 39]. This may be refuted using the Taylor Series. Let angle $AOM_2 = \theta$, angle $M_1OM_2 = \omega_1$, angle $M_2OM_3 = \omega_2$, $OM_2 = f(\theta)$, $OM_1 = f(\theta - \omega_1)$, $OM_3 = f(\theta + \omega_2)$. Then

$$NP = (1 - \cos\omega_1)f(\theta - \omega_1) = \frac{1}{2}\omega_1{}^2 f(\theta)$$

and

$$D_2T_2 = (1 - \cos\omega_2)f(\theta + \omega_2) = \frac{1}{2}\omega_2{}^2 f(\theta),$$

so that

$$NP - D_2T_2 = \frac{1}{2}(\omega_1{}^2 - \omega_2{}^2)f(\theta).$$

By the harmonic circulation

$$\frac{1}{2}f(\theta)f(\theta - \omega_1)\sin\omega_1 = \frac{1}{2}f(\theta)f(\theta + \omega_2)\sin\omega_2$$

so that

$$\omega_1 f(\theta - \omega_1) = \omega_2 f(\theta + \omega_2).$$

Expanding by the Taylor Series, this becomes

$$\omega_1[f(\theta) - \omega_1 f'(\theta)] = \omega_2[f(\theta) + \omega_2 f'(\theta)].$$

Hence $(\omega_1 - \omega_2)f(\theta) = (\omega_1{}^2 + \omega_2{}^2)f'(\theta)$, so that

$$NP - D_2T_2 = \frac{1}{2}(\omega_1^2 - \omega_2^2)f(\theta) = \frac{1}{2}(\omega_1 + \omega_2)(\omega_1^2 + \omega_2^2)f'(\theta),$$

which is a third-order infinitesimal.

Varignon's Alternative Model for Planetary Motion

In a memoir of 1701 (published in 1704) [2, p. 89], Varignon found an alternative formula for the centripetal force by equating its normal component to v^2/ρ, where v is the velocity in the orbit (supposed constant during an infinitesimal element of time) and ρ is the radius of the osculating circle.

Figure 7 is a simplified form of Varignon's diagram in which only the lines essential to the argument presented here are retained. Let C be the centre of force and AL the radius of the osculating circle normal to the tangent LE common to the circle and the orbit. Taking triangles

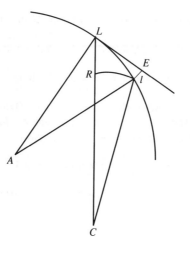

FIGURE 7

$AL\ell$ and $EL\ell$ to be similar, Vargignon deduced that $\ell E = ds^2/\rho$, where $L\ell = ds$. It will be noted that, if LE is taken to be the actual tangent, then the triangles are not in fact similar. But, as we have seen, at the time of publication of this memoir, Varignon conceived the tangent to be the prolongation of a first-order infinitesimal chord or side of a polygon, in which case the triangles may be taken as similar. The normal component of the centripetal force f is $f\,dz/ds$, where $dz = R\ell$. This component causes the deflection $E\ell$, which Varignon describes as uniformly accelerated, but he takes $E\ell$ to be equal to $(f\,dz/ds)dt^2$, so that $f = ds^3/(dxdt^2)$. Substituting the formula for $\rho = -ds^2 dr/(dz\,dds)$, where $r = CL$, he demonstrated the equivalence of his new formula for the centripetal force with his original formula $f = -v\,dv/dr$.

Varignon's analysis introduces another model in which the curve is represented, in a first-order infinitesimal element of time, by a first-order infinitesimal circular arc, or alternatively by a second-order infinitesimal polygon. In the second edition of the *Principia* (1713), Newton also developed the application of this model using the osculating circle, though his exposition is none too clear. [9] During the first-order infinitesimal element of time, the velocity in the orbit is supposed to be constant, though neither Newton nor Varignon attempted to justify this assumption. Although reasonable, it is not immediately obvious, for the centripetal force has a tangential component which must accelerate the motion in the orbit.

Bibliography

1. Aiton, E.J.: "The celestial mechanics of Leibniz in the light of Newtonian criticism," *Annals of Science* 18 (1962), 31–41.
2. Aiton, E.J.: "The inverse problem of central forces," *Annals of Science* 20 (1964), 81–99.
3. Aiton, E.J.: "The celestial mechanics of Leibniz: a new interpretation," *Annals of Science* 20 (1964), 111–123.
4. Aiton, E.J.: *The Vortex Theory of Planetary Motions*. London: Macdonald, 1972.
5. Aiton, E.J.: "The mathematical basis of Leibniz's theory of planetary motion," *Studia Leibnitiana*, Sonderheft 13 (1984), 209–225.
6. Aiton, E.J.: *Leibniz, a Biography*. Bristol: A. Hilger, 1985.

7. Aiton, E.J.: "Polygons and parabolas: some problems concerning the dynamics of planetary orbits," *Centaurus* 31 (1988), 207–221.

8. Bertoloni Meli, D.: "Leibniz's excerpts from the *Principle Mathematic*," *Annals of Science* 45 (1988), 477–505.

9. Brackenridge, J. Bruce: "Newton's mature dynamics; revolutionary or reactionary?" *Annals of Science* 45 (1988), 451–476.

10. Herivel, J.: *The Background to Newton's Principia*. Oxford: Oxford University Press, 1965.

11. Kues, Nikolaus von: *Die Mathmetischen Schriften*, (Joseph A. Hofmann, trans.). Hamburg, 1979.

12. Turnbull, H. W. (ed.): *The Correspondence of Isaac Newton*. Cambridge: Cambridge University Press, 1959–1977, vol. 6.

13. Whiteside, D.T.: "Newtonian Dynamics," *History of Science* 5 (1966), 104–117.

Mathematical Thinking and History of Mathematics

Man-Keung Siu

It seems most appropriate to begin with an exhortation by Niels Henrik Abel, the most distinguished mathematician of this land in which we gather for this workshop;

> It appears to me that if one wants to make progress in mathematics one should study the masters and not the pupils. [14, p. 138]

Harold M. Edwards has elaborated on this piece of advice in [9], giving no less than five reasons for its wisdom. He also cited Siegel's discovery of what are now known as the Riemann-Siegel formulas in the theory of the zeta function, through a study of Riemann's "Nachlass," as the greatest example of a combination of historical scholarship and mathematical research. I can add to this the story told by André Weil in a Ritt Lecture delivered at Columbia University in March, 1972 (later published in [19]). He described how, in 1947, he felt "bored and depressed" and "not knowing what to do" started to read Gauss's two memoirs on biquadratic residues. He noticed that the principles used could be applied to equations of a more general type and that this implied the truth of the so-called "Riemann hypothesis" for certain equations over finite fields. This discovery led to his classic paper "Number of solutions of equations in a finite field" [18] in which he made several conjectures about varieties over finite fields. One year after his lecture, these conjectures were completely settled by Pierre Deligne, a piece of work that has been hailed as one of the most remarkable achievements in mathematics of this century. But a Riemann, a Siegel, a Gauss, a Weil, a Deligne are exceptional people. For most people (like myself), Abel's exhortation sounds somewhat out of reach. However, can one still benefit from reading the masters? Will it help in the teaching of mathematics?

There are a number of difficulties involved in attempting to read the masters:

- change in terminology, in style, or even in approach,
- difference in mathematical environment,
- background knowledge on history of mathematics,
- proficiency in foreign languages.

I try to resolve these difficulties by aiming at a much more modest goal, viz. to study selected excerpts of the works of great mathematicians from the past with my students in order to:

- see how specific mathematical concepts evolve and crystallize,
- acquire enlightened understanding of specific topics,
- note important aspects of mathematical thinking, which includes good working habits in problem solving and also the nature and meaning of mathematics,
- experience the dynamic life of mathematics.

With these goals in mind I can select those excerpts that have been translated into English or those that were originally written in Chinese as my students are fluent in both languages.

Of course, by so doing I miss a lot, but even with our modest university library, we can get by with enough material from books such as [5, 8, 10, 13, 16, 17, 21]. Along with the discussions of these (translated) primary texts, I augment presentation with supplementary lectures on the relevant history of mathematics in order to put the readings in perspective. At suitable places I insert discussions of the nature of mathematics and expose students to various views. (A good source is [6] and its bibliography.) In particular, I like to impart to the class a regard for learning and ideas which form part of our cultural heritage. (See also [4, 12] for the description of two courses along allied but separate broader lines.) As to the difficulty arising from changes in terminology, style and approach, not too much can be done and indeed that is history! However, this situation can also have its merits. To cite one example gleaned from personal experience, I recall the time when I was a beginning graduate student at Columbia University and was advised by a senior graduate student who was completing his doctoral thesis to read Deuring's *Algebren* (1934) and to translate the section on "Faktorensysteme" into the modern language of cohomology. Although I never did carry out that project, that piece of advice helped to make me aware of this connection in subsequent study.

At the University of Hong Kong I offer a course titled "Development of Mathematical Ideas" which I developed over a period of years. Since intended topics, or even the teaching approach, can change from year to year, it provides me with an opportunity to try out new things and experiment with new teaching material. During 1986–88 I tried the aforementioned approach. Below is a list of some selections:

- Al-Khwarizmi, *Hisab Al-Jabr w'al Muqabala* [10, pp. 228—231; 17, pp. 55–60]
- R. Dedekind, *Stetigkeit und irrationale Zahlen* [7, pp. 1–27]
- Euclid, *Elements* [11]
- L. Euler, "Solutio problematis ad geometriam situs pertinentis" [1, pp. 1–8]
- D. Hilbert, "Mathematical Problems: Lecture delivered before the International Congress of Mathematicians at Paris in 1900" [3]
- Liu Hui, Commentary on *Jiu Zhang Suan Shu* [21]
- H. Lebesgue, "Sur le developpement de la notion d'intégrale" [5, pp. 721–725]
- N. Lobachevski, "Geometrical Researches on the Theory of Parallells" [2].

I shall illustrate with one "case study" in more detail. Let me choose Euler's memoir on the Seven Bridges of Königsberg [1, pp. 1–8] because the mathematics of it is so well known that we can concentrate on its pedagogical aspect. (For an interesting account of this memoir, see [20].) Let me go to the problem right away: Can one walk through all seven bridges, each exactly once, and come back to where one starts? (See Figure 1a.) The usual explanation given in textbooks is to apply the theorem on (semi) Eulerian graphs to the graph shown in Figure

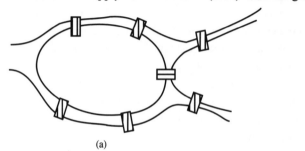
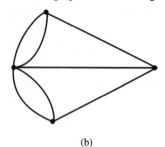

(a)   (b)

FIGURE 1

1b and conclude that it is impossible to do so. In the original memoir of Euler, presented to the St. Petersburg Academy on August 26, 1735, he solved the problem section by section (21 in all) in an illuminating way.

From the very beginning, in section 2, Euler did not confine his attention to the particular case of the Seven Bridges of Königsberg but looked at a general problem. This viewpoint helps in subsequent discussion. However, Euler did keep the particular case in mind and came back to it more than once to interpret and verify his new findings in its context. This illustrates how generalization and specialization complement each other in mathematical research.

Euler then introduced a good notation in section 4 ("the particularly convenient way in which the crossing of a bridge can be represented"), and from thence developed it into a useful device in later sections. Indeed, he had essentially focused on the notion of degree of a vertex in a graph, although throughout the memoir there is no mention of "graph" or "degree," and no record of any picture that resembles our modern graph. (The graph in Figure 1b was not due to Euler at all. It seems to appear for the first time in the first edition of *Mathematical Recreations and Problems* by Walter William Rouse Ball in 1892, almost 150 years after Euler's memoir!) This is a good illustration of how good notations, besides providing convenience, can breed new notions or concepts.

Euler then transformed the problem in section 7 and broke down the problem into sub-problems in section 8 by looking at one single region, and assembled subproblems to give a solution to the Problem of the Seven Bridges of Königsberg in section 9. He then outlined an algorithm for the general problem in section 14. These procedures, though commonplace for any working mathematician, are worth pointing out to students [15]. Careful readers will note a slip on Euler's part, for he had only explained the part on necessity but stated the condition as being both necessary and sufficient. But let us go on with Euler. In section 16 he stated that "I shall, however, describe a much simpler method of determining this which is not difficult to derive from the present method after I have first made a few preliminary observations." His observation is the result we usually refer to nowadays as the "Handshaking Lemma," and with it, he established in section 20 the result now commonly referred to as "Euler's Theorem" on (semi) Eulerian graphs. Again, he had only proved the part on necessity. Only in the last section did he mention, in passing, the construction of such a walk if it exists. Depending on how one interprets this section, one either says that Euler did not prove the part on sufficiency but merely reduced the problem to the case of a simple graph, or at best he hinted at a proof using mathematical induction. The first complete proof was given by Carl Hierholzer in 1873 (posthumously) over 130 years later [1, pp. 11–12]. Furthermore, Hierholzer was discussing the problem in connection with figure-tracing at one stroke and apparently was totally unaware of Euler's result in 1735! This illustrates how a mathematical proof evolves with time. In hindsight, the proof in modern textbooks appears much simpler and much neater, and is complete. But do not forget that we can avail ourselves of the language of graph theory together with its useful results, for which we have Euler to thank! What the first solution lacked in completeness and polish, it made up for in clarity, wealth of ideas, and revelation of the author's train of thoughts.

I shall conclude by giving a sample examination question (used by me in May of 1988):

Write a short account on ONE of the following.

(A) Book I of Euclid's *Elements* (about fourth century B.C.),

(B) Euler's "Solutio problematis ad geometriam situs pertinentis" (1736)

(C) Dedekind's *Stetigkeit und irrationale Zahlen* (1872).

You should give an outline of the content; explain the main mathematical ideas embodied therein; comment on how these ideas evolve; point out any features concerning mathematical thinking in general that you learned from reading it.

Bibliography

1. Biggs, N. L., Lloyd, E. K., and Wilson, R. J.: *Graph Theory*. Oxford: Clarendon Press, 1976.
2. Bonola, R.: *Non-Euclidean Geometry*. New York: Dover, 1955 (original edition published in 1912).
3. Browder, F. E. (ed.): *Mathematical Developments Arising From Hilbert Problems*. Providence: American Mathematical Society, 1976.
4. Brown, R. and Porter, T.: "Mathematics in Context: A New Course," *For the Learning of Mathematics* 10 (1990), 10–15.
5. Calinger, R. (ed.): *Classics of Mathematics*. Oak Park: Moore Publications, 1982.
6. Davis, P. J. and Hersh, R.: *The Mathematical Experience*. Boston: Birkhäuser, 1980.
7. Dedekind, R.: *Essays on the Theory of Numbers*. New York: Dover, 1963 (original English translation published in 1901).
8. Dedron, P. and Itard, J.: *Mathematics and Mathematicians*, Vol. 1 and 2, (J. V. Field, trans.). Milton Keynes: Open University Press, 1973 (original edition published in 1959).
9. Edwards, H. M.: "Read the masters!" in Steen, L. A. (ed.): *Mathematics Tomorrow*. New York: Springer-Verlag, 1981, 105–110.
10. Fauvel, J. and Gray, J. (ed.): *The History of Mathematics: A Reader*. London: Macmillan Education, 1987.
11. Heath, T. L.: *The Thirteen Books of the Elements*. New York: Dover, 1956 (original 2nd edition published in 1925).
12. Hersh, R.: "Let's Teach Philosophy of Math!" *College Mathematics Journal* 21 (1990), 105–111.
13. Midonick, H. (ed.): *A Treasury of Mathematics*. Harmondsworth: Penguin, 1968.
14. Ore, O.: *Niels Henrik Abel, Mathematician Extraordinary*. New York: Chelsea, 1974.
15. Pólya, G.: *How To Solve It*, 2nd edition. Princeton: Princeton University Press, 1957.
16. Smith, D. E.: *A Source Book in Mathematics*. New York: Dover, 1959 (original edition published in 1929).
17. Struik, D. J.: *A Source Book in Mathematics: 1200–1800*. Cambridge: Harvard University Press, 1969.
18. Weil, A.: "Number of solutions of equations in a finite field," *Bulletin of the American Mathematical Society* 55 (1949), 497–508.
19. Weil, A.: "Two Lectures on Number Theory, Past and Present," *L'Enseignement Mathématique* 20 (1974), 87–110.
20. Wilson, R. J.: "An Eulerian Trail Through Königsberg," preprint, Open University, 1985.
21. *Jiu Zhang Suan Shu (Nine Chapters on the Mathematical Art)* (in Chinese). Compiled at about 1st century, collected in *Si Ku Quan Shu (Complete Library of the Four Branches of Literature)*, 1774.

A Topics Course in Mathematics

Abe Shenitzer

In 1984 I published an article in the German journal, *Der Mathematikunterricht* describing my topics course for teachers. A virtually unaltered version of this article, entitled "A Topics Course in Mathematics," appeared in *The Mathematical Intelligencer* 9 (3) (1987). The present essay is an update of that article.

I described the aim and format of the course in these words.

The course covers a number of independent topics, each of which explores a significant mathematical problem or idea. While the treatment of each topic includes a description of the historical evolution of the relevant ideas, the course is not a course in the history of mathematics. Rather, it is an issues course intended to convince the participants that mathematics is meaningful, that some of its problems are profound, and that the evolution of some of its ideas is an exciting chapter of intellectual history.

The criterion for selection of topics is their mathematical and cultural significance and technical accessibility. While I try to make the material accessible, I include a minimum of technical details without which the course might degenerate into a facile talk show. On the other hand, technicalities are merely the means, and the end is the appreciation of ideas and their significance.

Neither the topics nor their treatment is fixed. Students are encouraged to suggest topics and to present them. Written accounts submitted by the students must not be paraphrases of encyclopedia articles but must deal with each topic as presented in class.

The question of aim requires some elaboration. I see my role as that of someone who opens a door to the riches of the world of mathematics and encourages the students (teachers) to take responsibility for their own education. The first step in this direction is the recognition of the evil of classroom paternalism, and the realization that the standard pattern of the teacher as the sole source of initiative, and of the teacher-and-textbook combination as the sole source of wisdom, induces a condition of intellectual bondage; an umbilical cord is indispensable at the fetal stage of growth, but it must be cut if the next stage of growth is to take place.

I am well aware that the need to grade creates a war zone atmosphere that is inimical to the habit of reflection and am glad to say that this evil is virtually eliminated when I give the course I am describing to teachers.

I conclude this introduction with two splendid and relevant quotations.

The great contribution of mathematics, pure or applied, is not rigor. It is ideas. Those are what our teaching should explain, and our own research should look for, and our writing should express. (From Gilbert Strang's letter in *The Mathematical Intelligencer* 10 (1988), p. 3)

Out of the interaction of form and content in mathematics grows an acquaintance with methods which enable the student to produce independently, albeit within certain moderate limits, and to extend his knowledge through his own reflection. The deepening of the consciousness of the intellectual powers connected with this kind of activity, and the gradual awakening of the feeling of intellectual self-reliance may well be considered the most beautiful and highest result of mathematical training. (Alfred Pringsheim in "über Wert und angeblichen Unwert der Mathematik," *Jahresbericht der DMV* (1904), 374. Quoted in *The Mathematical Intelligencer* 10 (1988), 49.)

And now from the general to the particular. The following topics were covered in the 1982–83 version of the course:

1. The evolution of the number system. (The discussion included a comparison of the contributions of Eudoxus and Dedekind, and consideration of major shifts of viewpoint.)
2. *The Method* of Archimedes.
3. Some Greek construction problems and their modern algebraic solutions.
4. Kepler's laws and Newton's Law of Universal Gravitation.
5. Huygens' cycloidal clock. (This was intended as a simple demonstration of the power of the calculus and of the importance of the idea of curvature.)
6. Minima and maxima. From the isoperimetric problem to the calculus of variations.
7. The discovery of hyperbolic geometry and its intellectual implications.
8. Geometry, geometries, and Klein's Erlangen program.
9. Fourier's series—its genesis and impact on mathematics.
10. Hilbert's third problem: the impossibility of an elementary theory of volume.
11. Uncertainty as progress: comments inspired by some of the work of Kurt Gödel.

I still regard some of these topics as musts, and the "miniessays" in *The Mathematical Intelligencer* article in 1987 as material that provides the generally insecure and disoriented would-be lecturer with a new view of the mathematical enterprise, which is, that mathematical skills are the tools to be used in the presentation of mathematical issues in a critical and historically grounded manner.

The first version of the course was a shoestring operation. By now I have accumulated a number of "minimonographs" and realistic lists of references, and would make these available to the students at the outset with the aim of providing them with "seed material," of exposing them to the variety of mathematics, and of encouraging them to explore and to make choices. I would also encourage them to look for accessible articles in the journals. The primary advantage of my handouts is their brevity.

A New Description of the Topics Course

The technical issues one constantly runs into are axiomatics, invariants, and, to a minor extent, cardinal numbers. A recurrent theme throughout the course is the uses of history. Hence the need for discussing these matters first. (The order in which other blocks of material are taken up is largely immaterial. A possible exception is the evolution of the number system.)

Axiomatics.   Lectures 7 and 35 in Eves' book [2] provide a splendid description of the patterns of material axiomatics and formal axiomatics, respectively. In material axiomatics meanings are attached to the basic terms, and the axioms "concerning the basic terms ... are felt to be acceptable to the reader as true on the basis of the properties suggested by the initial explanations ..." (of the basic terms). In formal axiomatics the "technical terms ... are deliberately chosen as unproved statements."

Examples of Invariants.   Euclidean invariants: Consider a unit square in a co-ordinatized Euclidean plane. The coordinates of its vertices vary with the rectangular coordinate system, but the distance between two successive vertices remains 1 and the distance between two opposite vertices remains $\sqrt{2}$; the distances in question are invariant with respect to allowable coordinate transformations.

Affine invariants: If instead of allowing only rectangular coordinates we allowed arbitrary skew coordinate systems, then the initial square would at all times remain a parallelogram. An invariant feature of the initial figure would be the parallelism of pairs of its opposite sides. The "affine setting" is the setting of a great deal of linear algebra. Thus let A be a linear transformation of a vector space V. Two matrices represent A (with respect to different bases) if they are similar. The "affine essence" of A is what all of these matrices have in common. One such common feature is the value of the determinant of each of the matrices in question. Another is the set of its eigenvalues, etc.

Needless to say, there are untold numbers of other sets of invariants, such as projective invariants, algebraic invariants, topological invariants, and so on; see [1] and [6].

Cardinal numbers and transfinite arithmetic.   Cantor introduced a hierarchy of infinities and extended to them the operations of addition, multiplication and exponentiation of the positive integers. See [3] and [4].

The uses of history.   For information on this topic, see [2].

Bibliography

1. Aleksandrov, A. D., Komogorov, A. N., and Lavrent'ev, M. A., eds.: *Mathematics. Its Content, Methods, and Meaning.* Cambridge: MIT Press, 1963. Invariant, I, pp. 238–241; topological III, pp. 310–311.
2. Eves, Howard: *Great Moments in Mathematics* (2 vols.), lectures 7 and 35. Washington: Mathematical Association of America, 1981.
3. Kaplansky, I.: *Set Theory and Metric Spaces.* Boston: Allyn and Bacon, 1972. (Try to read the material on set theory up to p. 66, ignoring difficult proofs.)
4. Vilenkin, N. Ya.: *Stories about Sets.* New York: Academic Press, 1968.
5. Weil, A.: "History of mathematics: Why and How," in Weil, *Oeuvres scientifiques.* New York: Springer-Verlag, 1979, 434–442.
6. Yaglom, I. M.: *A Simple non-Euclidean Geometry and its Physical Basis.* New York: Springer-Verlag, 1979. (Read pp. 1–15 on "What is Geometry?")

Topic A. The evolution of the number system

Theme 1. Evolution of Greek number concepts.   Until about 500 B.C. the Pythagoreans assumed that all magnitudes were commensurable. This meant that the (positive) integers and their ratios—our (positive) rational numbers—were thought to suffice for all quantitative studies (including metric geometry). The discovery of incommensurable magnitudes resulted in the collapse of the early theory of proportions, the adoption of the view that numbers are unequal to the tasks of geometry, and the rise of geometric algebra (ruler and compass constructions were the tool for the geometric solution of algebraic equations).

The theory of proportions was restructured by Eudoxus (about 350 B.C.), one of the most sophisticated thinkers of all time. After more than two millennia, Eudoxus's theory of proportions became the basis of Dedekind's theory of real numbers (dating to the 1870s). Roughly speaking, Eudoxus's theory of proportions corresponds to the theory of (positive) real numbers under multiplication.

Theme 2. Two effective number systems.   The widespread adoption of the decimal system in Europe (about 1600) was due largely to the support it received from Viète and Stevin.

The decimal system displaced the sexagesimal system first used by the Babylonians and then by Greek, Arab, and West European mathematicians. Both are positional systems; they use the bases 10 and 60, respectively.

However desirable, the (gradual) transition from ratios to numbers represented a retreat from earlier standards of rigor.

Theme 3. Negative and complex numbers.   For centuries the absence of good "models" of negative and complex numbers and of operations on them militated against their unreserved acceptance and use. Thus Descartes restricted himself to the positive quadrant of the plane. "Most authors [of eighteenth-century textbooks] felt it necessary to dwell at length on the rules governing multiplication of negative numbers and some rejected categorically the possibility of multiplication of two negative numbers." [1, p. 501] Leibniz thought of complex numbers "as a sort of amphibian, halfway between existence and nonexistence." Gauss's identification of complex numbers with the points of the plane at the beginning of the nineteenth century paved the way for their general acceptance. (While many eighteenth-century mathematicians used complex numbers extensively and even went from $a + b\sqrt{-1}$ to the point with coordinates a, b and conversely, they did not make the necessary identification, and did not interpret operations on complex numbers geometrically [3, vol. 2, p. 56]. This was first done in 1798 by C. Wessel, but his paper became widely known only at the end of the nineteenth century. It was done to greater effect by J. Argand and others in the beginning of the nineteenth century.

Theme 4. The rigorous development of the number system in the nineteenth century. Hamilton's definition of complex numbers as ordered pairs of reals (with suitable rules of addition and multiplication) came in 1837. Dedekind's and Cantor's definitions of reals in terms of rationals appeared in 1872 and 1871, respectively. An axiomatic basis for the natural numbers—and thus for the number system as well as for analysis—was provided by Peano (1889). It was not until 1894 that J. Tannery introduced the arithmetic of rationals as pairs of integers. It should be noted that the "logical" development of the number system, as presented, say in Landau's *Foundations of Analysis*, is almost the reverse of the historical process.

Bibliography

1. Boyer, Carl B.: *A History of Mathematics*. New York: Wiley, 1968.
2. Remmert, R.: "Genesis of the Complex Numbers," (A. Shenitzer, trans.) in J. H. Ewing, ed.: *Numbers*. New York: Springer-Verlag, 1991, 56–65.
3. Yushkevich, A. P., ed.: *History of Mathematics* (in Russian). Moscow, 1972.

Topic B. Vectors and vector spaces, Quaternions, Algebras

Composition of forces as well as composition of displacements readily lead to the notion of a vector space. Vector spaces and cosets of vector spaces arise naturally in the description of the solutions of systems of linear equations as well as in other contexts.

Hamilton's search for a system of "numbers" applicable to rotations and to the composition of rotations in space led to his discovery of quaternions. The use of quaternions for this purpose is described in [2].

Roughly speaking, algebras are vector spaces with multiplication. [2] provides a royal road to algebras (beginning with the algebra of quaternions) and their historical evolution.

Bibliography

1. Davis, H. A. and Snider, A. D.: *Introduction to Vector Analysis*. Boston: Allyn and Bacon, 1979, Appendix A.
2. Kantor, I. L. and Solodovnikov, A. S.: *Hypercomplex Numbers*, (A. Shenitzer, trans.). New York: Springer-Verlag, 1989. (In particular, consult the sections on quaternions and division algebras.)
3. May, K. O.: "The impossibility of a division algebra of vectors in three-dimensional space," *American Mathematical Monthly* 73 (1966), 289–291.

Topic C. The calculus and its uses

(a) A few phases in the evolution of the calculus. After a lapse of nineteen centuries the analytical problems first tackled by the Greeks—primarily Archimedes—were again worked on by the analysts of the seventeenth century. The new elements introduced at that time were greatly aided by the revival of algebra in the sixteenth century and the introduction of analytic geometry in the seventeenth. One could now rely on algebraic manipulation and adopt a "wholesome" approach to the kinds of geometric problems first solved by Archimedes. There was a gradual development of techniques of differentiation and integration. Greek rigor was set aside. Problems were solved using many imaginative and intuitive approaches. The use of infinitesimals (like much of what was done in mathematics at that time) was more showy than sound, but it yielded results that were largely correct. In a sense, this stream culminated in the work of Newton and Leibniz.

Newton and Leibniz replaced the special techniques of their predecessors by the general algorithmic procedures of the calculus and were fully aware of the potential of the new apparatus. (Thus they are rightly credited with invention of the calculus.) Newton must also be given credit for the creation of mathematical physics and the introduction of differential equations as the core of the calculus. It is of interest to note that calculus and its physical successes boosted the philosophy of determinism.

The introduction of the study of functions as an important concern of analysis is due to Euler.

The theoretical foundations of the calculus are due largely to Cauchy, Riemann and Weierstrass. Lebesgue provided what may be called a "best possible" concept of the integral.

(b) The uses of the calculus. "Analysis has been the dominant branch of mathematics for 300 years, and differential equations is the heart of analysis. This subject is the natural goal of elementary calculus and the most important part of mathematics for understanding the physical sciences. Also, in the deeper questions it generates, it is the source of most of the ideas and theories which constitute higher analysis." [11, preface] For the first glimpse of the uses of differential equations see [3]. [11] is also a splendid introduction to the subject.

Beginning students of the calculus learn to apply it to problems on minima and maxima. What these students see is the tip of an iceberg. Problems on minima and maxima have had a long history. The isoperimetric problem, that is, the assertion that of all simple, closed curves with a given perimeter the circle bounds the largest area, dates back to antiquity. (For a proof that uses virtually no analysis see [8, chap. 12].) Euclid considered the restriction of that problem to rectangles and proved that among all rectangles of fixed perimeter the square has the largest area. In analytical terms, both problems involve the search for stationary values of numerical-valued functionals (that is, functions of functions), but the second problem can be simplified and reduced to the search for stationary values of a function of a numerical variable.

Fermat (seventeenth century) considered Euclid's problem in the analytical form in which the variable area A is given by the expression $A(x) = x(a-x)$, where a is the fixed semiperimeter of each rectangle. He linked the stationary points on the curve $(x, A(x))$ to the points with horizontal tangent, that is, to the points at which $A'(x) = 0$.

Euler (eighteenth century) formulated the general problem of stationary values of functionals given by integrals and deduced the analog of the condition $A'(x) = 0$, the so-called Euler differential equation that must be satisfied by a solution of a "variational problem." The relevant analytical machinery is known as the calculus of variations.

To make Euler's great contribution comprehensible, we state a few typical problems of the calculus of variations.

To find the shortest curve joining (x_1, y_1) and (x_2, y_2), we must minimize the arc length integral

$$I = \int_{x_1}^{x_2} \sqrt{1 + y'^2} \, dx$$

subject to the conditions $y(x_1) = y_1$ and $y(x_2) = y_2$.

To find the curve joining (x_1, y_1) and (x_2, y_2) that yields a surface of revolution of minimum area when revolved about the x-axis, we must minimize

$$I = \int_{x_1}^{x_2} 2\pi y \sqrt{1 + y'^2} \, dx$$

subject to the conditions $y(x_1) = y_1$ and $y(x_2) = y_2$.

To find the curve of quickest descent from a point $P(x_1, y_1)$ to a lower point $Q(x_2, y_2)$ we must minimize the integral

$$I = \int_{x_1}^{x_2} \frac{\sqrt{1 + y}}{\sqrt{2gy}} \, dx$$

subject to the conditions that $y(x_1) = y_1$ and $y(x_2) = y_2$.

To solve the isoperimetric problem we must maximize the integral

$$A = \frac{1}{2} \int_{t_1}^{t_2} \left[x \frac{dy}{dt} - y \frac{dx}{dt} \right] dt$$

subject to the condition that

$$L = \int_{t_1}^{t_2} \sqrt{\left(\frac{dx}{dt} \right)^2 + \left(\frac{dy}{dt} \right)^2} \, dt$$

is constant.

To find the geodesics on a surface $G(x, y, z) = 0$ we must minimize the arc length integral

$$\int_{t_1}^{t_2} \sqrt{\left(\frac{dx}{dt} \right)^2 + \left(\frac{dy}{dt} \right)^2 + \left(\frac{dz}{dt} \right)^2} \, dt$$

subject to the side condition $G(x, y, z) = 0$.

In the first three examples the problem is that of finding the stationary values y of

$$\int_{x_1}^{x_2} f(x, y, y') = 0.$$

It turns out that such inputs y must satisfy the Euler differential equation

$$\frac{d}{dx} \left(\frac{df}{dy'} \right) - \frac{df}{dy} = 0.$$

This is an analog of the condition $\frac{d}{dx}(g(x)) = 0$ that must be satisfied by an input x at which a function g has a stationary value. In the remaining examples the stationary values must satisfy appropriate modifications of this equation.

The calculus of variations has been one of the major branches of analysis for more than two centuries. It is a tool of great power that can be applied to a wide variety of problems in pure mathematics. It can also be used to express the basic principles of mathematical physics in forms of the utmost simplicity and elegance. [11, p. 353]

One of the most impressive illustrations of the last sentence is Hamilton's principle of least action—the variational equivalent of Newton's second law of motion [11, p. 377]. That law is just Euler's differential equation for the action

$$A = \int_{t_1}^{t_2} (T - V) \, dt,$$

where T and V are, respectively, the kinetic and potential energies of a particle moving in a conservative force field. The following quotation conveys its great significance.

Hamilton's principle can be made to yield the basic laws of electricity and magnetism, quantum theory, and relativity. Its influence is so profound and far-reaching that many scientists regard it as the most powerful single principle in mathematical physics and place it at the pinnacle of physical science: Max Planck, the founder of quantum theory, expressed this view as follows: 'The highest and most coveted aim of physical science is to condense all natural phenomena which have been observed and are still to be observed into one simple principle ... Amid the more or less general laws which mark the achievements of physical science during the course of the last centuries, the principle of least action is perhaps that which, as regards form and content, may claim to come nearest to this ideal final aim of theoretical research.' [11, pp. 379–380]

Two unusual calculus-related items: We wish to call the reader's attention to two unusual issues involving the calculus. One has to do with Hilbert's third problem [2]. Its solution showed that in building up a theory of volume we must, from the very beginning, rely on the calculus. The first step in the development of a theory of area (volume) is to develop a theory of area (volume) of polygons (polyhedra). The two tasks—the development of theories of area of polygons and of volume of polyhedra—look similar but turn out to be radically different. Any two polygons with equal areas are congruent by addition, that is, one of them can be cut up into polygonal pieces which can be rearranged to form the second polygon. Dehn (1902) showed that the corresponding statement for polyhedra is false. This shows that while there is an "elementary" (that is, not requiring the use of limits) theory of areas of polygons, there is no simplification of Euclid's approach to these theories.

The other has to do with definitions in the calculus. Specifically, the Schwartz "paradox" [4] shows that the usual definition of arclength cannot be adapted in the "natural way" to yield a viable definition of area to a surface.

Bibliography

1. Alexandrov, P.: "Curves and Surfaces," in Aleksandrov, A. D. *et al.*, ed. *Mathematics: Its Content, Methods and Meaning*, vol. 2. Cambridge: MIT Press, 1963. (This is a splendid work, and Alexandrov's article is an expository masterpiece. Differential geometry uses the calculus in the study of curves and surfaces.)
2. Boltyanskii, V. G.: *Equivalent and Equidecomposable Figures*. New York: D. C. Heath, 1963.
3. Boltyanskii, V. G.: *Differentiation Explained* (in Russian). Moscow: Mir, 1964.
4. Dubnov, Ya. S.: *Mistakes in Geometric Proofs*. New York: D. C. Heath, 1963.
5. Edwards, C. H., Jr.: *The Historical Development of the Calculus*. New York: Springer, 1979.
6. Harnik, V.: "Infinitesimals from Leibniz to Robinson," *The Mathematical Intelligencer* 8 (2) (1986), 41–47.
7. Hildebrandt, S.: "The calculus of variations today" (A. Shenitzer, trans.), *The Mathematical Intelligencer* 11 (4) (1989), 50–60.
8. Niven, I.: *Maxima and Minima without Calculus*. Washington: Mathematical Association of America, 1981.
9. Pollard, H.: *Celestial Mechanics*. Washington: Mathematical Association of America, 1976. (See chapter 1, the central force problem.)
10. Robinson, A.: *Nonstandard Analysis*. Amsterdam: North Holland, 1966.
11. Simmons, G. F.: *Differential Equations*. New York: McGraw-Hill, 1972.
12. Tikhomirov, V. M.: *Stories about Maxima and Minima* (A. Shenitzer, trans.). Providence: American Mathematical Society, 1990. (See the eighth story, Newton's Aerodynamical Problem.)
13. Toeplitz, O.: *The Calculus: A Genetic Approach*. Chicago: University of Chicago Press, 1963. (See sections 40–42 on Kepler's Laws and Newton's Law of Universal Gravitation.)

Topic D. Geometries

Like numbers, geometry was at first viewed as an external datum. At its dawn—in the last decades of the eighteenth century — hyperbolic geometry was granted neither consistency nor physical significance. Lobachevsky and Gauss were inclined to grant it both. Bolyai was primarily concerned with the consequences of the axioms of hyperbolic geometry. He felt that he had "created a universe ex nihilo." This is an early expression of the notion of mathematics as a human creation.

Once Euclidean models of hyperbolic geometry were created, the problem of the (relative) consistency of hyperbolic geometry was settled and, like other mathematical systems, it was accepted as a branch of mathematics, a human creation whose utility—in the broadest sense of the word—was a function of human inventiveness.

The transition from one to two admissible geometries dethroned Euclidean geometry and paved the way for the transition to an infinity of geometries. (One such infinity was introduced in 1854 by Riemann. Another was introduced in 1872 by Klein as an answer to the question of what is a geometry. The needs of physics prompted Cartan to unify the views of Klein and Riemann by introducing [in 1922] Riemannian geometries that were locally Kleinian.)

Bibliography

1. Gould, S. H.: "Origins and development of concepts of geometry." *NCTM Yearbook 23*. Reston, Va.: National Council of Teachers of Mathematics, 1957.
2. Gray, J.: "The discovery of non-Euclidean geometry," in Phillips, Esther, ed., *Studies in the History of mathematics*. Washington: Mathematical Association of America, 1987.
3. Kelley, P. and Matthews, G.: *The Non-Euclidean Hyperbolic Plane*. New York: Springer-Verlag, 1981. (See chapter 1 for the historical background.)
4. Nikulin, V. V. and Shafarevich, I. R.: *Geometries and Groups*, (M. Reid, trans.). New York: Springer-Verlag, 1987.
5. Yaglom, I. M.: *A Simple Non-Euclidean Geometry and its Physical Basis* (A. Shenitzer, trans.). New York: Springer-Verlag, 1979.

Topic E. The function concept

To Leibniz the functional input was a curve and its output was some numerical characteristic of a curve. To Newton functions were tied to physics (e.g., distance is a function of time). D'Alembert identified "function" with "formula." Euler gave "function" a wider sense (including, for example, functions drawn free-hand). Both Euler and d'Alembert identified formula-given functions with what we now call analytic functions and thus greatly underestimated the power of "formulas."

Fourier claimed that all functions on an interval could be represented by a convergent Fourier series (but never defined "function"). (The closest modern equivalent of Fourier's claim is the theorem: Every function defined on an interval and belonging to L_p, $1 < p < \infty$, can be expanded in a Fourier series which converges almost everywhere.) This part of the story deserves elaboration.

In his paper "On vibrating strings" (1755), Daniel Bernoulli made the claim that every motion of a string, fastened at its endpoints and vibrating with zero initial velocity, is given by a formula

$$y = \sum_{p=1}^{\infty} a_p \sin p \frac{\pi x}{c} \cos pkt,$$

where c is the length of the string and k is a certain physical constant. This implied that if $y = f(x)$ is the initial shape of the string, then

$$f(x) = \sum_{p=1}^{\infty} a_p \sin p \frac{\pi x}{c};$$

that is, an "arbitrary" function $f(x)$ could be expanded in a sine series. This conclusion was regarded as inadmissible by such contemporaries of Bernoulli as Euler, d'Alembert, and (later) Lagrange for the following reasons.

In distinction to modern usage, Euler and his contemporaries divided curves into "continuous" curves, that is, curves given by "formulas," and "geometric" curves, that is, curves not given by single formulas. It seemed obvious that a "continuous" curve was determined throughout its domain of definition by its values on ever-so-small an interval of that domain. (That is, it seemed obvious that if two formulas had the same values on ever-so-small an interval, then they had the same values throughout their common domain of definition.) Thus no one doubted that the class of "geometric" curves was larger than the class of "continuous" curves. Now if an "arbitrary" function could be expanded in a sine series—and thus could be represented by a formula—then every "geometric" curve would be "continuous" and this was inadmissible.

Bernoulli's viewpoint was successfully defended half a century later by Fourier. In papers published in 1807 and 1822, Fourier claimed that an "arbitrary" function could be expanded in a series of sines and cosines, and determined the coefficients of the expansion (left undetermined by Bernoulli). While Fourier's computations were persuasive, few of them could be called proofs.

In 1829 Dirichlet lent substance to Fourier's claim by proving the convergence of the Fourier series of a large class of functions. The study of convergence of Fourier series continues to this day and has stimulated the development of whole areas of mathematics. In particular, in the nineteenth century, the study of Fourier series induced Riemann to improve the notion of the integral, stimulated Cantor's set-theoretic researches, made discontinuous functions respectable, and gave rise to the study of the subject now known as the theory of functions of a real variable.

It may be useful to explain briefly the references to Riemann and Cantor. The Fourier coefficients are definite integrals. If Fourier's claim that an arbitrary function could be expanded in a trigonometric series was to be substantiated, then one had to go beyond Cauchy's definition of an integral of a continuous function. The first step in this direction was taken by Riemann, who gave the familiar definition of an integral that involved no a priori restrictions on the integrand. (It is safe to say of the Riemann integral what Dorothy Sayers said about the sleuthing methods of Sherlock Holmes: "More showy than sound.") He also tried to find necessary conditions for the convergence of Fourier series. (The problem of what are necessary and sufficient conditions for the convergence of Fourier series is this: If a Fourier series vanishes on a period interval, then its coefficients are all zero. Cantor tried to eliminate the vanishing condition for infinite subsets of the period interval without affecting the truth of the conclusion and was led to the "derivative of a point set.")

What happened later is that Dirichlet's definition of function at first delighted all mathematicians, but eventually it was objected to on logical grounds. The Baire classification scheme represented a significant gain for the function-as-formula concept. The concept of generalized function was introduced as an extension of the function concept primarily as a response to the needs of science.

Bibliography

1. Langer, R. E.: "Fourier's Series," *The American Mathematical Monthly* 54 (Slaught Memorial Paper 1) (1947).
2. Luzin, N.: "Function," in *The Great Soviet Encyclopedia*. English translation published by Macmillan.
3. Van Vleck, E. B.: "The Influence of Fourier's Series upon the Development of Mathematics," *Science* 39 (1914), 113–124.

Topic F. Algebra and number theory

The solution in radicals of determinate polynomial equations was one of the key problems of algebra until well into the 19th century. The solution of indeterminate polynomial equations and of systems of such equations is an important component of number theory. But "diophantine analysis" has also profoundly influenced the evolution of algebra. (See [1].)

A beautiful branch of mathematics in which algebra and number theory are intertwined is algebraic number theory. (See [5].) Since the field of algebraic numbers is "of small size" (it is countable), it is only fair to say something about "the rest" of the field of complex numbers. What we do is demonstrate the existence of transcendental numbers. (See [4].)

The modern algebraic solutions of the three "unsolvable" classical Greek problems provide a natural opportunity for a discussion of significant issues in the history of mathematics. (See [3].)

Bibliography

1. Bashmakova, I. G.: "Diophantine equations and the evolution of algebra," (in Russian), *Proceedings of the International Congress of Mathematicians, 1986*. Providence: American Mathematical Society, 1987, 1612–1628.
2. Clark, A.: *Elements of Abstract Algebra*. New York: Wadsworth, 1971. (See the introdution, which includes material on Lagrange's analysis of solutions of quadratic, cubic, and quartic equations.)
3. Hadlock, C. R.: *Field Theory and its Classical Problems*. Washington: Mathematical Association of America, 1978. (See sections 1.1–1.4.)
4. Niven, I.: *Numbers: rational and irrational*. Washington: Mathematical Association of America, 1964. (See chapter 7 on the existence of transcendental numbers.)
5. Pollard, H. and Diamond, H. G.: *The theory of algebraic numbers*, 2nd edition. Washington: Mathematical Association of America, 1975.

Topic G. Uncertainty as progress: some of Gödel's discoveries

Kant claimed that "[Euclidean] geometry is a science which determines the properties of space synthetically and yet a priori," that is, without relying on experience. The retreat from this position began in the nineteenth century with the emergence of hyperbolic geometry which, while in one respect the logical opposite of Euclidean geometry, turned out to be its legitimate equal. The effect of this was that axioms in general, and the axioms of Euclidean geometry in particular, ceased to be "obvious truths" and gradually came to be viewed as intuitively motivated initial assumptions. By now

> there is no longer any question of defending the ancient and long-recurring rationalist hope that geometry gives us knowledge which is both synthetic and a priori. There is no one who can on this be quoted more aptly than Einstein. For he was the first

to give a physical application to a non-Euclidean geometry, that of Riemann: 'As far as the laws of mathematics refer to reality they are not certain; and as far as they are certain, they do not refer to reality.' [2, p. 427]

A different retreat from certainty was a consequence of the profound intellectual discovery made in 1931 by Kurt Gödel. Roughly speaking, Gödel showed that any axiomatic system that includes ordinary arithmetic includes undecidable propositions. Also—and this is of crucial importance—through an ingenious encoding one of the undecidable propositions can be interpreted as asserting the consistency of the system. This makes arithmetic, and thus all of its extensions, "a branch of theology" [1, lecture 38].

A century separates the discovery of hyperbolic geometry by Lobachevski, Bolyai, and Gauss from Gödel's discovery. These are the most dramatic of a long list of developments that have reshaped our view of mathematics. Any such list must include the following:

1. Elimination of the special role of Euclidean geometry.
2. Axiomatization of arithmetic (and also of the number system and of analysis).
3. Improvement of the logical basis of Euclidean geometry and insight into the logical consequences of the various groups of axioms comprising the system of axioms of Euclidean geometry.
4. Discovery of paradoxes in set theory and efforts aimed at their elimination.
5. Reemergence of the ancient debate on the potential infinite and the actual infinite.
6. Insight into the role of the axiom of choice. The Hausdorff paradox.
7. Hilbert's program to prove the consistency of arithmetic and its termination by the discoveries of Gödel.

In addition to changing our view of mathematics, these developments have also changed our view of the nature of mathematical activity. In the words of H. Weyl, "'Mathematizing' may well be a creative activity of man, like language or music, of primary originality, whose historical decisions defy complete objective rationalization." [*Obituary Notices of Fellows of the Royal Society* 4 (1944), 547–553.]

Bibliography

1. Eves, H.: *Great Moments of Mathematics*, vol. 2. Washington: Mathematical Association of America, 1981.
2. Flew, A.: *An Introduction to Western Philosophy*. Indianapolis: Bobbs-Merrill, 1971.
3. Nagel, E. and Newman, J. R.: *Gödel's Proof*. New York: NYU Press, 1958.
4. Quine, W. V.: "Paradox," *Scientific American* 206 (April, 1962), 84–94.
5. Smorynski, C.: "The incompleteness theorems," in J. Barwise, ed., *Handbook of Mathematical Logic*. Amsterdam: North Holland, 1978.

Topic H. The Method of Archimedes

The mathematicians of the seventeenth and eighteenth centuries relied heavily on the heuristic method of indivisibles without being aware of its antiquity. Heilberg's discovery of *The Method* in 1906 showed that rigor alone did not make classical Greek mathematics. Archimedes' (c. 250 B.C.) brilliant use of physical reasoning and of the method of indivisibles as a means of arriving at some of his results on area and volume of figures with curved boundaries is a demonstration of the continuity of approach of creative minds separated—in this case—by 20 centuries. [This material is not part of the Calculus topic since it has practically nothing to do with the calculus(!).]

Bibliography

1. Aaboe, A.: *Episodes from the Early History of Mathematics*. Washington: Mathematical Association of America, 1964. (See especially section 3.6.)
2. Toeplitz, O.: *The Calculus. A genetic approach*. Chicago: University of Chicago Press, 1963. (See sections 13 and 14.)

Niels Henrik Abel

Niels Henrik Abel (1802–1829)
A Tribute

My eyes have been opened in the most surprising manner. If you disregard the very simplest cases, there is in all of mathematics not a single infinite series whose sum has been stringently determined. In other words, the most important parts of mathematics stand without foundation. It is true that most of it is valid, but that is very surprising. I struggle to find the reason for it, an exceedingly interesting problem. (January 16, 1826)

So wrote Niels Henrik Abel to his lifelong friend and teacher, Bernt Holmboe. During his short life, Abel found many such interesting problems to investigate and resolve. His concern with the status of infinite series resulted in the publication of several classical theorems on the convergence of power series and the "stringently determined" sum of a binominal series with either real or complex exponents. At the age of nineteen, Abel had already determined that a general equation of fifth or higher degree could not be solved by radicals and settled a problem that had lingered from the seventeenth century. Today, this result is known as the Abel–Ruffini Theorem. Further work with equations led Abel to seek out solutions for equations that admitted commutative groups (Abelian equations). In travels to Germany, he made the acquaintance of August Leopold Crelle, a civil engineer and amateur mathematician. Together they worked to publish Germany's first mathematics journal, *Journal für die reine und angewandte Mathematik (Journal for Pure and Applied Mathematics)* which became commonly known as *Crelle's Journal.* Abel was one of its most prolific early contributors.

In the spring of 1826 Niels Abel completed his masterpiece, *Mémoire sur une propriété générale d'une classe très-éntendue de fonctions transcendentes (Memoir on a General Property of a Very Extensive Class of Transcendental Functions)* in which he presented a theory of integrals of algebraic functions and enunicated Abel's Theorem: the sum of integrals of a given algebraic function can be expressed as a fixed number of such integrals, with integration arguments that are algebraic functions of the original arguments. Later this work gave rise to the theory of Abelian integrals and Abelian functions. By 1827, in friendly competition with K. G. J. Jacobi, Abel had produced a theory of elliptic functions.

Neils Abel was just beginning to win recognition for the scope and power of his mathematical insights; his work had extended the theory of equations, strengthened group theory and greatly enlarged the theory of functions, when his life was cut short by tuberculosis. He died on his fiancée's estate at Froland on April 6, 1829 and is buried in the churchyard there.

August Crelle eulogized his lost friend as follows:

All of Abel's works carry the imprint of an ingenuity and force of thought which is unusual and sometimes amazing, even if the youth of the author is not taken into consideration. One may say that he was able to penetrate all obstacles down to the very foundations of the problems, with a force which appeared irresistible; he attacked the problems with an extraordinary energy; he regarded them from above and was able to soar so high over their present state that all the difficulties seemed to vanish under the victorious onslaught of his genius.... But it was not only his great talent which created the respect for Abel and made his loss infinitely regrettable. He distinguished himself equally by the purity and nobility of his character and by a rare modesty which made his person cherished to the same unusual degree as was his genius. [*Crelle's Journal* 4 (1829), 402]

On a visit to his grave site on August 10, 1988, another eulogy was delivered by Man-Keung Siu on behalf of the Kristiansand Conference participants, who although separated from Niels Henrik Abel by both time and distance, still sense his mathematical greatness and lament his lost potential.

Two days ago Otto asked me to say a few words on this occasion of making our pilgrimage to Froland, where exactly 160 years ago Niels Henrik Abel spent the summer and where a year later he died and was buried at the age of 26. I was told that I was chosen to speak because I come from afar. Since I keep a travel log, I can even tell you the exact distance I have covered so far, starting from Hong Kong. It is 16,919 km. Mathematically speaking, the only connection I can make of this and Abel is that a group of order 16,919 ($= 7 \times 2417$) is abelian (actually cyclic), the proof of which will be left to the reader. But in this wonderful gathering at Kristiansand we share the conviction that mathematics is not just a collection of theorems and formulae; there is a cultural and human element to it. So, we come here today not just for a technical mathematical connection. For myself, and I believe it also true for others, I harbour a deep admiration and respect for Abel, not just for the contributions he made in mathematics, but also for his qualities as a human being. I am glad and I feel honored to be given this chance to speak.

I first read of the life of Abel in the popular account "Genius and Poverty" by E. T. Bell when I was a first year undergraduate. At that time, young and ignorant as I was, I was already moved and inspired by this story of a young man who gave so much of himself despite the adversity he had to face, so much so that he was robbed of his life in his prime. Schopenhauer maintained that a strong motive that led people to art and science was flight from the harshness of everyday life. I think there is another, perhaps even stronger, motive: namely, an inner call of intellectual curiosity and quest for learning.

If I may now be allowed to indulge in a more personal note, I would add that 25 years after I read this story, now that I am no longer young but only slightly less ignorant, this feeling and admiration can only be strengthened, especially after I learned of the painful sufferings most of my mathematician friends in China had gone through in those infamous years labelled "the Cultural Revolution." Of course, that dreadful experience is a nightmare behind them for over a decade by now, and hopefully history will not repeat itself at that, but they are still working under difficult conditions, this time in financial terms. Many things we take for granted are to them ·luxuries. In some sense their situation is not unlike what Abel had been facing. Like Abel, they continue to work hard despite that, and many literally give their lives for science in their motherland. I know of quite a number of Chinese mathematicians who have died in their forties, some of whom had made noted contributions in their own fields.

History will surely reward scholars with such devotion, just like Abel is remembered today. Even without such recognition and honor, they would, I believe, gladly do the same at their inner calls. But would it not be infinitely better if they can work and contribute in a more agreeable environment which they more than deserve when they were living? A regard for learning is what we need in the society of today. With this wish I shall end my humble speech.

For readers who may wish to learn more about the life and work of Niels Abel, his collected works have appeared in two editions: *Oeuvre complètes N. H. Abel, mathématicien,* B. Holmboe, ed. (Oslo, 1839) and *Nouvelle édition,* M. M. L. Sylow and S. Lie, eds., 2 vols., (Oslo, 1881); and an English language biography has been written by Øystein Øre, *Niels Henrik Abel, Mathematician Extraordinary* (Minneapolis: University of Minnesota Press, 1957), republished by Chelsea, 1974.

About the Authors

E. J. Aiton was formerly a Senior Lecturer in Mathematics in the Didsbury School of Education, Manchester Polytechnic, England. He held the degrees of PhD and DSc from the University of London and was a member of the International Academy of the History of Science. Among his publications are a book on the vortex theory of planetary motions, a biography of Leibniz and a translation of Peurbach's *Theoricae novae planetarum.* Having collaborated on an edition of Kepler's *Mysterium cosmographicum,* he was working on a translation of Kepler's *Harmonice mundi* when he died in 1991.

Shmuel Avital is Professor of Mathematics at the Technion, Israel Institute of Technology, Haifa. He is interested in incorporating the history of mathematics into secondary school teaching. He works closely with Israeli teachers to accomplish this goal.

Otto Bekken is Professor of Mathematics at Adger College, Kristiansand, Norway. His major research interests focus on the life and work of Niels Henrik Abel. He is particularly active in the incorporation of the history of mathematics into teacher training programs. He has written materials for this purpose for use in Scandinavia and Peru. He served as the chief organizer and director of the Kristiansand Conference.

David Burton is Professor of Mathematics at the University of New Hampshire, where he has been for the past 29 years. He earned his doctorate from the University of Rochester in 1961, writing his PhD thesis on double Laplace transforms under the tutelage of Dorothy Bernstein. He has written a number of books, including *A First Course in Rings and Ideals* (Addison-Wesley), *Elementary Number Theory* (Allyn and Bacon), *A History of Mathematics* (Allyn and Bacon), and *Abstract Algebra* (W. C. Brown).

John Fauvel is Lecturer in Mathematics at the Open University, England, and chaired the Open University course, "Topics in the History of Mathematics." He has been an editor of several books, including *Darwin to Einstein: Historical studies on science and belief* (1980), *Conceptions of Inquiry* (1981), *The History of mathematics: A Reader* (1987) and *Let Newton Be!* (1988). He is the chair of the International Study Group on the Relations between History and Pedagogy of Mathematics.

Tony Gardiner is Reader in Mathematics and Mathematics Education at the University of Birmingham (UK). Alongside research in group theory and combinatorics, he uses the history of mathematics to bring home the way "real" mathematics develops; his book *Infinite Processes: Background to Analysis* (Springer, 1982) was much praised but little read. *Discovering Mathematics* (Oxford, 1987) introduces "real" mathematics in a different way—getting students to tackle and solve unfamiliar problems. He is deeply involved in mathematics competitions—both popular and select; his OKSMC has expanded within five years to almost 100,000 entries.

Michel Helfgott   is Professor of Mathematics at the Universidad de San Marcos (Lima, Peru). His fields of interest are the relations between mathematics and the natural sciences, especially chemistry, and the use of history as a pedagogical device in the classroom. He is co-author of a textbook on differential equations intended for students of science and engineering, and the author of a secondary school textbook on plane geometry.

Phillip Jones   is Professor Emeritus of Mathematics and Education at the University of Michigan, Ann Arbor. He is an "elder statesman" in the movement to combine the history of mathematics with its pedagogy. For many years, he worked with the National Council of Teachers of Mathematics in the United States in bringing historical material to mathematics teachers. In particular, he edited the feature "Historically Speaking" in the NCTM journal *Mathematics Teacher*.

Victor J. Katz   is Professor of Mathematics at the University of the District of Columbia in Washington, D.C. He has been interested in the history of mathematics for many years, and in particular, in ways in which it can be used in the mathematics classroom. He has published several papers on both history and on its use in teaching. His textbook *A History of Mathematics* (Harper Collins, 1993) contains many suggestions as to how the material can be used in teaching on both the secondary and the undergraduate levels.

Israel Kleiner   received his PhD in ring theory at McGill University. He is presently Professor of Mathematics at York University, Toronto, where he has taught for over twenty years. He has been involved in teacher education at the undergraduate and graduate levels and has given numerous talks to high-school students and teachers. One of his major interests is the history of mathematics and its relation to the learning and teaching of mathematics.

Joel P. Lehmann   is an Associate Professor of Mathematics and Computer Science at Valparaiso University in Valparaiso, Indiana and was formerly the Director of Academic Computing. His doctorate was in applied mathematics, and he later earned a master's degree in computer science. Current activity includes teaching a course in the history of mathematics, and developing a program to incorporate microcomputers and computer algebra systems into the calculus sequence.

Jan A. van Maanen   studied mathematics at the University of Utrecht where he concentrated on number theory and on the history of mathematics. He wrote his PhD thesis on "Facets of seventeenth century mathematics in The Netherlands" (1987), and has published several articles on the history of mathematics, mathematics education and the application of history in teaching mathematics. Presently he teaches mathematics at a secondary school and the history of mathematics at Central Netherlands Polytechnic, Utrecht (to prospective mathematics teachers) and the University of Groningen (to students of mathematics).

Lars Mejlbo   studied at the University of Copenhagen, completing a masters degree in mathematics in 1957, and undertook further study at Princeton University in the United States. Presently, he is Lecturer in Mathematics at Aarhus University, Denmark. Before 1968, most of his work involved functional analysis, and he published several papers in that field. From 1968 onwards, he has concentrated on the history and teaching of mathematics and has published books and papers in Danish on these subjects.

Donovan H. Van Osdol   is Professor of Mathematics and Chairman of the Department of Mathematics at the University of New Hampshire, where he has been since 1970. His dissertation, written under the direction of Michael Barr at the University of Illinois in 1969, was concerned with sheaf theory and homological algebra. He has held visiting positions at Oslo, Utrecht, Sussex, Aarhus, McGill, and Granada. He is currently interested in semi-simplicial topology and calculus curriculum reform.

Karin Reich has been Professor of the History of Science and Technology at the Fach-hochschule für Bibliothekswesen and Professor of the History of Mathematics at the University of Stuttgart since 1980. She has also lectured at the universities of Hamburg and Heidelberg. Since 1987 she has organized, in collaboration with the Oberschulamt in Stuttgart, a course in advanced teacher training on the topic "History of Mathematics in High School Teaching."

V. Frederick Rickey is Professor of Mathematics at the Bowling Green State University, Ohio. He was initially trained in logic and has served as the editor of the *Notre Dame Journal of Formal Logic*. Over the years his interests have changed to the history of mathematics, which is now his primary research interest, especially the history of the calculus. He has given numerous short courses around the U.S. on using history in the teaching of calculus.

Abe Shenitzer is Professor of Mathematics at York University in Canada. Born in Warsaw in 1921, he attended Brooklyn College and received his PhD from New York University in 1954. He has translated many Russian and German papers and books. He is interested in the history of mathematics and in the psychological and intellectual preconditions for effective teaching and learning.

Man-Keung Siu completed his BSc at the University of Hong Kong and received his PhD from Columbia University in 1972. After three years at the University of Miami, he returned to teach at the University of Hong Kong. He has published in the fields of algebra, combinatorics, applied probability, mathematics education, and the history of mathematics. Besides teaching and doing research, he likes popularizing mathematics among school pupils. For the making of a slide show for that purpose he received (jointly with N. K. Tsing) a 1981 CASME Award from the Commonwealth Association of Science and Mathematics Educators.

Frank Swetz is Professor of Mathematics and Education at the Pennsylvania State University at Harrisburg. His research concerns societal impact on the learning and teaching of mathematics. These interests have led him into studies on the history of mathematics and ethnomathematics. His findings are incorporated into teaching strategies which humanize mathematics teaching. His most recent books are: *The Sea Island Mathematical Manual: Surveying and Mathematics in Ancient China* (1992) and *From Five Fingers to Infinity: A Journey Through the History of Mathematics* (1993).